HYDROCARBON BIOTECHNOLOGY

Challenges and Future Trends

HYDROCARBON BIOTECHNOLOGY

Challenges and Future Trends

Edited by

Wael Ahmed Ismail, PhD

Jonathan Van Hamme, PhD

AAP | APPLE ACADEMIC PRESS

First edition published 2023

Apple Academic Press Inc.
1265 Goldenrod Circle, NE,
Palm Bay, FL 32905 USA

760 Laurentian Drive, Unit 19,
Burlington, ON L7N 0A4, CANADA

CRC Press
6000 Broken Sound Parkway NW,
Suite 300, Boca Raton, FL 33487-2742 USA

4 Park Square, Milton Park,
Abingdon, Oxon, OX14 4RN UK

Library and Archives Canada Cataloguing in Publication

Title: Hydrocarbon biotechnology : challenges and future trends / edited by Wael Ahmed Ismail, PhD, Jonathan Van Hamme, PhD.
Names: Ismail, Wael Ahmed, editor. | Van Hamme, Jonathan, editor.
Description: First edition. | Includes bibliographical references and index.
Identifiers: Canadiana (print) 20220213690 | Canadiana (ebook) 20220213747 | ISBN 9781774639894 (hardcover) | ISBN 9781774639900 (softcover) | ISBN 9781003277354 (ebook)
Subjects: LCSH: Hydrocarbons—Biotechnology. | LCSH: Microbiology. | LCSH: Bioremediation. | LCSH: Fossil fuels.
Classification: LCC QD305.H5 H93 2023 | DDC 547/.01—dc23

Library of Congress Cataloging-in-Publication Data

CIP data on file with US Library of Congress

ISBN: 978-1-77463-989-4 (hbk)
ISBN: 978-1-77463-990-0 (pbk)
ISBN: 978-1-00327-735-4 (ebk)

About the Editors

Wael Ahmed Ismail, PhD
Environmental Biotechnology Program,
Life Sciences Dept., College of Graduate Studies,
Arabian Gulf University, Manama, Bahrain

Wael Ahmed Ismail, PhD, is a Professor of Microbiology and Environmental Biotechnology, and the former director of the Biotechnology Program at the Arabian Gulf University. Having more than 25 years of research experience in the field of microbial biodegradation of hydrocarbons (HCs), Dr. Ismail received his BSc in Microbiology and Chemistry from Ain Shams University (Cairo-Egypt) and joined the Egyptian Petroleum Research Institute in 1994 to pursue his MSc in Petroleum Microbiology. He then moved to the Department of Microbiology at Freiburg University, Germany, to earn his PhD (funded by the German Research Foundation) under the supervision of Prof. Georg Fuchs. His PhD research was dedicated to the biochemistry and molecular biology of microbial degradation of aromatic compounds, with particular focus on the elucidation of unprecedented catabolic pathways (the hybrid pathways) for aromatic compounds degradation. During his postdoctoral training in the same group, he continued research on the aerobic and anaerobic degradation of aromatic compounds. In addition, he joined projects on the anaerobic degradation of cholesterol, where he contributed to the molecular and biochemical elucidation of previously uncharacterized catabolic pathways. In 2008, Dr. Wael was appointed as a Research Assistant Professor at the Center of Environmental Biotechnology at the University of Massachusetts, Amherst (USA), where he worked with Prof. Derek Lovley. During that time, he conducted molecular biology studies on Geobacter sulfurreducens and Rhodoferax ferrireducens, key players in microbial fuel cells (MFCs). Currently, his research focuses on biodegradation of crude oil and steroids, biosurfactants production and applications, biodesulfurization (BDS) of diesel fuel, as well as biocatalytic upgrading of heavy oil and refinery residues. A major part of his research deals with the application of systems biology to advance the understanding of biochemistry and physiology of bacteria adopted in petroleum biotechnological processes.

He works in close association with the oil and gas authorities and research institutes in Bahrain and Kuwait. He was recently appointed as an associate editor at Frontiers in Microbiology and was also the lead guest editor for two special issues on Petroleum Biotechnology and Biodesulfurization in the same journal.

Jonathan Van Hamme, PhD
Department of Biological Sciences, Thompson Rivers University, Kamloops, British Columbia, Canada

Jonathan Van Hamme, PhD, is a Professor of Microbiology and founding Director of the TRUGen Applied Genomics Laboratory, the only high-throughput sequencing facility in central British Columbia. He is an applied environmental microbiologist with over 25 years of experience in petroleum microbiology, industrial waste treatment, contaminant biotransformation, phytoremediation, analytical chemistry, proteomics, and genomics, including bacterial genome sequencing, microbial community characterization, and eDNA detection of insects, plants, and animals. He received an Honors BSc in Biology with a Chemistry Minor from the University of Waterloo in 1996, and a PhD in petroleum microbiology from the University of Waterloo in 2000. After three years with the National Resources Canada at their National Center for Upgrading Technology, where he was leading the work on microbial biotechnologies for heavy oil upgrading, he joined Thompson Rivers University and established his own research program. Since that time, he has built a world-class molecular microbiology laboratory that includes equipment for aerobic and anaerobic microbiology, fermentations up to 40 liters, proteomics, high-throughput sequencing (Ion S5, Oxford Nanopore, Sanger), and stable isotope probing (SIP). With a major focus on environmental remediation of both organic and inorganic pollutants, he has a long history of academic and industrial collaborations in a wide range of fields (petroleum, agriculture, food, brewing, chemical). He is currently focused on uncovering the biochemical basis of bacterial per- and poly-fluoroalkyl substance (PFAS) metabolism, as well as tracking SARS-CoV-2 in sewage. He served for 10 years as an editor for *Biodegradation* and is currently an editor for the *Canadian Journal of Microbiology.* He collaborated with Dr. Ismail and others as a guest associate editor for a special issue on Petroleum Microbiology in *Frontiers in Microbiology.*

Contents

Contributors

R. M. M. Abed

Sultan Qaboos University, College of Science, Biology Department, P.O. Box – 36, Al Khoud, Postal Code 123, Muscat, Sultanate of Oman, Tel.: +968-24141450, E-mail: rabed@mpi-bremen.de

Saif Al-Bahry

Department of Biology, College of Science, Sultan Qaboos University, Oman;
Oil and Gas Research Center, Sultan Qaboos University, Oman

Ali Al-Bemani

Petroleum and Chemical Engineering Department, College of Engineering, Sultan Qaboos University, Oman

Yahya Al-Wahaibi

Petroleum and Chemical Engineering Department, College of Engineering, Sultan Qaboos University, Oman

Subramania Angaiah

Electro-Materials Research Lab, Center for Nanoscience and Technology, Pondicherry University, Puducherry – 605014, Tamil Nadu, India

Vanessa Ayala-Espinoza

Molecular Microbiology and Environmental Biotechnology Laboratory,
Department of Chemistry and Center of Biotechnology Daniel Alkalay Lowitt,
Universidad Técnica Federico Santa María, Valparaiso, Chile

Ximena Báez-Matus

Molecular Microbiology and Environmental Biotechnology Laboratory,
Department of Chemistry and Center of Biotechnology Daniel Alkalay Lowitt,
Universidad Técnica Federico Santa María, Valparaiso, Chile

Dawoud Bahzad

Petroleum Research Center, Kuwait Institute for Scientific Research, P.O. Box – 24885,
Safat – 13109, Kuwait

Abhijeet P. Borole

Bredesen Center for Interdisciplinary Research and Education, The University of Tennessee,
Knoxville – 37996, Tennessee, US; Chemical and Biomolecular Engineering,
The University of Tennessee, Knoxville – 37996, Tennessee, US

Georgina Corti-Monzón

Institute of Science and Technology in Food & Environment , National University of Mar del Plata,
CONICET. J.B. Justo – 4302, Mar del Plata, Bs. As., Argentina

Heather D. Dettman

Natural Resources Canada, CanmetENERGY Devon, 1 Oil Patch Drive, Devon, AB T9G 1A8, Canada,
E-mail: heather.dettman@canada.ca

Flavia Dorochesi
Molecular Microbiology and Environmental Biotechnology Laboratory,
Department of Chemistry and Center of Biotechnology Daniel Alkalay Lowitt,
Universidad Técnica Federico Santa María, Valparaiso, Chile

Roberto E. Durán
Molecular Microbiology and Environmental Biotechnology Laboratory,
Department of Chemistry and Center of Biotechnology Daniel Alkalay Lowitt,
Universidad Técnica Federico Santa María, Valparaiso, Chile

Nour Sh. El-Gendy
Department of Process Design and Development, Egyptian Petroleum Research Institute (EPRI),
Nasr City, Cairo PO – 11727, Egypt; Center of Excellence,
October University for Modern Sciences and Arts (MSA), 6th of October City, Giza, PO – 12566,
Egypt, E-mail: nourepri@yahoo.com

Abdulkadir Elshafie
Department of Biology, College of Science, Sultan Qaboos University, Oman

Lisette Hernández
Molecular Microbiology and Environmental Biotechnology Laboratory,
Department of Chemistry and Center of Biotechnology Daniel Alkalay Lowitt,
Universidad Técnica Federico Santa María, Valparaiso, Chile

Sanket J. Joshi
Oil and Gas Research Center, Sultan Qaboos University, Oman;
Central Analytical and Applied Research Unit, College of Science, Sultan Qaboos University,
Oman, Phone: +96824142456, Fax: +96824414120,
E-mails: sanket@squ.edu.om; sanketjj@gmail.com

Constanza C. Macaya
Molecular Microbiology and Environmental Biotechnology Laboratory,
Department of Chemistry and Center of Biotechnology Daniel Alkalay Lowitt,
Universidad Técnica Federico Santa María, Valparaiso, Chile

H. Mahmoud
Kuwait University, Department of Biological Sciences, P.O. Box – 5969, Safat – 13060, Kuwait

Nisenbaum Melina
Institute of Science and Technology in Food & Environment , National University of Mar del Plata,
CONICET. J.B. Justo – 4302, Mar del Plata, Bs. As., Argentina

Fernanda Lopes Motta
Natural Resources Canada, CanmetENERGY Devon, 1 Oil Patch Drive, Devon, AB T9G 1A8, Canada

Silvia E. Murialdo
Institute of Science and Technology in Food & Environment , National University of Mar del Plata,
CIC. J.B. Justo – 4302, Mar del Plata, Bs. As., Argentina

Kadarkarai Murugan
Division of Entomology, Department of Zoology, School of Life Sciences, Bharathiar University,
Coimbatore – 641046, Tamil Nadu, India

Punniyakotti Parthipan
Environmental Molecular Microbiology Research Laboratory, Department of Biotechnology, Thiruvalluvar University, Serkkadu, Vellore – 632115, Tamil Nadu, India;
Electro-Materials Research Lab, Center for Nanoscience and Technology, Pondicherry University, Puducherry – 605014, Tamil Nadu, India

Pattanathu K. S. M. Rahman
Pharmacy and Biomolecular Sciences, Liverpool John Moores University, Liverpool, UK;
TeeGene Biotech and Tara Biologics, Wilton Center, UK

Aruliah Rajasekar
Environmental Molecular Microbiology Research Laboratory, Department of Biotechnology, Thiruvalluvar University, Serkkadu, Vellore – 632115, Tamil Nadu, India,
E-mails: rajasekargood@gmail.com; rajasekargood@tvu.edu.in; rajasekar.aruliah@gmail.com

M. M. Ramírez-Corredores
Energy and Environment S&T Division, Idaho National Laboratory, Idaho Falls, ID – 83415, USA

Mohan S. Rana
Petroleum Research Center, Kuwait Institute for Scientific Research, P.O. Box – 24885, Safat – 13109, Kuwait, E-mail: mrana@kisr.edu.kw

Laura Rodríguez-Castro
Molecular Microbiology and Environmental Biotechnology Laboratory, Department of Chemistry and Center of Biotechnology Daniel Alkalay Lowitt, Universidad Técnica Federico Santa María, Valparaiso, Chile

Geetha S. J.
Department of Biology, College of Science, Sultan Qaboos University, Oman

Nayereh Saborimanesh
Natural Resources Canada, CanmetENERGY Devon, 1 Oil Patch Drive, Devon, AB T9G 1A8, Canada

Felipe Salazar-Tapia
Molecular Microbiology and Environmental Biotechnology Laboratory, Department of Chemistry and Center of Biotechnology Daniel Alkalay Lowitt, Universidad Técnica Federico Santa María, Valparaiso, Chile

Patricio Santis-Cortés
Molecular Microbiology and Environmental Biotechnology Laboratory, Department of Chemistry and Center of Biotechnology Daniel Alkalay Lowitt, Universidad Técnica Federico Santa María, Valparaiso, Chile

Scott J. Satinover
Bredesen Center for Interdisciplinary Research and Education, The University of Tennessee, Knoxville – 37996, Tennessee, US

Michael Seeger
Molecular Microbiology and Environmental Biotechnology Laboratory, Department of Chemistry and Center of Biotechnology Daniel Alkalay Lowitt, Universidad Técnica Federico Santa María, Valparaiso, Chile, E-mail: michael.seeger@gmail.com

N. Sivakumar
Sultan Qaboos University, College of Science, Biology Department, P.O. Box – 36, Postal Code 123, Al Khoud, Sultanate of Oman

Matthew L. Smith
Pharmacy and Biomolecular Sciences, Liverpool John Moores University, Liverpool, UK

James G. Speight
CD&W Inc., Laramie, Wyoming, USA

Yen-Peng Ting
Department of Chemical and Biomolecular Engineering, National University of Singapore, 4 Engineering Drive 4, Singapore – 117585

Abbreviations

2DE	two-dimensional gel electrophoresis
2-HBP	2-hydroxybiphenyl
ADH1	alcohol dehydrogenases
ALDH	aldehyde dehydrogenase
AMO	alkene monooxygenase
API	American Petroleum Institute
AR	atmospheric residue
BAC	bacterial artificial chromosomes
BDC	bioconversion
BDN	biodenitrogenation
BDO	benzene 1,2-dioxygenase
BDS	biodesulfurization
BDS-ODS-RADS	biodesulfurization/oxidative desulfurization/reactive adsorptive desulfurization
BES	bioelectrochemical systems
BMMs	bacterial multicomponent monooxygenases
BMO	benzene monooxygenase
BotB	bottom of the barrel
bpd	barrels of oil per day
BTEX	benzene, toluene, ethylbenzene, and xylene
C	carbon
CAPP	Canadian Association of Petroleum Producers
CC	closed circuit
CCE	cathodic conversion efficiency
CCR	carbon Conradson residue
CD	current density
CFU	colony-forming unit
CH_4	methane
CMC	critical micelle concentration
COD	chemical oxygen demand
DBT	dibenzothiophene
DGGE	denaturing gradient gel electrophoresis
DRO	diesel range organics
DST	Department of Science and Technology

EBC	Enchira Biotechnology Corporation
EEOR	enzyme-enhanced oil recovery
EIA	Energy Information Administration
EOR	enhanced oil recovery
FBP	final boiling point
FCC	fluid catalytic cracking
FMN	flavin mononucleotide
FTIR	Fourier Transform Infrared Spectroscopy
GMOs	genetically modified microorganisms
GRO	gasoline range organics
GWPC	Ground Water Protection Council
H	hydrogen
H_2S	hydrogen sulfide
HAC	high acid crudes
HC	hydrocarbon
HCK	hydrocracking
HDC	hydroconversion
HDP	hydroprocessing
HDS	hydrodesulfurization
HDT	hydrotreatment
IBP	initial boiling point
IOR	improved oil recovery
LC	liquid chromatography
MAHs	monocyclic aromatic hydrocarbons
MALDI-ToF	matrix-assisted laser desorption ionization time-of-flight mass spectrometry
MCDC	microbial capacitive deionization cell
MCR	microcarbon residue
MDCs	microbial desalination cells
MECs	microbial electrolysis cells
MELs	mannosylerythritol lipids
MEOR	microbial enhanced oil recovery
MFC	microbial fuel cell
MIC	microbial induced corrosion
MMO	monooxygenase-like enzymes
MOFC	microbial osmotic fuel cell
MS	mass spectrometry
MWS	midway sunset
N	nitrogen

Ni	nickel
NMR	nuclear magnetic resonance
NOx	nitrogen oxides
O	oxygen
OC	open circuit
OPEC	Organization of Petroleum Exporting Countries
OSC	offshore California
OUT	operational taxonomic unit
P	phosphorus
PAH	polycyclic aromatic hydrocarbon
PCT	patent cooperation treaty
PDMS	polydimethylsiloxane
PEDEC	Petroleum Engineering Development Company
PEP	phosphoenolpyruvate
PFPS	bis-(3-pentafluorophenylpropyl)-sulfide
pMMO	particulate MMO
PTS	phosphotransferase system
PVT	pressure-volume-temperature
RACDC	repetitive alternating carboxylation-decarboxylation cycle
RHDs	ring-hydroxylating dioxygenases
RT-qPCR	real time-quantitative polymerase chain reaction
S	sulfur
SARA	saturates-aromatics-resins-asphaltenes
SCMFC	sediment column MFC
SCO	synthetic crude oil
SDS	sodium dodecyl sulfate
SERB	Science and Engineering Research Board
SIP	stable isotope probing
sMMO	soluble MMO
SO_2	sulfur dioxide
SOx	sulfur oxides
SRB	sulfate-reducing bacteria
SWDs	saltwater disposal wells
TAN	total acid number
TB2MO	toluene/benzene 2-monooxygenase
TCA	tricarboxylic acid
TDO	toluene 2,3-dioxygenase
TDS	total dissolved solids

TMO	toluene monooxygenase
ToMO	toluene/o-xylene monooxygenase
TPH	total petroleum hydrocarbon
TSS	total suspended solids
ULSF	ultra-low sulfur fuels
UNITAR	United Nations Institute for Training and Research
V	vanadium
VGO	vacuum gas oil
VOOEP	vanadium oxide octaethyl porphyrin
VR	vacuum residue
WMS	whole metagenome shotgun sequencing

Acknowledgments

The editors thank Prof. Ibrahim Banat (University of Ulster-Northern Ireland, UK) and Dr. Ki-Hyouk Choi (Saudi Aramco-Kingdom of Saudi Arabia) for their critical review of some chapters.

Preface

In the foreword to the inaugural issue of the *Journal of Bacteriology*, published in February of 1916, Dr. W. T. Sedgwick, the first president of the Society of American Bacteriologists, noted that "Bacteriology must henceforward be recognized as a broad and fundamental branch of the science of immense theoretical and practical importance, and it bids fair to become vastly greater and more important tomorrow" (Sedgwick, 1916). By that time, theories on the role of microorganisms in petroleum formation had permeated the literature (Dake, 1913), and in 1946, Claude E. ZoBell published a 49-page review, *Action of Microorganisms on Hydrocarbons* (Zobell, 1946), which included 182 citations dating back to Miyoshi's observations in 1895 of *Botrytis cinerea* penetrating thin layers of paraffin.

ZoBell's review was wide-ranging, discussing: hydrocarbon (HC)-containing culture media and effects of growth conditions; nearly one hundred species of aerobic and anaerobic hydrocarbonoclastic bacteria, yeasts, and molds able to exploit gaseous, liquid, and solid aliphatic, olefinic, and naphthenic hydrocarbons; the broad environmental range of hydrocarbon-utilizing microorganisms; the activities of these organisms on petroleum, and rubber products; and even the use of microorganisms as indicators of oil deposits. ZoBell presciently noted that "Obtaining intermediate products of economic value such as fatty acids, for example, from the microbial decomposition of hydrocarbons, or employing microorganisms for the elimination of industrial wastes around refineries are almost unexplored possibilities."

The themes discussed by ZoBell continued to be explored, thanks to the enduring willingness of microbiologists interested in hydrocarbon biotechnology to embrace new technologies in efforts to better understand the microbial physiology and ecology of hydrocarbon-metabolizing microorganisms, and how those microorganisms may be exploited to the benefit of the world. A simple search for "hydrocarbon bacteria" in the Web of Science will yield more than 10,000 results up to the present day, evidence that the field of hydrocarbon microbiology continues to endure into the third century. Although research on bioremediation, the classical application of hydrocarbon biotechnology/microbiology, has retained momentum over these long decades, novel applications have emerged, particularly in the fields of oil recovery and refining, in addition to the valorization of hydrocarbon wastes.

This volume represents an up-to-date view of hydrocarbon microbiology and biotechnology, presented by experts around the world with interests in how our expanding understanding of hydrocarbonoclast ecology and physiology can translate to better tools for bioremediation, oil recovery, bioupgrading of unconventional crudes, the development of biorefining technologies, and the production of hydrogen and electricity from hydrocarbon wastes. A common theme across the chapters in this book is an interest in how developing hydrocarbon biotechnologies may reduce our impact on the global environment.

While classical and modern microbiological approaches are major drivers of hydrocarbon biotechnology, the field moves forward through collaboration with chemists, engineers, mathematicians, computer scientists, and others. We are taken back to wise words from W. T. Sedgwick in 1916 when he noted, "It was an acute observation of the late C. S. Peirce that some of the most fruitful of modern sciences have been bred by the crossing of older ones." We hope that this volume will contribute to catalyzing continued collaboration and innovation among the next generation of hydrocarbon biotechnologists.

—Wael Ahmed Ismail
Manama, Kingdom of Bahrain

Jonathan Douglas Van Hamme
Kamloops, Canada

REFERENCES

Dake, C.L., 1913. Petroleum. J. Geography 12, 9–12.

Sedgwick, W.T., 1916. Foreword the Genesis of a New Science,-Bacteriology. J. Bacteriol. 1, 1–4.

Zobell, C.E., 1946. Action of microorganisms on hydrocarbons 1. Bacteriol. Rev. 10, 1–49.

CHAPTER 1

Petroleum Chemistry

DAWOUD BAHZAD and MOHAN S. RANA

Petroleum Research Center, Kuwait Institute for Scientific Research, P.O. Box–24885, Safat–13109, Kuwait,
E-mail: mrana@kisr.edu.kw (M. S. Rana)

1.1 INTRODUCTION

Petroleum, also called "rock oil," forms naturally in the deep subsurface and may be recovered in a solid (coal), liquid (crude oil), or gas (LPG) form. It is estimated that more than 80% of the world's energy comes from fossil fuels (Riazi et al., 2013; Rana and Faisal, 2016; Rana et al., 2017). Understanding petroleum chemistry in the refining industry is complex in terms of a large number of molecular components and their reactivity, particularly in heavier fractions for which analytical techniques are not available. Gaps in understanding these complex mixtures and their interactions are a major challenge for refining clean fuels from diverse petroleum sources. Apart from refining, heavy oil reservoirs are usually challenging to explore and have short production lives due to the asphaltenes and other heavy components, causing formation damage. As such, heavy oil wells are often candidates for secondary (enhanced) and tertiary recovery processes, which require significant capital investment, labor, infrastructure, and time. Heavy and light crudes are classified based on density, and most often the American Petroleum Institute (API) gravity, where the higher the density, the more complex is the crude, which is mainly due to the heavier resin and asphaltene fractions (Rana et al. 2007). The API gravity is an estimate of flow properties of crude oil; the higher the API gravity, the less dense is the liquid. The API gravity can be calculated using the crude oil's specific gravity, which is the unitless ratio of the density of a liquid to the density of water, using the following formula:

$$\text{API} = \left|\frac{141.5}{\text{Specific gravity at }25°\text{C}}\right| - 131.5 \quad (1)$$

API gravities range between 10° to 70° and can be used to broadly classify crude oils (Table 1.1).

TABLE 1.1 Crude Oil Classification and Compositional Properties (Ramirez et al., 2007)

	API Gravity	Asphaltene, wt.%	Resin, wt.%	Oils, wt.%	S, wt.%	N, wt.%	Metals (Ni+V), ppm
Extra light	>50°	0 to <2	0.05–3	–	0.02–0.2	0.0–0.01	<10
Light crude	22°–32°	<0.1–12	3–22	67–97	0.05–4.0	0.02–0.5	10–200
Heavy crude	10°–22°	10–25	14–39	24–64	0.1–5.0	0.2–0.8	50–500
Extra heavy crude oil	<10°	15–40	–	–	0.8–6.0	0.1–1.3	200–600
Residue	-	15–30	25–40	<49	–	–	100 to >1,000

The world's reserves of light crude oil resources are declining while heavy crude production is increasing, which contains a lower amount of distillates and more residues. Therefore, it is foreseen that in the future, some refineries will increasingly replace light crudes with heavier oils (Riazi et al., 2013; Marafi et al., 2019). The price differential between light and heavy oils is mainly based on their composition and blending properties. Globally, crudes are specified with their properties, mainly density and sulfur content, which vary with the origin of crude oil, region to region, and country to country. Recently, Marafi et al. (2019) showed a relationship between the API gravity, naphthenic acid, and sulfur content, where lower API gravity is correlated with higher sulfur content. Figure 1.1 is an example of the relationship between sulfur content and API gravity for various types of crude oils. A lower API gravity indicates that crude oil has a higher average molecular weight for chemical constituents, lower flow properties, lower hydrogen and higher carbon content, and inferior quality as a feedstock.

Considering the chemical makeup of heavy and extra heavy oil, it is important to understand the composition of underground fluids (upstream) and reactor feedstock (downstream) in detail, so that the oil industry can tailor their transport, handling, and refining practices appropriately (Benjamin, 1997; Moore and Richmond 2008; Ancheyta et al., 2010).

There is a consensus that cross-linked asphaltene molecules destabilize oil matrices leading to processing difficulties, such as pore and transport plugging (Mansoori, 1997), facility corrosion (Kilpatrick, 2012), and increased oil viscosity (Pena et al., 2005; Kilpatrick, 2012) during oil production and refining (Ancheyta et al., 2010; Riazi et al., 2013). Therefore, a broader chemical understanding of these components is critical for the development of mitigating technologies.

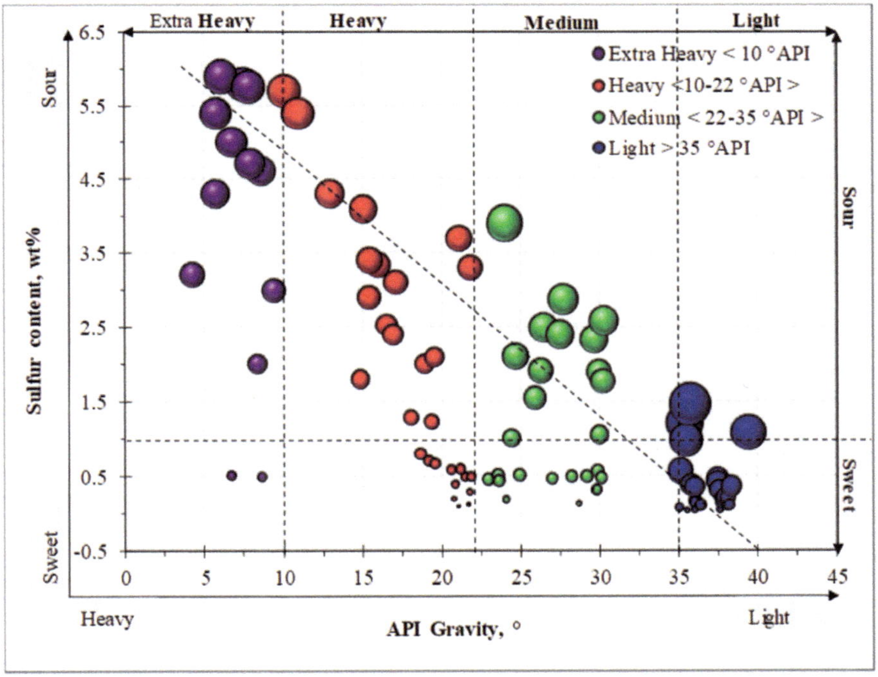

FIGURE 1.1 Relationship between API gravity (heavy, medium, light) and sulfur (sour, sweet) content in various crude oils.

Source: Reprinted with permission from Marafi et al., 2019. © 2018 Elsevier.

1.2 COMPOSITION AND CLASSIFICATION OF CRUDE OIL

Crude oil is made up of diverse hydrocarbon (HC) compounds, many of which are associated with heteroatoms and metals, such as sulfur (S), nitrogen (N), oxygen (O), vanadium (V), and nickel (Ni), that have yet to be fully classified. Two of the most common methods for broadly classifying the chemical makeup of crudes are illustrated in Figure 1.2. The first one is based on the physical separation of the crude components into four broad classes, namely; saturates, aromatics, resins, and asphaltenes (collectively known as the SARA fractions), while the second one is based on boiling points, that is useful for guiding distillation practices in complex refineries. For the latter, fractional distillation cuts are normally classified as naphtha, gasoline, kerosene, diesel, and atmospheric residue (AR). The AR fraction is further divided into vacuum gas oil (VGO) and vacuum residue (VR). All fractions may be further classified into narrower subgroups as needed (Bissada et al., 2016).

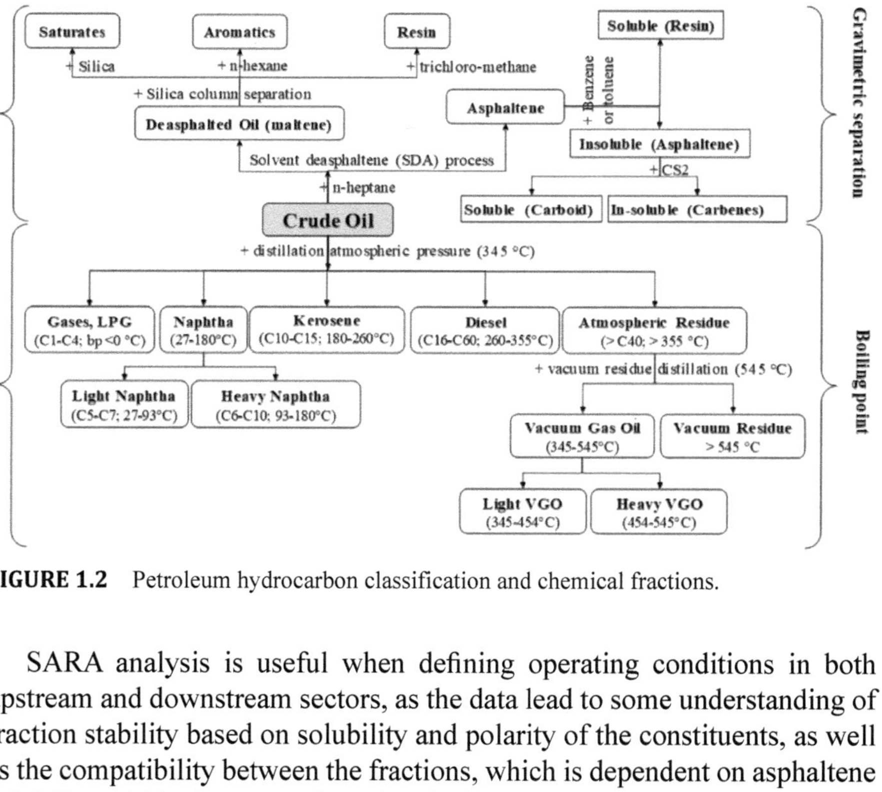

FIGURE 1.2 Petroleum hydrocarbon classification and chemical fractions.

SARA analysis is useful when defining operating conditions in both upstream and downstream sectors, as the data lead to some understanding of fraction stability based on solubility and polarity of the constituents, as well as the compatibility between the fractions, which is dependent on asphaltene solubility within the aromatic and resin fractions (Figure 1.3) (Takatsuka et al., 1989; Wiehe, 1992; Speight et al., 1996, 2004).

The saturate (S) fraction (i.e., paraffins) consists of saturated nonpolar compounds that are rich in hydrogen in the form of straight-chain, branched, or cyclic alkanes, chemically represented as C_nH_{2n+2}. This fraction is usually light, boils at a low temperature, and promotes the destabilization of asphaltenes resulting in precipitation (Ancheyta et al., 2010). The aromatic (A) fraction contains molecules with at least one benzene ring, which can be

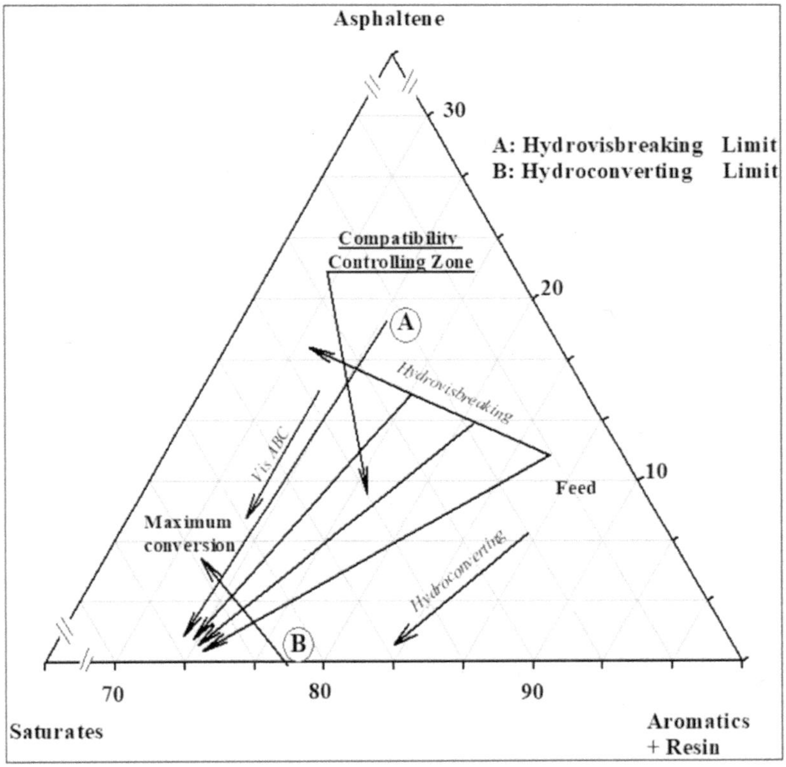

FIGURE 1.3 Compatibility reaction chemistry diagram for SARA fractions and their relationships with refining processes.

Source: Reprinted from Takatsuka et al., 1989.

fused into di-aromatic and polyaromatic compounds with or without aliphatic substituents. The resin (R) fraction is similar to the aromatic fraction, but the constituent molecules have higher molecular weights and contain at least one heteroatom in their aromatic ring structure, which makes them more polar. Resins are soluble in *n*-pentane and *n*-heptane but insoluble in propane (Speight, 2014). The asphaltene (A) fraction contains high-molecular weight polar molecules that are thought to include layered aromatic sheets connected via intermolecular bridges and non-covalent attractive forces. Asphaltenes are defined as being soluble in toluene but insoluble in paraffin solvents such as *n*-pentane and *n*-heptane. This is the first fraction to be separated from the crude oil or residue during refining as a solid, and being the heaviest and most polar fraction, it has a high content of heteroatoms and metal complexes. The proposed structures of asphaltenes in the literature are diverse, leading to misunderstandings about this complex fraction. This may be due to asphaltenes varying with origin and the fact that researchers interrogate this fraction with a diversity of analytical techniques, each providing different datasets. Asphaltene remains one of the crude oil components that cannot yet be thoroughly analyzed. Nevertheless, a detailed account of the microstructure parameters, such as hypothetical molecular determination, average molecular weight, quantitative analysis, aliphatic and aromatic composition, and so forth have been reported (Mullins, 1985; Mansoori, 2005; Ancheyta et al., 2010; Alhumaidan et al., 2015–2017). Recently, Alhumaidan et al. (2015, 2017) reported asphaltene structural changes as a function of temperature, indicating that it is a coke precursor substance.

SARA analysis is a simple yet powerful tool for classifying crude oils into unique categories. Figure 1.4 (inset) shows that heavy crude properties are very similar to the residue (355°C+). Therefore, the processability of heavy oils is expected to be similar to that of the residue. However, catalytic processes are more vulnerable to the presence of metals and asphaltenes (Ancheyta et al., 2005). Generally, asphaltene is the smallest fraction of the crude, but it has a more severe impact on the refining process. The density or viscosity and aromatic content significantly vary with increasing asphaltene content, as shown in Figure 1.5. The asphaltene not only affects the density, but also is responsible for instability of the crude oil and residue, particularly critical for blended feedstocks (Ancheyta et al., 2010). Thus, working in heavy crude and residue, one must understand the chemistry between the feedstock composition and catalyst properties to obtain adequate conversions, catalyst life, product stability, and the final marketable products.

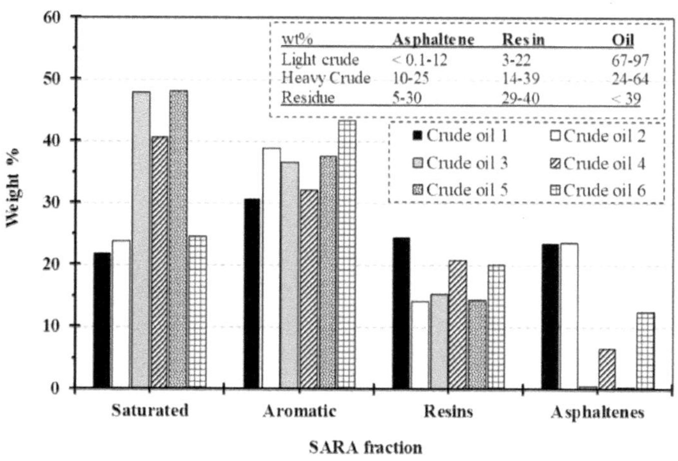

FIGURE 1.4 Various crude oil SARA fractions and their compositional (wt.%) analysis.

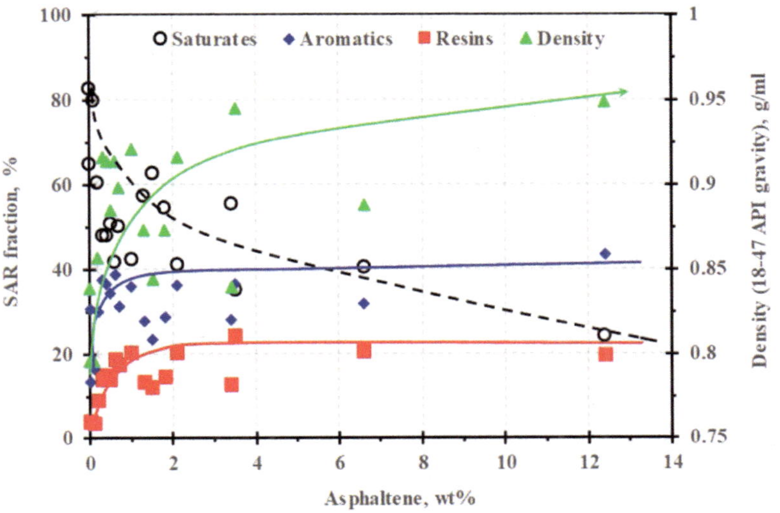

FIGURE 1.5 SARA distribution and density of crudes from different regions (API gravity between 18 to 47).

Based on the boiling point or distillation cuts, the individual fractions can be further identified at a molecular level, such as naphtha (light and heavy), middle distillates (kerosene, diesel, and jet fuel), and VGO (light, medium, and heavy). The AR (bottom of an atmospheric distillation tower, 350°C+) and the VR (bottom of a vacuum distillation tower, 545C+) are still an issue to scientists as well as refiners to separate at the molecular level (Marafi et al., 2019). Naphtha is a mixture of many HC compounds, which has an initial boiling point (IBP) of 30°C and a final boiling point (FBP) about 180°C. Naphtha fraction consists of a large amount of paraffin and cycloparaffins (naphthene) but a limited amount of aromatic HCs (< 5 wt.%) ranging from C4–C11 carbon atoms and is further classified as light naphtha (27–93°C) and heavy naphtha (93–180°C) (Ancheyta et al., 2010).

The various HC components of crude oil play a significant role in fuel or energy production. The lighter the HC fraction, the better is the fuel efficiency or the higher is the energy content, which leads to cleaner burning mainly due to the higher hydrogen content. Fossil fuels such as crude oil constitute the most exploitable source of energy and are also a source of raw materials as a feedstock for the chemical industries (Rana et al., 2017). This is particularly obvious for the transportation sector where all refined fractions of crude oil are used as transportation fuels, such as naphtha (gasoline), aviation fuel (kerosene), middle distillate (diesel), and bunker oil (fuel oils), which are used for light vehicles, jet, heavy transportation (trucks), and marine transport, respectively (Rana et al., 2017; Marafi et al., 2019). Nonetheless, the share of fossil fuels in the world energy matrix is steadily decreasing and is expected to further decrease in the future. On the other hand, the share of alternative sources of energy slowly increases, but their contribution remains less than 20% (Rana et al., 2017). The main constraints of renewable energy implementation are availability as well as economic feasibility compared to fossil fuels. The new technological developments may eliminate those economic constraints if clean energy is produced under the guidance of environmental policies and regulations, which impose many limitations on fossil fuel utilization.

It is a worldwide environmental apprehension that the fossil fuel-based energy sectors are facing prevalent challenges that remain difficult to tackle despite the enormous technological developments since extensive damage has occurred, such as climatic change and pollution. Not only that, but also only 50% of fossil fuel is expected to be replaced by cleaner fuels within the next 30 years, so further damage is expected in the next 30 years (McCarthy and Copeland, 2012; Venkatesh et al., 2012; Rana et al., 2013). Today, about 85% of the global energy demand is being satisfied with fossil fuels, which is releasing enormous CO_2 emissions into the atmosphere. Apart from CO_2, fossil fuels are the primary source of other atmospheric emissions, such as particulate matter, benzene emission, sulfur dioxide (SO_2), and nitrogen oxides (NOx). These emissions are not only environmental pollutants, but also carcinogenic to humankind (Rana and Humaidan, 2016). To accommodate environmental legislation, fossil fuels require significant refinery processing that would remove contaminants (heteroatoms) and enhance the fuel efficiency by improving its properties (high octane and cetane numbers). The production of clean fuels from crude oil depends mainly on the composition of the crude (Rana and Faisal, 2016). Hydrogen content plays a crucial role where the presence of high hydrogen amount enhances clean and efficient burning. Moreover, with higher H/C ratio, the fuel combustion will release lower CO_2 emission, which is about 30% for crude oil and 45% for coal (Huggins, 2015). Natural gas contains a large amount of methane (CH_4) with a high hydrogen to carbon ratio (H/C=4) that leads to significantly higher combustion energy of about 812 kJ/mol and relatively lower CO_2 emission (Tseng, 2002; Huggins, 2015).

1.3 ELEMENTAL AND TRACE METAL COMPOSITION

Crude oil is composed mainly of carbon, hydrogen, sulfur, nitrogen, and oxygen, along with traces of metals as described in the following:

a. **Carbon:** The content and nature of the carbon play a significant role in the density, viscosity, and other physical properties of the crude, which becomes more complex and heavier with an increasing carbon content. The HC compounds are grouped as alkanes (paraffin), cycloalkanes (naphthenes), olefins (alkenes), and aromatics (benzene) (Hayes and Anderson, 1986). Heteroatoms, such as S, N, O, and metals (Ni, V, and Fe) are often associated with the HC components where they constitute heterocompounds (Table 1.2).
b. **Hydrogen (H):** It is the simplest component of crude oil, and its content can define the burning quality of the fuel. Hydrogen is a high-quality energy source and is used for many power generation applications (Alhumaidan et al., 2011; Gianotti et al., 2018; Modisha et al., 2019), particularly for fuel cell vehicles. It is produced mainly during reforming processes in the refinery. Hydrogen-fueled vehicles emit only water vapor during combustion, so it is considered one of the cleanest fuels.

TABLE 1.2 Hydrocarbon and Heterocompound Components of Crude Oil and Molecular Classification

Name of Compounds	Group	Formula/Structure
Alkanes	Paraffins	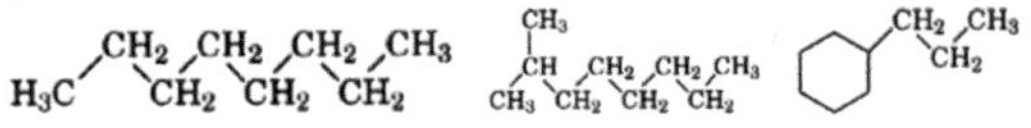
Cycloalkanes	Naphthenes	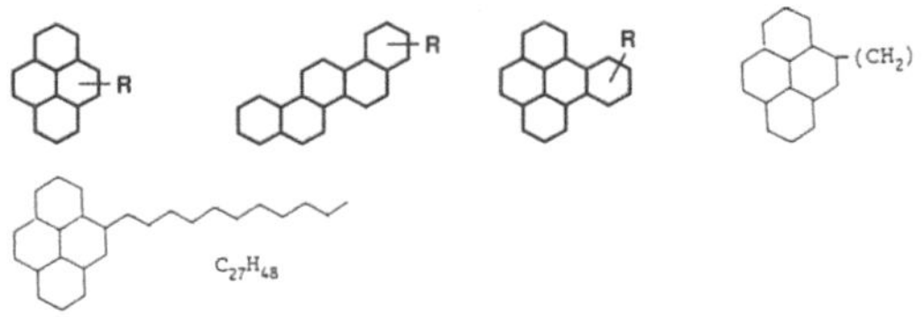
Olefins	Alkenes	
Aromatics	Benzene	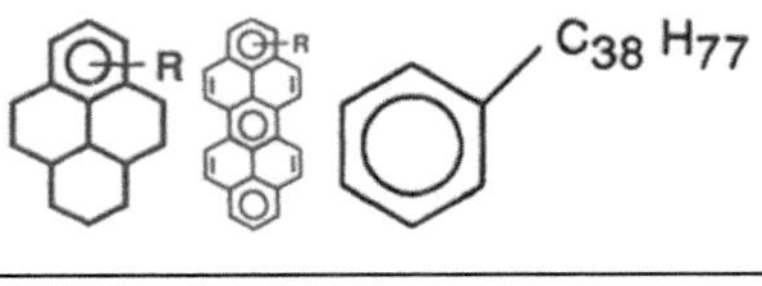

TABLE 1.2 *(Continued)*

Name of Compounds	Group	Formula/Structure
Heterocompound 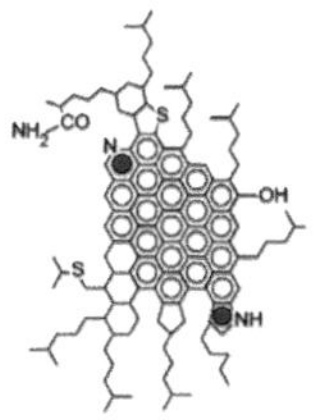	Sulfur	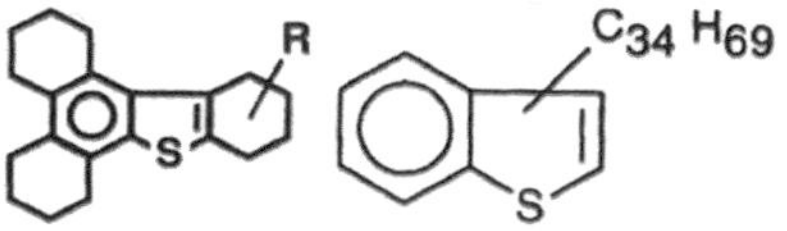
	Nitrogen	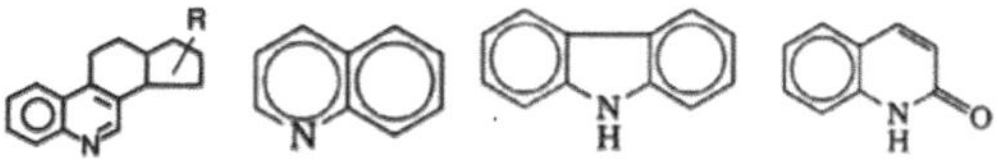
	Oxygen	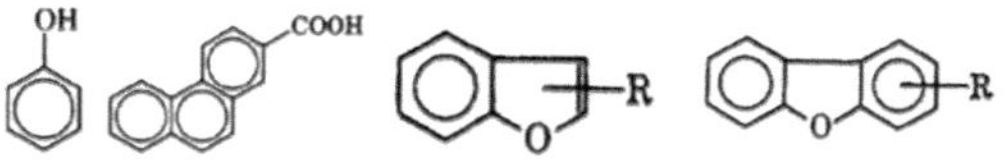
	Metals	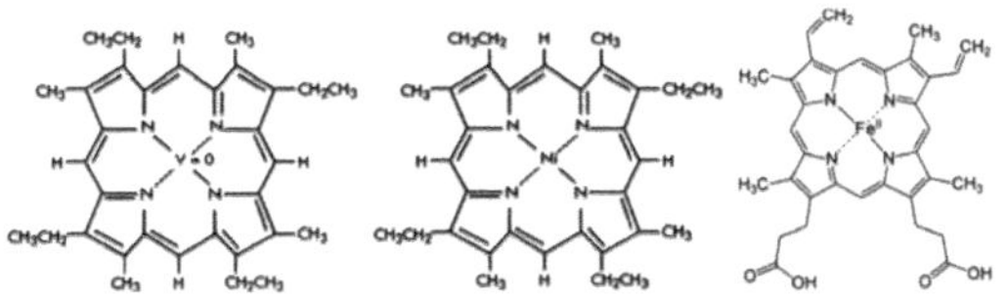

c. **Sulfur:** It is a main component of crude oil regardless of its source, and it originates essentially from constituents of living organisms (animals and plants). A part of sulfur may be available in the rock or core that yields hydrogen sulfide (H_2S) and even free sulfur. H_2S is usually produced by the decomposition of organic compounds under reservoir conditions as well as after the refinery processing (hydroprocessing). However, the majority of sulfur in crude oil usually exists as organic sulfur compounds, which vary in structure and molecular size; hence they are classified as thiols, organic sulfides, disulfides, polysulfides, as well as aliphatic and aromatic sulfur compounds (Stanislaus et al., 2010). The aromatic sulfur compounds are present mainly as benzothiophene, dibenzothiophene (DBT), and their alkylated derivatives. Organic sulfur compounds are available in all crudes and their abundance is based on the crude origin (Lobodin et al., 2014). Usually, aliphatic sulfur compounds are highly reactive in catalytic processes, while aromatic compounds have limited reactivity without a catalyst even at 400°C (Beck et al., 1997). The presence of sulfur in crude oil has drawn attention not only due to the offensive odor and corrosive nature, but also because of its impact on the environment. Hence, refineries play a crucial role in removing sulfur compounds from crude oil and refined fuels, which is compulsory to produce clean fuels (Stanislaus et al., 2010).

d. **Nitrogen and Oxygen:** Nitrogen compounds are naturally present in crude oil, Usually, nitrogen compounds are both basic and non-basic in nature, and they are known for their sticky nature and instability. They tend to exert competitive adsorption on the catalytic sites of the refining catalysts, and thus contribute to the difficulty in meeting clean fuel demands (Zeuthen et al., 2001; Kwak et al., 2001, Rana et al., 2018). Hence, nitrogen compounds are known for their detrimental role in the refining processes (Mills et al., 1950; Niquille-and Prins et al., 2007; Bauserman et al., 2014; Raghuveer et al., 2016). Nitrogen compounds, particularly pyrrole, make complexes with the transition metals, which act as a building block for the metal complexes (heterocyclic systems), as shown in Figure 1.6. Usually, basic nitrogen compounds (i.e., quinoline and acridine) are stronger inhibitors of the hydrodesulfurization (HDS) reactions than the non-basic moieties, i.e., carbazole and indole (Moses et al., 2007; Temel et al., 2010). However, the detrimental effect due to non-basic nitrogen species cannot be overlooked entirely (Stanislaus et al., 2010). Oxygen content in crude oil is relatively lower than nitrogen and sulfur. Oxygen compounds are relatively less polluting and easier to remove during the refining processes, mainly HDP, where oxygen is removed as water vapor under hydrodeoxygenation. Oxygen usually appears in the form of organic acids, alcohols, and basic aromatic molecules, such as furan. The N and O compounds increase with increasing distillation temperature (Figure 1.7), which indicates that a considerable amount of refining effort will be required to eliminate contaminants from a heavy fraction (i.e., 540 to 700°C) in order to make them marketable products (Snyder, 1970).

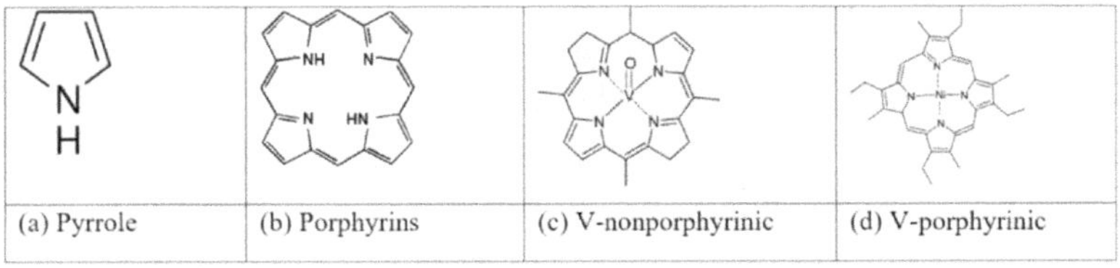

FIGURE 1.6 Pyrrole and its association with complex porphyrin structures.

Source: Reprinted with permission from Rana et al., 2013. © 2013 ASTM International.

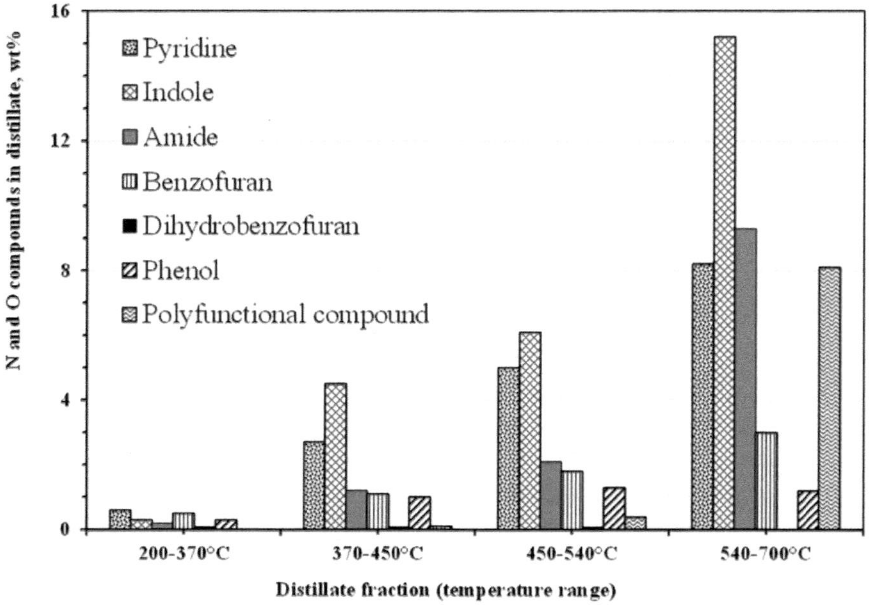

FIGURE 1.7 Nitrogen and oxygen compound types in distillates from Wilmington, California.

e. **Trace Metals:** All crude oils contain some inorganic constituents in the form of the metal complexes known as metalloporphyrins (Figure 1.6) (Yen, 1975). Nickel and vanadium are normally the most abundant metals in crude oil with Ni content ranging from 1 to 250 ppm and V from 10 to 1,500 (Dunning et al., 1960). Other transition metals are present in crude oils at much lower concentrations such as Na, Mg, Ca, Si, Al, P, Cu, Cd, and Pb (Rana et al., 2007; Rana et al., 2014). Generally, metals are concentrating more in the heavier fraction (asphaltenes and resins) than in saturated and aromatic fractions, as presented in Figure 1.8 (Ancheyta et al., 2007). Dunning and co-workers (1960) reported that from 3 to 10% of the vanadium content in most metal-rich oils is present as vanadium porphyrin complexes. They have also studied the nature of vanadium compounds along with the phase transformation (decomposition), chemical treatment (acid-base), variation in oxidation state, irradiation, and volatilities. The major analysis problem is due to the separation of metals compounds from the complex crude and residues, once the separation is carried out more precise analysis has been reported via spectroscopic techniques (Tynan and Yen, 1969; Yen, 1975; Mullins, 1985; Speight, 2014).

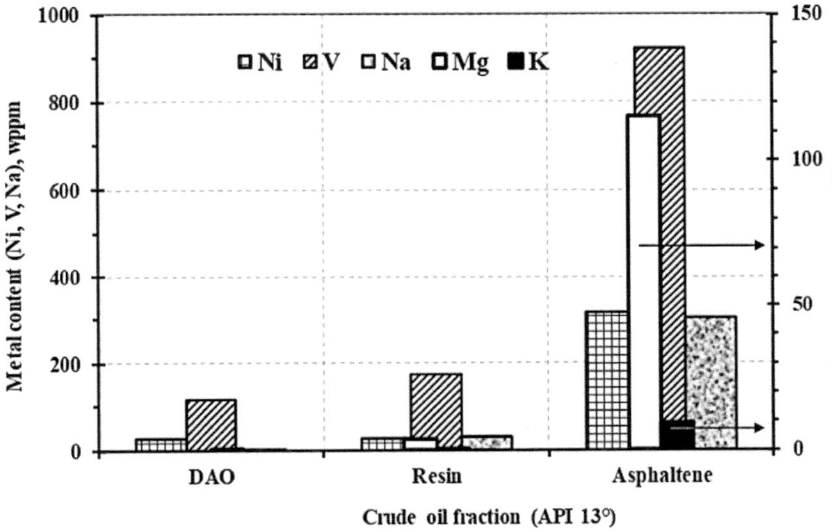

FIGURE 1.8 Content of metals in different fractions of a heavy crude (DAO, deasphalted oil).
Source: Reprinted with permission from Ancheyta et al., 2010. © Taylor & Francis.

1.4 CRUDE OIL COMPOSITION AND ITS PROPERTIES

Based on the chemical composition, crude oils possess different properties. The heavier the crude oil, the more complex is the composition, and the higher is the asphaltene content. In general, the relative amounts of the SARA fractions vary significantly in different crude types. For instance, as saturates decrease, aromatics, and asphaltenes increase. However, the resin content remains almost constant, which is related to the stability of the crude oil (Figure 1.9) (Ancheyta et al., 2010). As described above, the major heteroatoms in crude oil, S and N, exist as organic compounds, which are distributed based on their boiling point in each distillate fraction. The S and N content increases significantly with increasing fraction boiling point (Rana et al., 2007; Marafi et al., 2019). Usually, the lower the API gravity, the higher the boiling point of crude oil, hence API gravity is an indirect measurement of boiling point. Figure 1.10(a) shows that both S and N content is significantly high for low API gravity crude oil. On the other hand, both heteroatoms show an increasing tendency against each other, as shown in Figure 1.10(b). The scattered relationship between the heteroatoms is due to the wide range of data collected from extra light to extra heavy crude oils (65–5° API gravity) (Marafi et al., 2019).

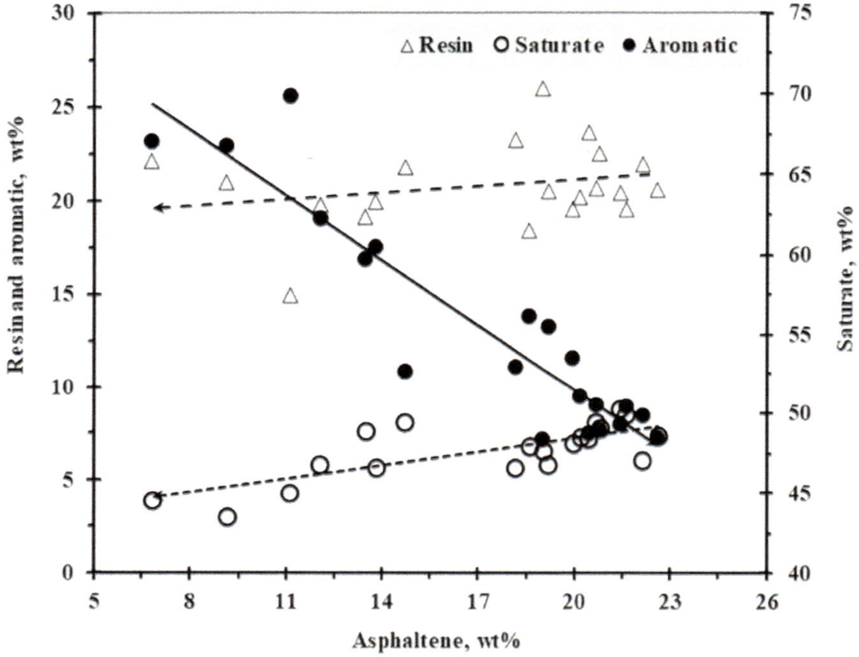

FIGURE 1.9 Relationship between hydrocarbon composition and asphaltenes.

Source: Reprinted with permission from Leyva et al., 2007. © American Chemical Society.

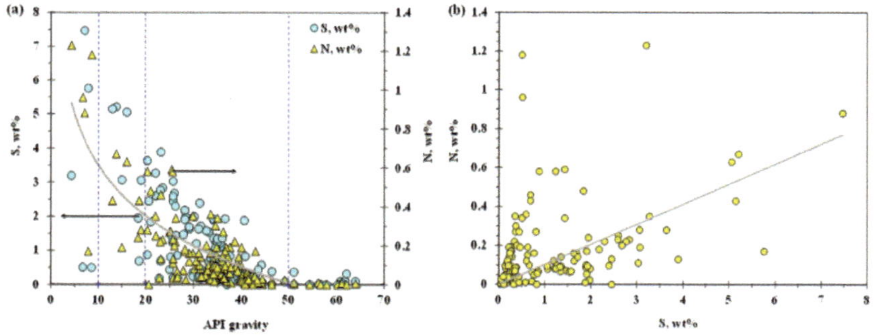

FIGURE 1.10 Effect of API gravity on S and N heteroatom content (a); and the relationship between S and N content (b) (Rana et al., 2013).

Source: Reprinted with permission from Rana et al., 2013. © ASTM International.

According to the API gravity, crude oils can be classified as super light (API > 50°), light (API = 50–35°), medium-light (API= 35–20°), heavy (API = 10–20), and extra heavy (API < 10°) (Ramirez et al., 2007). Usually, super light and light crude oils only contain S, N, and O heteroatoms, but have a negligible amount of metals (Ni and V) (Marafi et al., 2019). The medium-light and heavy crude oils have a considerable amount of Ni and V, which must be removed in the guard bed reactor in order to protect subsequent catalytic processes. The metal complexes are mainly associated with the asphaltene sheets, which enhances the polar nature of the asphaltene (Ancheyta et al., 2010) (Figure 1.11).

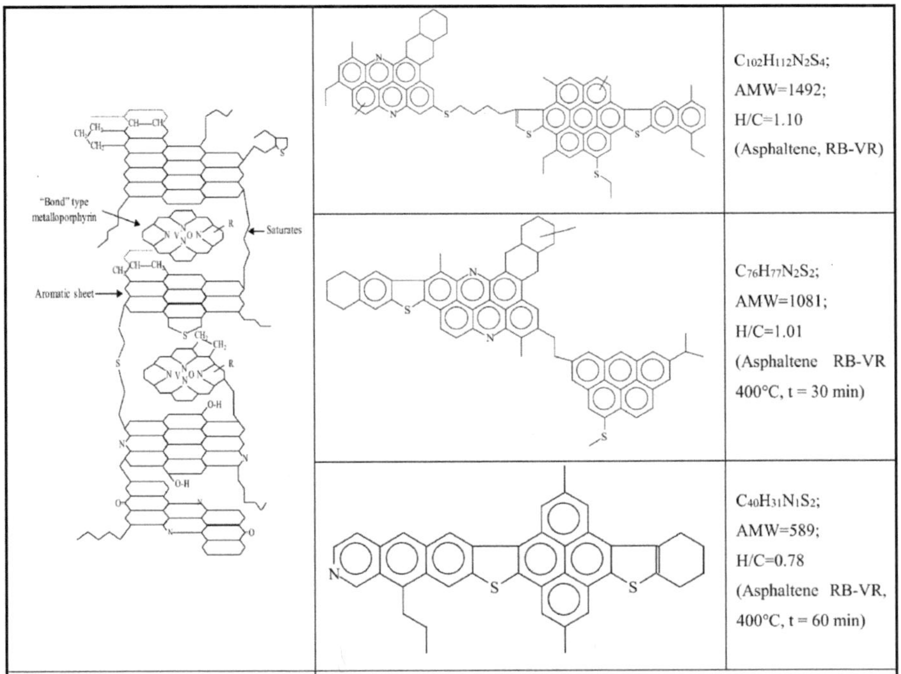

(a) Hypothetical molecular structure (Tynan and Yen, 1969; Rana et al., 2007)

(b) Molecules derived from NMR (Alhumaidan et al., 2017)

FIGURE 1.11 Hypothetical molecular structure of asphaltene derived from a Kuwaiti crude (RB-VR: Ratawi-Burgan vacuum residue).

A major challenge facing the utilization of crude oil remains to be the characterization of the feedstock at a molecular level. Sincere efforts and significant success have been reported by using an advanced analytical technique like Fourier Transform-Ion Cyclotron Resonance-Mass Spectrometry (FT-ICR-MS) for composition analysis (Kekalainen et al., 2013; McKenna et al., 2014; Witt and Timm, 2016; Rogel and Witt, 2017), sulfur speciation in heavy fractions (Purcell et al., 2007; Lobodin et al., 2014, 2015), identification of sulfur and nitrogen species (Chakravarthy et al., 2017), identification of reactive and non-reactive sulfur species (Lobodin et al., 2015), and detection of vanadium and nickel species (McKenna et al., 2014). Asphaltene, the heaviest fraction in heavy crude oil (Rogel and Witt, 2017), acts as a carrier for most heteroatoms and metals (Ni and V). An asphaltene molecule may be 4 to 5 nm in diameter, which is difficult to process in a refinery by the conventional catalytic systems (Marafi et al., 2019). Metals in the asphaltene aggregates are believed to be associated with the asphaltene sheets, making the asphaltene molecule heavier than its original structure (Speight, 2013), which requires catalysts that are more resistant to deactivation and have a balance between textural properties and number of active sites.

1.5 CONCLUSIONS AND FUTURE RESEARCH DIRECTIONS

Sustainable use of crude oil is a challenging task for scientists as well as refiners and the comprehensive molecular characterization of crude oil can be helpful in understanding its chemistry and help design proper and efficient refining catalysts. Compositional analysis of fossil fuels plays an important role in improving refinery operations and mitigating environmental problems. Although compositional analysis of crude oil is possible, the role of its components remains partially explainable, which may further limit the effective crude oil processing, that is indispensable for the production of clean fuels and accommodating environmental legislations and economic shortcomings. This is particularly important in view of the fact that refinery feedstock is gradually changing to low-quality heavy crudes, which require cost-efficient and green unconventional technologies for production, transport, and refining.

KEYWORDS

- **asphaltenes**
- **atmospheric residue**
- **final boiling point**
- **heavy oil**
- **hydrocarbon**
- **initial boiling point**
- **petroleum composition**
- **sulfur**

REFERENCES

AlHumaidan, F. S., Hauser, A., Rana, M. S., & Lababidi, H. M. S., (2017). NMR characterization of asphaltene derived from residual oils and their thermal decomposition. *Energy and Fuels, 31*(4), 3812–3820.

Alhumaidan, F. S., Hauser, A., Rana, M. S., & Lababidi, H. M. S., (2016). Impact of thermal treatment on asphaltene functional groups. *Energy Fuels, 30*(4), 2892–2903.

Alhumaidan, F. S., Hauser, A., Rana, M. S., Lababidi, H. M. S., & Behbehani, M., (2015). Changes in asphaltene structure during thermal cracking of residual oils: XRD study. *Fuel, 150*, 558–564.

Alhumaidan, F., Cresswell, D., & Garforth, A., (2011). Hydrogen storage in liquid organic hydride: Producing hydrogen catalytically from methylcyclohexane. *Energy Fuels, 25*(10), 4217–4234.

Ali, M., Pernazowski, H., & Haji, A., (1993). Nickel and vanadyl porphyrins in Saudi Arabian crude oils. *Energy Fuels, 7*, 179–183.

Ancheyta, J., Rana, M. S., & Furimsky, E., (2005). Hydroprocessing of heavy petroleum feeds: Tutorial. *Catal. Today, 109*(1–4), 3–15.

Ancheyta, J., Trejo, F., & Rana, M. S., (2010). Asphaltenes: Chemical transformation during hydroprocessing of heavy oils, CRC, New York.

Bauserman, J. W., Mushrush, G. W., Hardy, D. R., Willauer, H. D., Laskoski, M., & Keller, T. M., (2014). The removal of organic nitrogen compounds from unstable diesel fuels. *Petrol. Sci. Tech., 32*, 638–645.

Beck, J. S., Haag, W. O., Buonomo, F., Sanfilippo, D., Trifir, F., Arnold, H., et al., (1997). Organic reactions: Sections 4.1– 4.4. In: Ertl, G., Knozinger, H., & Weitkamp, J., (eds.), *Handbook of Heterogeneous Catalysis* (pp. 2123–2231).

Benjamin, I., (1997). Molecular structure and dynamics at liquid-liquid interfaces. *Annu. Rev. Phys. Chem., 48*(1), 407–451.

Biofuels Market-Segmented by Type, Feedstock, and Geography-Growth, Trends, and Forecast 2020–2025. https://www.mordorintelligence.com/industry-reports/biofuels-market (accessed on 25 October 2021).

Bissada, K. K., Tan, J., Szymczyk, E., Darnell, M., & Mei, M., (2016). Group-type characterization of crude oil and bitumen. Part I: Enhanced separation and quantification of saturates, aromatics, resins and asphaltenes (SARA). *Org. Geochem., 95*, 21–28.

BP. (2020). *Statistical Review of World Energy*. 69th edition. https://www.bp.com/content/dam/bp/business-sites/en/global/corporate/pdfs/energy-economics/statistical-review/bp-stats-review-2020-full-report.pdf (accessed on 25 October 2021).

Chakravarthy, R., Naik, G. N., Savalia, A., Kedia, J., Saravanan, C., Das, A. K., Sreedharan, U., & Gudasi, K. B., **(2017).** Simultaneous determination of boiling range distribution of hydrocarbon, sulfur, and nitrogen in petroleum crude oil by gas chromatography with flame ionization and chemiluminescence detections. *Energy Fuels, 31(*3), 3101–3110.

Dunning, H. N., Moore, J. W., Bieber, H., & Williams, R. B., (1960). Porphyrin, nickel, vanadium, and nitrogen in petroleum. *J. Chem. Eng. Data, 5*(4), 546–549.

Fish, R. H., & Komlenic, J. J., (1984a). Molecular characterization and profile identifications of vanadyl compounds in heavy crude petroleums by liquid chromatography/graphite furnace atomic absorption spectrometry. *J. Anal. Chem., 56*, 510–517.

Fish, R. H., Komlenic, J. J., & Wines, B. K., (1984b). Characterization and comparison of vanadyl and nickel compounds in heavy crude petroleums and asphaltenes by reverse-phase and size-exclusion liquid chromatography/graphite furnace atomic absorption spectrometry. *Anal. Chem., 56*(13), 2452–2460.

Fleischer, E. B., (1963). The structure of Ni etioporphyrine: I. *J. Am. Chem. Soc., 85*, 146–148.

Gianotti, E., Taillades-Jacquin, M., Roziere, J., & Jones, D. J., **(2018).** High-purity hydrogen generation via dehydrogenation of organic carriers: A review on the catalytic process. *ACS Catal., 8(*5), 4660–4680.

Gray, M. R., (1994). *Upgrading Petroleum Residue and Heavy Oils*. Marcel Dekker, New York, (1994).

Hauser, A., Bahzad, D., Stanislaus, A., & Behbahani, M., (2008). Thermogravimetric analysis studies on the thermal stability of asphaltenes: Pyrolysis behavior of heavy oil asphaltenes. *Energy and Fuels, 22*(1), 449–454.

Hayes, P. C., & Anderson, S. D., (1986). Hydrocarbon group type analyzer system for the rapid determination of saturates, olefins, and aromatics in hydrocarbon distillate products. *Anal. Chem., 58*, 2384–2388.

Huggins, R. A., (2016). *Energy Storage: Fundamentals, Materials and Applications*. Springer, New York.

Kekalainen, T., Pakarinen, J. M. H., Wickström, K., Lobodin, V. V., McKenna, A. M., & Janis, J., **(2013).** Compositional analysis of oil residues by ultrahigh-resolution Fourier transform ion cyclotron resonance mass spectrometry. *Energy Fuels, 27(*4), 2002–2009.

Kilpatrick, P. K., (2012). Water-in-crude oil emulsion stabilization: Review and unanswered questions. *Energy Fuels, 26*(7), 4017–4026.

Kwak, C., Lee, J. J., Bae, J. S., & Moon, S. H., (2001). Poisoning effect of nitrogen compounds on the performance of $CoMoS/Al_2O_3$ catalyst in the hydrodesulfurization of DBT, 4-MDBT and 4. 6-DMDBT. *Appl. Catal. B: Env., 35*, 59–68.

Leyva, C., Rana, M. S., Trejo, F., & Ancheyta, J., (2007). On the use of acid-base-supported catalysts for hydroprocessing of heavy petroleum. *Ind. Eng. Chem. Res., 46*, 7448.

Lobodin, V. V., Juyal, P., McKenna, A. M., Rodgers, R. P., & Marshall, A. G., **(2014).** Silver cationization for rapid speciation of sulfur-containing species in crude oils by positive electrospray ionization Fourier transform ion cyclotron resonance mass spectrometry. *Energy Fuels, 28(*1), 447–452.

Lobodin, V. V., Robbins, W. K., Lu, J., & Rodgers, R. P., (**2015**). Separation and characterization of reactive and non-reactive sulfur in petroleum and its fractions. *Energy Fuels, 29(*10), 6177–6186.

Mansoori, G. A., (1997). Modeling of asphaltene and other heavy organic depositions. *J. Pet. Sci. Eng., 17*(1, 2), 101–111.

Mansoori, G. A., (2005). *Principles of Nanotechnology: Molecular-Based Study of Condensed Matter in Small Systems.* World Scientific, NJ.

Marafi, A., Albazzaz, H., & Rana, M. S., (2019). Hydroprocessing of heavy residual oil: Opportunities and challenges. *Catal. Today, 329*, 125–134.

McCarthy, J. E., & Copeland, C., (2011). *EPA's Regulation of Coal-Fired Power: Is a "Train Wreck" Coming?, CRS Report for Congress R41914.* https://www3.epa.gov/region1/npdes/merrimackstation/pdfs/ar/AR-1162.pdf (accessed on 25 October 2021).

McKenna, A. M., Williams, J. T., Putman, J. C., Aeppli, C., Reddy, C. M., Valentine, D. L., Lemkau, K. L., et al., **(2014).** Unprecedented ultrahigh-resolution FT-ICR mass spectrometry and parts-per-billion mass accuracy enable direct characterization of nickel and vanadyl porphyrins in petroleum from natural seeps. *Energy Fuels, 28(*4), 2454–2464.

Mills, G. A., Boedeker, E. R., & Oblad, A. G., (1950). Chemical characterization of catalysts. I. Poisoning of cracking catalysts by nitrogen compounds and potassium ion. *J. Am. Chem. Soc., 72*(4), 1554–1560.

Modisha, P. M., Ouma, C. N. M., Garidzirai, R., Wasserscheid, P., & Bessarabov, D., (**2019**). The prospect of hydrogen storage using liquid organic hydrogen carriers. *Energy Fuels, 33*(4), 2778–2796.

Moore, F., & Richmond, G., (2008). Integration or segregation: How do molecules behave at oil/water interfaces? *Accounts Chem. Res., 41*(6), 739–748.

Moses, P. G., Hinnemann, B., Topsøe, H., & Nørskov, J. K., (2007). The hydrogenation and direct desulfurization reaction pathway in thiophene hydrodesulfurization over MoS_2 catalysts at realistic conditions: A density functional study. *J. Catal., 248*, 188–203.

Mullins, O. C., (1985). Sulfur and nitrogen in asphaltenes and related material quantified by XANES spectroscopy. In: *Asphaltene*. Plenum, New York.

Niquille-R, A., & Prins, R., (2007). Influence of nitrogen-containing compounds on the hydrodesulfurization of 4,6-dimethyl dibenzothiophene over Pt, Pd, and Pt-Pd on amorphous silica-alumina catalysts. *Ind. Eng. Chem. Res., 46*(12), 4124–4133.

O'Connell, A., Kousoulidou, M., Lonza, L., & Weindorf, W., (2019). Considerations on GHG emissions and energy balances of promising aviation biofuel pathways. *Renew. Sust. Energy Rev., 101*, 504–515.

Pena, A. A., Hirasaki, G. J., & Miller, C. A., (2005). Chemically induced destabilization of water-in-crude oil emulsions. *Ind. Eng. Chem. Res., 44*, 1139–1149.

Purcell, J. M., Juyal, P., Kim, D. G., Rodgers, R. P., Hendrickson, C. L., & Marshall, A. G., (2007). Sulfur speciation in petroleum: Atmospheric pressure photoionization or chemical derivatization and electrospray ionization Fourier transform ion cyclotron resonance mass spectrometry. *Energy Fuels, 21*(5), 2869–2874.

Raghuveer, C. S., Thybaut, J. W., & Marin, G. B., (2016). Pyridine hydrodenitrogenation kinetics over a sulphided NiMo/γ-Al_2O_3 catalyst. *Fuel, 171*, 253–262.

Ramírez, J., Rana, M. S., & Ancheyta, J., (2007). Characteristics of heavy oil hydroprocessing catalysts. In: Ancheyta, J., & Speight, J. G., (eds.), *Hydroprocessing of Heavy Oils and Residua* (pp. 121–190). Taylor & Francis, New York.

Rana, M. S., & Al Humaidan, F. S., (2016). Statistical data on worldwide coal reserves, production, consumption, and future demand. In: Raizi, M. R., & Gupta, R., (eds.), *Coal Production and Processing Technology* (pp. 31–50). CRC Press, Boca Raton, FL.

Rana, M. S., Al-Barood, A., Brousli, R., Al-Hendi, A. W., & Mustafa, N., (2018). Effect of organic nitrogen compounds on deep hydrodesulfurization of middle distillate. *Fuel Process. Technol., 177*, 170–178.

Rana, M. S., Ancheyta, J., Riazi, M., & Marafi, M., (2013). Future direction in petroleum and natural gas refining. In: Riazi, M. R., et al., (eds.), *American Society for Testing and Materials (ASTM) Manual Series MNL 58* (pp. 769–799). ASTM International, West Conshohocken, PA. https://doi.org/10.1520/MNL5820131214533.

Rana, M. S., Ancheyta, J., Sahoo, S. K., & Rayo, P., (2014). Carbon and metal deposition during the hydroprocessing of Maya crude oil. *Catal. Today, 220–222*, 97–105.

Rana, M. S., Samano, V., Ancheyta, J., & Diaz, J. A. I., (2007). Review of recent advances on process technologies for upgrading of heavy oils and residua. *Fuel, 86*(9), 1216–1223.

Rana, M. S., Vinoba, M., & AlHumaidan, F. S., (2017). Sustainability Challenges in oil and gas development in the middle east and north Africa. *Curr. Sustain. Energy Rep., 4*, 232–244.

Riazi, M. R., Rana, M. S., & Pena, D. J. L., (2013). Worldwide statistical data on proved reserves, production and refining capacities of crude oil and natural gas. In: Riazi, M. R., et al., (eds.), *American Society for Testing and Materials (ASTM) Manual Series MNL 58* (pp. 33–78). ASTM International, West Conshohocken, PA.

Rogel, E., & Witt, M., (2017). Asphaltene characterization during hydroprocessing by ultrahigh-resolution Fourier transform ion cyclotron resonance mass spectrometry. *Energy Fuels, 31*(4), 3409–3416.

Snyder, L. R., (1970). Petroleum nitrogen compounds and oxygen compounds. *Acc. Chem. Res, 3*(9), 290–299.

Speight, J. G., & Long, R. B., (1996). The concept of asphaltenes revisited. *Fuel Sci. and Technol. Int., 14*, 1–12.

Speight, J. G., (2004). Petroleum asphaltenes - part 1: Asphaltenes, resins and the structure of petroleum. *Oil Gas Sci. Technol. Rev. IFP, 59*(5), 467–477.

Speight, J. G., (2013). *Heavy and Extra-Heavy Oil Upgrading Technologies*. Elsevier.

Speight, J. G., (2014). *The Chemistry and Technology of Petroleum* (5th edn.). CRC Press, Boca Raton.

Speight, J. G., Long, R. B., & Trowbridge, T. D., (1984). Factors influencing the separation of asphaltenes from heavy petroleum feedstocks. *Fuel, 63*, 616–620.

Stanislaus, A., Marafi, A., & Rana, M. S., (2010). Recent advances in the science and technology of ultra-low sulfur diesel (ULSD) production. *Catal. Today, 153*(1, 2), 1–68.

Takatsuka, T., Wada, Y., Hirohama, S., & Fukui, Y., (1989). A prediction model for dry sludge formation in residue hydroconversion. *J. Chem. Eng. Jpn., 22*, 298–303.

Temel, B., Tuxenl, A. K., Kibsgaard, J., Topsøe, N. Y., Hinneman, B., & Knudsen, K. G., (2010). Atomic-scale insight into the origin of pyridine inhibition of MoS_2-based hydrotreating catalysts. *J. Catal., 271*, 280–289.

Tseng, C., (2002). Effects of hydrogen addition on methane combustion in a porous medium burner. *Int. J. Hydrog Energy, 27*, 699–707.

Tynan, E. C., & Yen, T. F., (1969). Association of vanadium chelates in petroleum asphaltenes as studied by ESR. *Fuel, 43*, 191.

US Energy Consumption, (2009). Maxwell School, Syracuse University, Syracuse, NY. http://wilcoxen.maxwell.insightworks.com/pages/804.html (accessed on 25 October 2021).

Venkatesh, A., Jaramillo, P., Griffn, W. M., & Matthews, H. S., (2012). Implications of near-term coal power plant retirement for SOx and NOx and life cycle GHG emissions. *Environ. Sci. Technol., 46*(18), 9838–9845.

Wiehe, I. A., & Liang, K. S., (1996). Asphaltenes, resins, and other petroleum macromolecules. *Fluid Phase Equilibr., 117*, 201–210.

Wiehe, I. A., (1992). A solvent-resid phase diagram for tracking resid conversion. *Ind. Eng. Chem. Res., 31*(2), 530–536.

Witt, M., & Timm, W., **(2016).** Determination of simulated crude oil mixtures from the North Sea using atmospheric pressure photoionization coupled to Fourier transform ion cyclotron resonance mass spectrometry. *Energy and Fuels, 30(*5), 3707–3713.

World Oil Outlook, (2009). *Organization of the Petroleum Exporting Countries.* Vienna, Austria, www.opec.org (accessed on 25 October 2021).

Yen, T. F., (1975). The role of trace metals in petroleum. *Ann. Arbor. Science*, USA.

Zeuthen, P., Knudsen, K. G., & Whitehurst, D. D., (2001). Organic nitrogen compounds in gas-oil blends and their importance to hydrotreatment. *Catal. Today, 65*, 307–314.

CHAPTER 2

Biotechnology in the Refinery

NOUR SH. EL-GENDY[1,2] and JAMES G. SPEIGHT[3]

[1]*Department of Process Design and Development, Egyptian Petroleum Research Institute (EPRI), Nasr City, Cairo PO–11727, Egypt*

[2]*Center of Excellence, October University for Modern Sciences and Arts (MSA), 6th of October City, Giza, PO–12566, Egypt,*
E-mail: nourepri@yahoo.com

[3]*CD&W Inc., Laramie, Wyoming, USA*

2.1 INTRODUCTION

The major sources of energy in the world are crude oil and natural gas, and this is likely to be the *status quo* for the next 50 years (Speight, 2011). In fact, the world still depends mainly on fossil fuels (crude oil, natural gas, and coal) for 81.1% of its total energy need, with crude oil being the main source of energy (31.3%). The United States Energy Information Administration (EIA) reported that the world demand for oil would exceed 123 Mbd (million barrels per day) by 2025. In addition, the organization of petroleum exporting countries (OPEC) expected that within 2025, the oil production would be less than half of the request, i.e., approximately 61 MMbd (Hutzler et al., 2003). Petroleum is universally classified into different types according to the American Petroleum Institute (API) gravity (Figure 2.1). Thus, there is light crude oil (API 36–38) and heavy crude oil (API $\leq$ 20) (Head et al., 2006; Speight, 2014). It has been estimated that; 2 trillion barrels (2×10^{12} bbls) of conventional light crude oil and 5 trillion barrels (5×10^{12} bbls) of heavy crude oil remain in reservoirs worldwide (Al-Sulaimani et al., 2011a, b; Lavania et al., 2012; Demirbas et al., 2016; Speight and Islam, 2016). The largest heavy and extra-heavy crude oil reserves are found in the Orinoco oil belt of Venezuela, whereas bitumen (API <10°) is found in the Athabasca oil sands (also called Athabasca tar sands) in Alberta-Canada, while Olenik oil sands are found in Siberia-Russia (BGR, 2015).

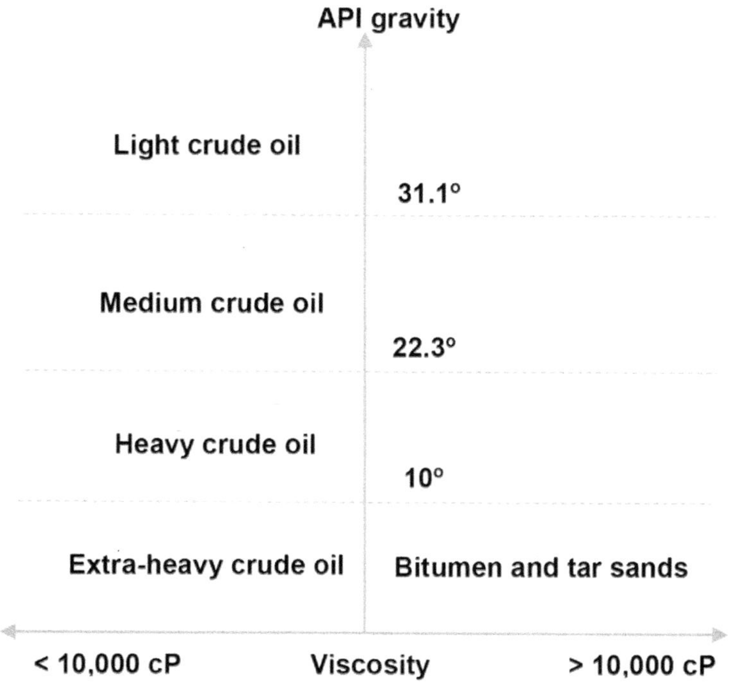

FIGURE 2.1 Classification of crude oils according to API gravity and viscosity (Speight and El-Gendy, 2017).

The increasing global energy demand will eventually lead to depletion of the light crude oil reserves and an increase in the production and refining of heavy crude oil. Unfortunately, heavy crude oil cannot be produced, transported, and refined by conventional methods that are suitable for light crude oil due to its challenging physicochemical properties (Speight, 2015, 2017). Heavy crude oils, on contrary to light ones, are thick, black, sticky fluids with higher viscosity and density and higher content of sulfur, nitrogen, and metals (predominantly nickel and vanadium). Moreover, heavy crudes have higher content of aromatic compounds and polar high molecular weight resins and asphaltenes (Figure 2.2) (Leòn and Kumar, 2005; Speight, 2014; Al-Bahry et al., 2016; El-Gendy and Nassar, 2018). The high viscosity together with the flocculation and deposition of asphaltenes make production and refining processes more difficult. High viscosity significantly hampers pumping, transportation, refining, and handling of heavy crude oil, and common methods used to overcome problems associated with the high viscosity (heating, dilution, and chemical additives) are expensive and require specialized equipment and/or safety procedures (Speight, 2017). The presence of sulfur, nitrogen, and metals leads to the formation of coke and poisons the refining and upgrading catalysts (Ancheyta and Speight, 2007; Speight, 2013a, b; Ancheyta, 2016), that comes with strict worldwide environmental regulations, which require significant decrease in the emissions of the environmentally problematic sulfur dioxide (SO_2) and nitrogen oxides (NOx) (Ayala et al., 2007; Nassar et al., 2021a). In addition, refining of heavy crudes produces lower yields of refined fuels as compared to light crudes. To remove those pollutants and to obtain cleaner fuels, the traditional refineries must work under severe conditions of higher temperature and pressure and must use more expensive catalysts, which would lead to an increase of the refining cost. Thus, heavy oils require deep conversion and substantial upgrading involving deep thermal and catalytic processes and significant hydrogen inputs to obtain fractions suitable to produce transportation fuels (Speight, 2013a, b, 2014; Ancheyta, 2016; Speight, 2017; Nassar et al., 2021b).

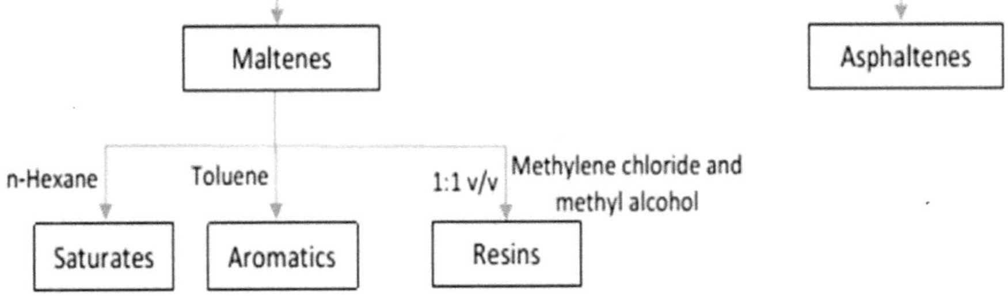

FIGURE 2.2 Major constituents of crude oil (i.e., SARA fractions) (Speight and El-Gendy, 2017).

The environmental, economic, and technical shortcomings associated with heavy crude oils production, urge for evolution of new technologies that can efficiently overcome these hurdles. Petroleum biotechnology has emerged as a green and economic approach to enable the oil industry to cope with environmental legislations and meet the increasing global demand for cleaner fuels. As compared to the conventional thermochemical treatments, biotechnological processes are environmentally friendly, cost-effective, and can be more efficient. In principle, biotechnology can be potentially applied throughout the value chain of the oil industry, starting with the exploration all the way through recovery, transportation, refining, and waste treatment (Montiel et al., 2009; Speight and El-Gendy, 2017).

2.2 MICROBIAL ENHANCED OIL RECOVERY (MEOR)

Microbial enhanced oil recovery (MEOR) is an example of the tertiary EOR techniques, which applies microorganisms and/or bioproducts to enhance oil production (Lazar et al., 2007; Al-Sulaimani 2011a, b; Sun et al., 2011). While MEOR is not specially a refinery process, it can play a role in refining insofar a change made to the heavy oil prior to and/or during recovery, can have a beneficial effect on the refinability of the heavy oil. Hence, it is worthy of consideration here. MEOR is considered as a non-hazardous and economically viable strategy. Other advantages have attracted attention, such as high efficiency, easy process, robust adaptableness, and eco-friendly (Jack, 1991; Al-Sulaimani et al., 2011a, b; Zhang et al., 2014; Gassara et al., 2015; Daryasafar et al., 2016). In addition, microorganisms can be maintained on low-cost renewable raw materials (Gao and Zekri, 2011; Karimi et al., 2012; Zhang et al., 2014).

The MEOR process utilizes reservoir microorganisms or other specially selected natural bacteria for the production of some specific metabolites which enhance oil recovery (Speight, 2014). From a microbiologist's viewpoint, MEOR is, in some respects, similar to *in situ* bioremediation practices (Speight and Arjoon, 2012). In the MEOR process, nutrients, together with indigenous or added microbes, are injected into the oil-bearing formation to promote (i) *in situ* microbial growth and/or product accumulation, which would mobilize additional oil to producing wells through reservoir re-pressurization, (ii) interfacial tension/oil viscosity reduction, and (iii) discriminating plugging of the most permeable zones (Bryant et al., 1989; Bryant and Lindsey, 1996; Bryant et al., 1998). Alternatively, *ex situ* production of oil-mobilizing bioproducts occurs via fermentation and is then injected into the reservoir.

Usually, essential microbial nutrients (i.e., the C, N, and P sources) are injected to stimulate microbial growth, some of which may be indigenous to the oil-bearing formation. The microbes generate biosurfactants and carbon dioxide that help to displace the oil and, since growth occurs at exponential rates, the process generates considerable amounts of biosurfactants that assist in the production of the oil.

The application of biosurfactants in MEOR is very promising because of low toxicity, biodegradability, and effectiveness at different conditions of temperature, pressure, salinity, and pH (Simpson et al., 2011; Speight and El-Gendy, 2017). Biosurfactants can alter the surface wettability of solids, reduce the interfacial tension between formation water and crude oil (Vaz et al., 2012), and increase the emulsification of crude oil (Zou et al., 2014). Other products, including the acids (fatty acids of low molecular weight), gases (carbon dioxide, methane, and hydrogen), and organics (polymers, alcohols, aldehydes), also play significant roles in decreasing oil viscosity. In addition, the heavy component of crude oil may be potentially biodegraded, with a concomitant decrease in oil viscosity and consequent enhancement in oil quality (Yang and Lou, 1997; Lazar et al., 2007; Liu et al., 2012; Daryasafar et al., 2016).

Abdel-Waly (1999) found that indigenous bacterial communities in reservoirs could influence the viscosity and pH of formation water, viscosity, and SARA fractions of crude oil and oil/water interfacial tension (Abdel-Waly, 1999). Examples for the various microbial strains that are capable of biosurfactants production and biodegradation of heavy oil fractions for enhancing oil recovery include; *Bacillus licheniformis*, *Bacillus subtilis*, *Bacillus sonorensis*, and *Bacillus mojavensis* (Jinfeng et al., 2005; McInerney et al., 2005). McInerney et al. (2005) explored the ability of various strains of *Bacillus licheniformis*, *Bacillus subtilis*, *Bacillus sonorensis* and *Bacillus mojavensis* to produce biosurfactants and degrade heavy oil fractions for enhancing oil recovery (Jinfeng et al., 2005). She et al. (2011) used biosurfactant-producing *Bacillus* strains to *in situ* degrade crude oil and enhance the production from a petroleum reservoir in the Daqing oil field, whereas the biosurfactants produced by *B. subtilis* W19 reduced the interfacial tension from 20.9 to 1.85 mN/m. In core flooding tests, a mixture of diluted biosurfactants (0.4 g/L) with synthetic surfactants (v/v, 75:25) was used for oil displacement and 28% oil recovery (Souayeh et al., 2014). Liu et al. (2015) found that surfactants produced by *Bacillus subtilis* BS-37 were stable in alkaline environments with high concentration of minerals (50 g/L NaCl and 110 g/L $CaCl_2$) at high temperature (120°C). Moreover, the crude oil displacement efficiency was 13.48% making the material attractive for enhancing oil recovery (Liu et al., 2015).

2.3 BIOTRANSFORMATION OF HEAVY CRUDE OIL

Despite the fact the structure of asphaltenes has not been fully elucidated, it is widely accepted that they consist of interacting systems of polyaromatic sheets bearing alkyl side-chains (Figure 2.3). The asphaltene constituents of heavy crude oils and tar sand bitumen has extremely complex and variable molecular structures containing sulfur (0.3–10.3%), oxygen (0.3–4.8%), nitrogen (0.6–3.3%), and metals such as Fe, Ni, and V in a small amount, with an average molecular weight between 600 and 2,000,000 (Tavassoli al., 2012). Asphaltenes are also considered as the product of complex heteroatomic aromatic macrocyclic structures polymerized through sulfide linkages (El-Gendy et al., 2006; Ali et al., 2012; El-Gendy et al., 2014a).

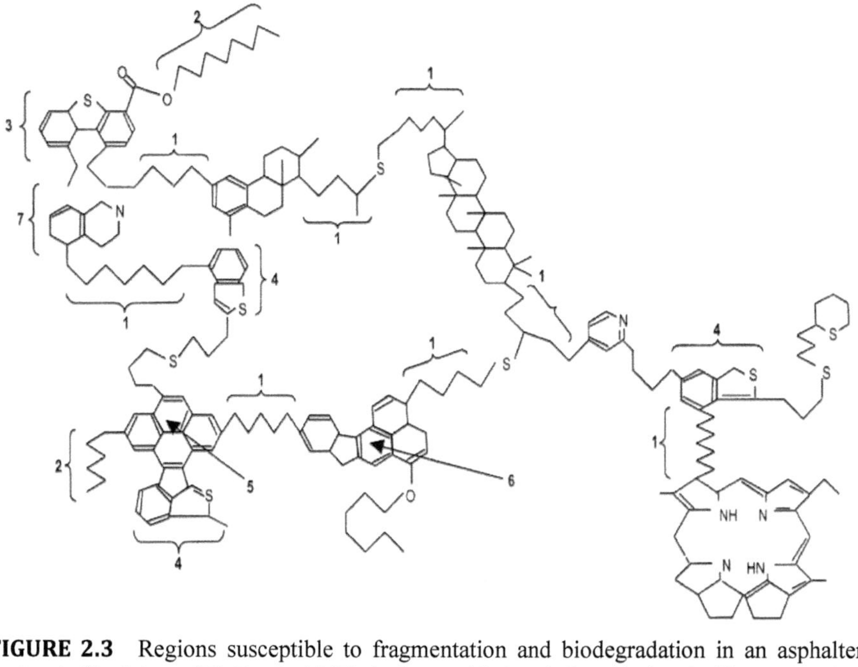

FIGURE 2.3 Regions susceptible to fragmentation and biodegradation in an asphaltene molecule (Speight and El-Gendy, 2017). 1: photooxidation; 2: β-oxidation; 3: dibenzothiophene pathway; 4: similar to dibenzothiophene pathway; 5: pyrene pathway; 6: similar to that of benzo(a)pyrene pathway; 7: similar to carbazole pathway.

The efficient recovery and processing of heavy crude oils and tar sand bitumen are hampered by the presence of high concentration of asphaltenes as they are one of the major contributors to crude oil viscosity (El-Gendy, 2001, 2004; El-Gendy et al., 2006; Bachman et al., 2014; Speight, 2014). Asphaltenes enhance the formation of emulsions, polymers, and coke (Vazquez-Duhalt et al., 2002; Hernández-López et al., 2015). The asphaltene constituents have drawn considerable attention due to problems caused by their detrimental effects in the extraction, transportation, and processing of residua because of their viscous and flocculating nature and their relative resistance to biodegradation following spills (Speight and Arjoon, 2012; Speight, 2014). Moreover, high asphaltene content can prevent the transfer of biodegradable components in oil droplets toward the oil-bacteria interface and reduce the rate and degree of degradation of biodegradable components (Uraizee et al., 1998). Thus, the biotransformation or upgrading of asphaltenes and other aromatic compounds in crude oil and other heavy hydrocarbon (HC) streams is of interest in the petroleum industry to maximize crude oil usage and minimize waste production (Gupta and Gera, 2015). The biodegradation of asphaltenes can markedly reduce the viscosity of crude oil by breaking asphaltenes into smaller molecules and cutting internal aliphatic linkages (sulfides, esters, and ethers) of an asphaltene molecule (Peng et al., 1997; El-Gendy, 2001, 2004). This will satisfy the need to extract heavier crude oils and to produce higher amounts of light fractions by thermal processes such as visbreaking and thermal cracking. Therefore, screening bacteria with the ability to degrade asphaltenes in crude oil is a component of research into MEOR in high-asphaltene heavy oil reservoirs.

Given their chemical properties, mass transfer limitations are expected in aqueous reactions and biotransformation rates are limited for enzymatic biocatalysts and, in the case of whole cells, transport across the cell wall may be impossible (Leòn and Kumar, 2005).

Despite these difficulties, there is a real potential for the bacterial transformation of these complex high molecular weight constituents of the crude oil. This is possible because the asphaltene constituents contain carbon, hydrogen, sulfur, nitrogen, and oxygen, which are necessary elements for the survival of microorganisms, besides, the complex structure of asphaltene offers several possible routes to biotransformation (Figure 2.3).

The asphaltene fraction of Castilla crude oil which is known to be rich in heavy metals, was reported to be treated with three different hemoproteins, namely chloroperoxidase, cytochrome C peroxidase, and lignin peroxidase in both aqueous buffers and organic solvents. Only chloroperoxidase was effective in eliminating the Soret peak and removed approximately 27% and 53% of V and Ni, respectively from the petroporphyrin-rich fractions and asphaltene constituents (Mogollón et al., 1998). According to Fourier transform infrared spectroscopy (FTIR), chemically modified *cytochrome c* catalyzed the oxidation of carbon and sulfur atoms in the asphaltene molecules (Garcia-Arellano et al., 2004). The enzymatic treatment of asphaltene constituents is an interesting alternative to heavy metal removal in order to reduce catalyst poisoning during hydrotreatment (HDT) and cracking processes. Five asphaltene-degrading bacterial strains were obtained from crude oil and polluted soil samples collected from Dorood oilfield in the south of Iran. Maximum degradation efficiency of 46% and 48% was achieved by *B. lentus* and a consortium of five selected bacterial isolates, respectively. Statistical optimization of asphaltene biodegradation was successfully carried out and the optimum values of pH-6.7, 76 g/L salinity and 22 g/L asphaltene concentration for asphaltene biodegradation at 40°C (104°F) were obtained for pure cultures of *B. lentus* (Tavassoli et al., 2012).

The naphthalene degradation pathway (Figure 2.4) serves as the basis to develop a biodearomatization process. It is suggested that if step 5 (Figure 2.4) is interrupted, a more linear oxidized HC with the same carbon content as naphthalene could be obtained. This would both reduce the aromaticity of the compound and generate HCs that could be hydrogenated under milder conditions. The naphthalene degradation pathway has been characterized mainly in *Pseudomonas*, *Sphingomonas*, *Burkholderia*, *Ralstonia*, *Rahnella*, *Polaromonas*, *Comamonas*, *Rhodococcus*, *Nocardioides*, and *Mycobacterium* species (Speight and El-Gendy, 2017).

However, caution is advised in the interpretation of the asphaltene-related data since most of the studies on asphaltene biodegradation depend on the gravimetrical determination of the asphaltene content by precipitation (using n-alkane precipitant) and, thus, the reported changes may be attributed to the disruption of the asphaltene-resin-oil matrix by the production of surfactants during bacterial growth, liberating the trapped HCs. There are also the claims that the constituents of the asphaltene fraction do not support bacterial growth, and no changes in asphaltene content are observed after bioconversion (BDC) of heavy oil and the asphaltene constituents (Lacotte et al., 1996; Thouand et al., 1999). However, Ali et al. (2012, 2014), El-Gendy et al. (2014), and El Mahdi et al. (2016) reported the biodegradation of asphaltenes.

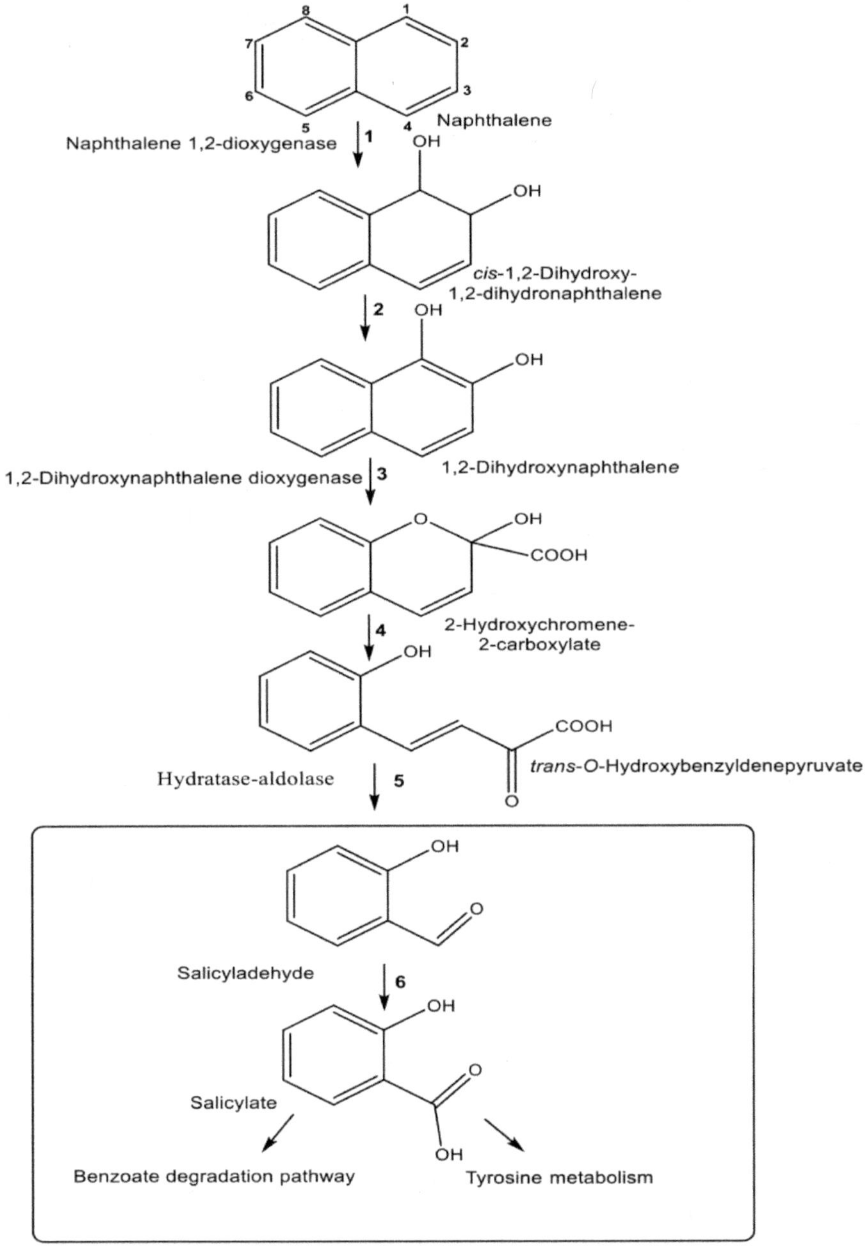

FIGURE 2.4 Naphthalene biodegradation pathway (Speight and El-Gendy, 2017).

2.4 BIOREFINING OF CRUDE OIL

Microorganisms and enzymes can be used to catalyze specific biochemical reactions on crude oil and refining streams that make the basis for the development of biorefining technology (Figure 2.5) (Le Borgne and Quintero, 2003). Due to the revolution in protein and genetic engineering, the study of extremophilic microorganisms, development of non-aqueous biocatalysts and nano-biocatalysts, biotechnology became interesting to companies developing petroleum biorefining technologies. The majority of research in the area has focused on biodesulfurization (BDS), biodenitrogenation (BDN), and biodemetallization (BDM) (Speight and El-Gendy, 2016, 2017).

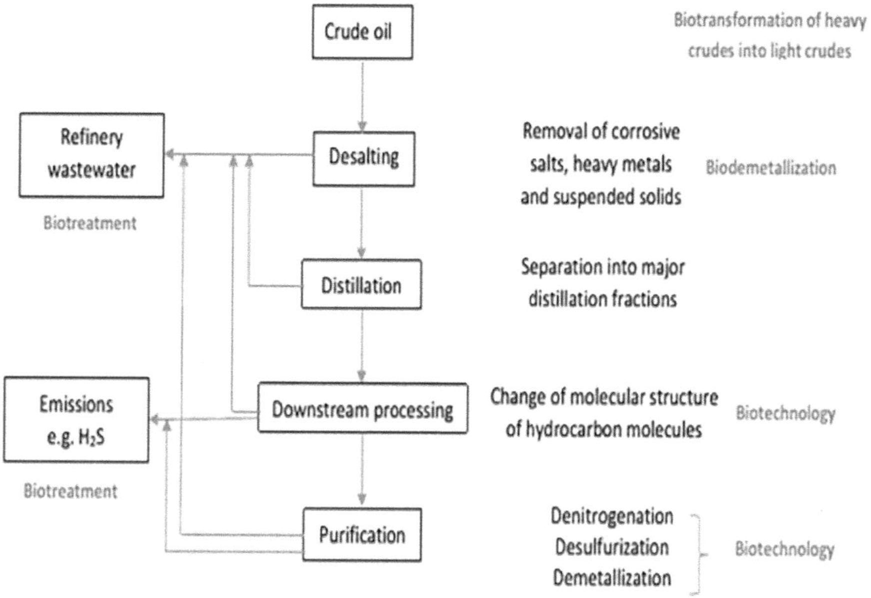

FIGURE 2.5 A simplified flow chart for potential applications of petroleum biotechnology during the processing of heavy crude oil (Speight and El-Gendy, 2017).

2.4.1 *BIODESULFURIZATION (BDS)*

Sulfur-which is the third most abundant element in crude oil after carbon and hydrogen-is a major concern for refiners and is used as a determinant of the value of crude oils. Heavy crude oil and bitumen contain 3 to 6% sulfur, and sulfur must be removed before oils are used as a refinery feedstock.

The combustion of sulfur-containing fuels leads to the increased emissions of sulfur oxides (SOx) which are major contributors to acid rain and also particulate matter. It has been reported that the total particulate matter emissions from diesel-fueled vehicles are proportional to the diesel sulfur content (Stanislaus et al., 2010; Mohebali and Ball, 2016). If desulfurization occurs on the crude feedstocks before they enter the refinery system, this would minimize the downstream desulfurization costs through reduced catalyst poisoning and reduced hydrogen usage (El-Gendy and Nassar, 2018). Viscous, high-density crude oils normally contain higher amounts of more complex sulfur compounds, with aliphatic acyclic sulfides (thioethers) and cyclic sulfides (thiolanes) being most amenable to hydrodesulfurization (HDS) and thermal treatments. Aromatic sulfur compounds, such as thiophenes and the benzo-derivatives (e.g., benzothiophenes, dibenzothiophenes (DBTs), and benzonaphthothiophenes) are considerably recalcitrant for HDS and thermal conversion (Gray et al. 1996; El-Gendy and Nassar, 2018; Nassar et al. 2021c).

The most challenging goal of biodesulfurization (BDS) technology development is the catalytic removal of recalcitrant organosulfur compounds, e.g., DBT, and its derivatives, using microbial cells or enzymes under aerobic or anaerobic conditions. The main disadvantage of BDS is lower reaction rates compared to HDS. There are three main routes for aerobic BDS (Figure 2.6): (i) complete conversion where the end products are carbon dioxide and water; (ii) the Kodama pathway, where C-C bonds are cleaved and some water-soluble by-products are produced (Kilbane and Jackowsky, 1992); and (iii) the 4S-pathway, in which sulfur is selectively removed leaving the carbon skeleton intact. The first two pathways are not recommended for desulfurization of fuels, as carbon skeleton oxidation reduces the fuel value (El-Gendy and Nassar, 2018).

The key challenge to address for commercially viable BDS processes is to establish a cost-effective process that utilizes a two-phase bioreactor system and optimizing de-emulsification for product recovery (Kaufman et al., 1998; Pacheco et al., 1999; McFarland, 1999). Multiple-stage air-lift reactors can reduce mixing costs, and centrifugation approaches facilitated de-emulsification, desulfurized oil recovery, and recycling of the cells (Ohshiro et al., 1996), but this has never been proven at operational scales. Having said that, while most of petroleum biotechnologies have only been tested in the laboratory, BDS pilot plants were built in the mid-1990s by Enchira Biotechnology Corporation (ENBC) (formerly Energy Biosystems Corporation).

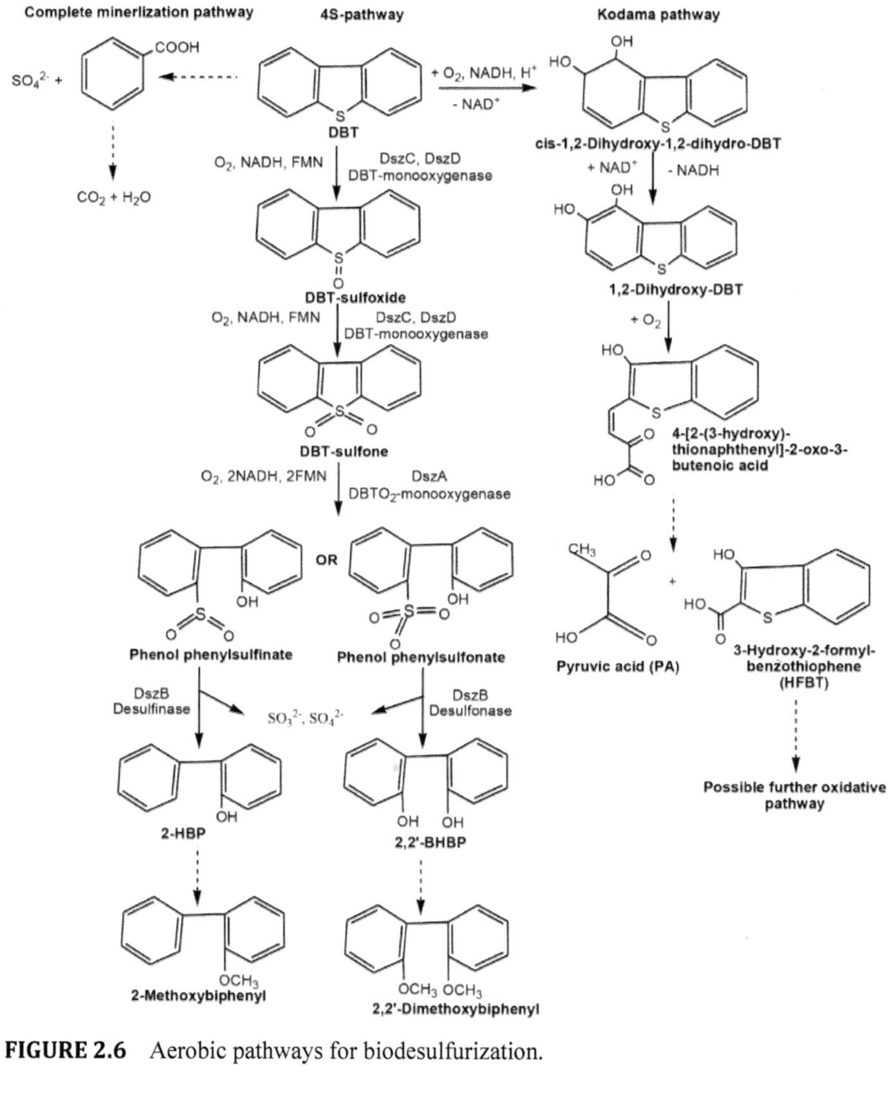

FIGURE 2.6 Aerobic pathways for biodesulfurization.

Zobell (1953) reported in a Texaco patent on the anaerobic BDS of crude oil using a hydrogenase from the sulfate-reducing bacteria (SRB) *Desulfovibrio desulfuricans*. Anaerobic BDS proceeds more slowly than aerobic reactions, but generates the same products as conventional HDS technology, namely H_2S and desulfurized oil (Figure 2.7). In 1961, ESSO Research and Engineering Co. patented an anaerobic process using a *Thiobacillus* strain (ESSO Research and Engineering Company, 1961). Microbial conversion of organic sulfur in crude petroleum slurry (10–20% water with added surfactant) was described with claims of more than 90% sulfur removal in 10 minutes.

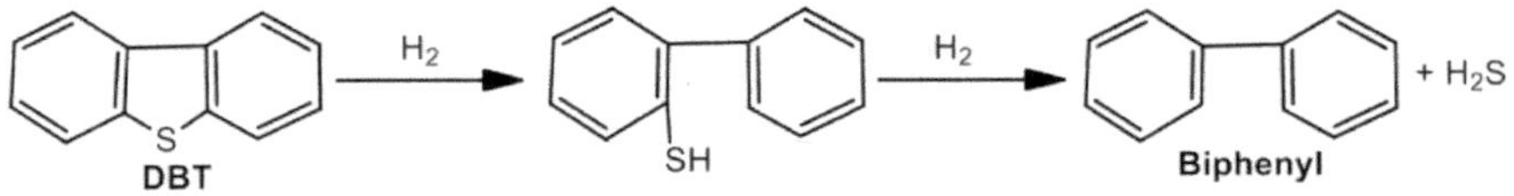

FIGURE 2.7 Anaerobic pathway for biodesulfurization of dibenzothiophene (DBT).

VEB Petrolchemisches Kombinat Schwedt in former East Germany investigated anaerobic routes for crude oil desulfurization in 1974 (Eckart et al., 1981, 1982). Uncharacterized mixed cultures obtained from oil-polluted soils and sludge, were contacted with various oils in a 4-dm^3 batch fermenter. Removal of sulfur was achieved but at the expense of considerable loss of HC; mainly through metabolism of n-alkanes. In 1997, VEB Petrochemisches Kombinat Schwedt patented an anaerobic fermentation process where 40% reduction in the organic sulfur content of Romaschkino crude oil was claimed after 2–3 days of anaerobic incubation. The sulfur was released as H_2S or mercaptan derivatives, which were detected in the waste gases. Some 4–6% of the HCs were lost, probably through volatilization. Other problems included foaming, increased crude oil viscosity and short biocatalyst life, rendering the process impractical (Kohler et al., 1984). In a further study, the anaerobic BDS of 1.8% sulfur Romashkino crude oil was examined (Eckart et al., 1986), and mixed cultures grown with lactate and sulfate were dominated by *Desulfovibrio* spp. Upon, the control of the pH maintains circumneutral conditions, desulfurization of 26 to 40% was noted in a few days. Pifferi et al. (1990) performed continuous anaerobic microbial desulfurization of crude oil, petroleum fractions, and petroleum products using sulfate-reducing microorganisms in the presence of H_2 at ambient temperature and atmospheric pressure. The culture medium contained ferrous ions that enhanced the formation of a colloidal iron sulfide precipitate in the reaction mixture. Desulfurization tests of fuel oils with different sulfur content were conducted in a 10 L bench-scale laboratory reactor. Desulfurization yield ranged from 25% to 90% for a fuel oil with a S-content of 3 wt.%. Kim et al. (1990a) investigated *Desulfovibrio desulfuricans* M6, a strain found to desulfurize 96% and 42% of benzothiophene and DBT, respectively, as well as other S-containing organic substrates, including phenyl sulfide, benzyl sulfide, benzyl disulfide, ethanethiol, and butanethiol. Metabolite analyzes confirmed the strain capability to metabolize DBT into biphenyl and H_2S. Kim et al. (1990b) reported that *Desulfovibrio desulfuricans* M6 removed up to 21% of the 3 wt.% sulfur in Kuwait crude oil within 6 days and up to 17 wt.% from other crude oils and their distillate products (Kim et al., 1995). Anaerobic BDS of crude oil and its distillates was also reported by the hyperthermophile, *Pyrococcus furiosus* (Tilstra et al., 1992), *Nocardioform actinomycete* FE9 (Finnerty 1993), *Desulfomicrobium scambium* and *Desulfovibrio longreachii* (Yamada et al., 2001). Agarwal and Sharma (2010) reported the desulfurization of both heavy crude oil (1.88% S) and light crude oil (0.4% S) by a three-step process; biodesulfurization/oxidative desulfurization/reactive adsorptive desulfurization (BDS-ODS-RADS), under both aerobic and anaerobic conditions. BDS of both crudes using *Pantoea agglomerans* D23W3 resulted in 61% and 63% S removal under aerobic conditions, while, under anaerobic conditions, 63% and 69% desulfurization were achieved with heavy oil. The second step of the process, oxidative desulfurization, resulted in a total of 94% desulfurization, while the third step, reactive adsorptive desulfurization, led to 95% S-removal. The use of the thermophilic *Klebsiella* sp. 13T resulted in 62% S removal from light crude oil and 68% S removal from heavy crude oil during the first treatment stage. In addition, Li and Jiang (2013) reported 51% sulfur removal from bunker C oil within 7 days of incubation using a mixed microbial culture. The anaerobic biodesulfurization of diesel oil (Kareem et al., 2014) and kerosene (Kreem et al., 2016) using *Desulfatiglans anilini* has been also reported.

Overall, anaerobic desulfurization of petroleum has the advantage of liberating sulfur as H_2S gas that can be treated with existing refinery desulfurization plants (e.g., Claus process), it eliminates the costs related to aeration in aerobic process and does not liberate sulfate as a waste product (Setti et al., 1997; Speight, 2014). Under anaerobic conditions, the oxidation of HC derivatives to produce undesirable compounds such as gum-forming products is minimal (McFarland, 1999). However, maintaining an anaerobic process is extremely difficult, and the specific activity of most of the isolated strains have been reported to be insignificant for DBT derivatives (Armstrong et al., 1995). Due to low reaction rates, safety, and cost concerns, and the lack of identification of specific enzymes responsible for anaerobic desulfurization and anaerobic microorganisms effective enough for practical petroleum desulfurization, anaerobic BDS processes have not been developed (El-Gendy and Nassar, 2018). Consequently, aerobic BDS has been the focus of most of the research in BDS (Le Borgne and Quintero, 2003).

Malik (1978) reported the 4S pathway (Figure 2.6), which is a metabolic pathway that removes sulfur without altering fuel octane value. By the 1990s, due to the depletion of light crude oil reserves, increasing world-wide fuel demand, stricter environmental regulations, and the exploration of new reservoirs of sulfur-rich oils (e.g., heavy crude oil, tar sands and oil shale), interest in the biotechnological utility of this pathway was high. The petroleum industry engaged with academia to study the applicability of BDS through the specific oxidative 4S pathway (Figure 2.6). The pathway is named for the four S-containing intermediates involved: sulfoxide, sulfone, sulfinate or sulfonate, and sulfate. Four enzymes are required for the 4S pathway: flavin-dependent dibenzothiophene mono-oxygenase (DszC), dibenzothiophene sulfone mono-oxygenase (DszA), a flavin reductase (DszD), and a desulfinase (DszB). Generally, the *dszABC* genes are found in an operon, with *dszD* found at another location in the genome. The end product of the pathway, 2-hydroxybiphenyl (2-HBP), is oil, conserving the calorific value of the fuel. Some microorganisms have been found to express an extended 4S pathway where 2-HBP is methylated to 2-methoxybiphenyl (Xu et al., 2007). Chen et al. (2009); and El-Gendy et al. (2014b) reported the production of 2-methoxybiphenyl and 2,2'-dimethoxy-1,1'-biphenyl (2,2'-DMBP) as the end products of BDS of DBT through the 4S pathway using *Mycobacterium* sp. and *Rhodococcus erythropolis* HN2, respectively.

The difference in desulfurization efficiency of the different microorganisms might be related to mass transfer limitations and metabolic regulation controlling reaction rates (Bustos-Jaimes et al., 2003). The presence of n-alkanes favors the removal of sulfur aromatic compounds by acting as a co-substrate and solubilizing the aromatic compounds. Microbial adsorption to the oil phase is the most likely mechanism for explaining *n*-alkane biodegradation (Setti et al., 1993, 1995). Substrate uptake apparently happens through the diffusion or dynamic transport at the point of contact. In addition, most of the aerobic microorganisms adhere to the low molecular weight n-alkanes (i.e., those below n-C_{16}) which are usually liquid at room temperature. Although some aerobic microorganisms can remove high amounts of organic sulfur, the sulfur percentage in the residual heavy oil may increase because of simultaneous aliphatic compounds biodegradation. It has been also recommended to apply BDS for crude oil because of its high-water content relative to diesel oil and gasoline (Zhou and Zhang, 2004).

Microorganisms have been used for the relatively controlled destruction of petroleum constituents with the result that the viscosity of the treated product was stabilized. BDS can be applied to decrease the viscosity of viscous heavy crude oil and bitumen. It is applicable for viscous oil containing sulfur-heterocyclic compounds, since the physicochemical properties of the heterocycles significantly contribute to the viscosity of crude oil and its fractions. Briefly, BDS contacts the viscous petroleum liquid with a biocatalyst that converts sulfur-heterocyclic compounds into compounds that lack the physicochemical characteristics contributing to viscosity. The biocatalyst works in a sulfur-specific manner, such that the sulfur-heterocyclic compound is selectively attacked at the sulfur heteroatom. Then the carbon-sulfur bonds are cleaved and/or polar substituents such as hydroxyl groups are joined to the sulfur heteroatom, the HC skeleton of the sulfur-bearing heterocycle, or both. Preferred biocatalysts for viscosity reduction include preparations of *Rhodochrous erythropolis* ATCC 53968 microorganism and enzymes obtained therefrom (Monticello et al., 1996; Monticello, 2000). BDS as a viscosity reducer process does not require specified equipment or safety processes and does not affect the HC skeleton of the treated oil and consequently keeps the caloric (fuel) value intact (Nassar et al., 2021c).

In 1982, the Atlantic Richfield Company patented a BDS process for crude oils using *Bacillus sulfasportare* ATCC 39909 (Atlantic Richfield Company, 1986). The sulfur removal ranged from 17–23 wt.% for vacuum residual oil to 45 wt.% removal for Maya crude oil without a recordable loss in carbon content.

In 1987, it was reported that an aerobic process has been developed by the Unocal Corporation. The process was based on hydroxylation of the sulfur heterocycles to water-soluble products by a *Pseudomonas* strain, which also produced a surfactant. The disadvantages of the process included low efficiency and loss of fuel value (Monticello and Finnerty, 1985).

Treatment of high-sulfur crude oil with the soil isolate *Pseudomonas alcaligenes* (Finnerty and Robinson, 1986), which efficiently oxidizes DBT to water-soluble products, resulted in the removal of more than 70% of organic sulfur from the pentane-soluble fraction of crude oil. However, this bacterium does not remove organic sulfur from the heavy fractions (asphaltenes and resins) of the crude oil, which contains the major concentrations of organic sulfur. Moreover, the biocatalyst causes significant fuel value loss and a zero-value water-soluble sulfur waste product is produced.

Kleshchev (1989) designed an integrated process for HDS of crude oil followed by BDS using *Rhodococcus rhodochrous* to remove the labile organic sulfides, then the recalcitrant sulfur-heterocyclic compounds, e.g., DBT, respectively. It was reported by Setti et al. (1997) that in an experiment of heavy oil desulfurization in which *R. erythropolis* IGTS8 was used, non-destructive desulfurization was followed by the complete degradation of the aliphatic fraction, which constituted 60% of the oil. A thermophilic bacterial strain, identified as *Paenibacillus*, can selectively desulfurize DBT without degrading its HC matrix. Thus, it is proposed for the development of a BDS process for crude oil at high temperatures where crude oil viscosity is lower and mass transfer limitations are reduced (Konishi et al., 2000). Bladi et al. (2003) reported the isolation of a yeast strain; *Rhodosporidium toruloides* DBVPG 6662, that can grow on a variety of sulfur compounds and can desulfurize Orimulsion (a bitumen amended with an emulsifying agent and water) by 68% in 15 days.

Egyptian crude oils are characterized by being dominated with mixtures of substituted benzothiophene and DBT derivatives. Salama et al. (2004) reported the BDS of Egyptian crude oil (Balayim Barry) by *Rhodococcus rhodochrous* ATCC 53968 (which was later identified as *R. erythropolis* IGTS8) that achieved 45.4% total sulfur removal. The optimized parameters affecting the BDS process were found to be 24 h, 30C, pH 7.2, 1% oil/water, 10 g/L glucose, and 3 g/L ammonium chloride. The desulfurized crude oil was characterized by a pronounced drop in sulfur content in asphaltenes, while no difference was reported for saturates, aromatics, and resins. Microbial desulfurization of Balayim crude oil was associated with a decrease in paraffins and iso-paraffins with C-numbers C13–C18, with a slight degradation of the biomarker phytane. In another study performed on Balayim crude oil by a *Bacillus subtilis* isolate, the same optimum conditions for maximum efficiency of BDS were found at pH 7, recording 38% total sulfur removal (Ibrahim et al., 2004). That was accompanied by a decrease in polynuclear aromatic, phytane, polynuclear naphthene compounds and asphaltenes. However, *B. subtilis* raised paraffins with higher carbon number (C20–30) and isoparaffins (C21–30). Both ATCC 53968 and isolated *B. subtilis* recorded 36.5% and 22.6%, respectively, reduction in asphaltene content during the BDS process. BDS of the aromatic fraction showed that IGTS8 removed up to 63% of sulfur content from that fraction (Moustafa et al., 2006). El-Gendy et al. (2006) reported BDS of crude oil by the halotolerant yeast, *Candida parapsilosis* NSh45, isolated from Egyptian HC-polluted sea water, in a batch process of 1/3 oil/water phase after 7 days at 30°C with mixing rate of 200 rpm. NSh45 reduced the sulfur content of Balayim Mix crude oil (2.74% w/w sulfur) by 75%, with a decrease in the average molecular weight of asphaltene constituents by approximately 28%, and dynamic viscosity reduction by approximately 70%, compared to that of *R. erythropolis* IGTS8, which removed approximately 64% of the sulfur, and a decrease in average molecular weight of asphaltenes and crude dynamic viscosity of approximately 24% and 64%, respectively, under the same conditions (El-Gendy et al., 2006). The reported maximum sulfur removal from heavy crude oil using different microorganisms ranged between 47–68% (Setti et al., 1993; EL-Gendy et al., 2006; Torkamani et al., 2008a, b; Agarwal and Sharma, 2010; Bhatia and Sharma, 2010, 2012; Adlakha et al., 2016).

Gundam et al. (2006, 2013) reported that BDS removed most of the sulfur (59%) from Liaoning crude oil within 72 hours using *Shingomonas subarctic* T7B. Yu et al. (2006) reported the efficient BDS of hydrodesulfurized diesel, Fushun crude oil and Sudanese crude oil by resting cells of *Rhodococcus erythropolis* XP, which recorded 94.5%, 62.3% and 47.2% within 24 h and 72 h for diesel and crude oil, respectively. *Mycobacterium goodie* X7B and *R. erythropolis* XP removed 47.2 to 62.3% of sulfur from crude oil after 72 h treatment at 30°C (Li et al., 2007). A subsidiary Petroleum Engineering Development Company (PEDEC) of National Iranian Oil Company isolated fungus *Stachybotrys* sp. which selectively removes sulfur and nitrogen from heavy crude oil at 30°C. It is reported to remove 76% and 64.8% of sulfur from heavy crude oil of Soroush oil field and the Kuhemond oil field within 72 and 144 h, respectively (Torkamani et al., 2009).

The effect of adding different surfactants (Tween 20, Triton X-100, and PEG 4000) on the BDS of fuel oil (MFO 380) has been investigated in an attempt to enhance the BDS of heavy oils (Li and Jiang, 2013). Upon the usage of soil or oil sludge as an inoculum, no desulfurization was recorded and that was attributed to the absence of the specific BDS microorganisms in the inoculated seed, besides, the low-to-no solubility of fuel oil (MFO 380) in the aqueous phase might, formations of large oil particles with the adherence of a large amount of heavy oil to the flask's walls. Consequently, low contact between the biocatalyst and oil occurred. For achieving a successful BDS process, the decrease in oil viscosity and the increase of the contact of oil with the biocatalyst are very essential. The results of bunker oil BDS showed that the sulfur removal efficiency was only 2.88% after 7 days of incubation, but this could be significantly improved by adding the surfactants Triton X-100 or Tween 20. That was ascribed to the reduced viscosity of the heavy oil and the increased mass transfer of the sulfur compounds from heavy oil into water. Adding Triton X-100 achieved the highest sulfur removal efficiency of up to 51.7% after 7 days of incubation. The optimal amount of Triton X-100 was 0.5 g/50 mL medium (Li and Jiang, 2013).

Jiang et al. (2014) studied the BDS of model thiophene derivatives and heavy oil by an enriched consortium from oil sludge where the BDS of bunker oil recorded 2.8% sulfur removal without de-asphalting within 7 days. After a de-asphalting process, the efficiency of the BDS was considerably increased recording 26.2–36.5%, due to the decrease in bunker oil viscosity which consequently enhanced the contact between the oil, water, and microbial consortium. Chauhan et al. (2015) reported batch BDS of crude oil (1:5 oil/water) using *Gordonia* sp. IITR 100, where GC-MS revealed the production of 2-HBP, 1-methylhydroxybiphenyl and benzonaphthothiophene-hydroxide, which possibly were produced from DBT, 1-methyldibenzothiophene and benzonaphthothiophene.

The desulfurization capability of *B. subtilis* Wb600 on light crude oil was examined by Nezammahalleh (2015). The Wb600 strain utilized the sulfur-containing compounds of the oil as a source of sulfur without degrading the HC skeleton and produced biosurfactants, which facilitated the transfer of the organic sulfur compounds into the aqueous phase. It decreased the total sulfur content of the light crude oil to about 40% during 35 h with an oil/water ratio of 0.2. Mohamed et al. (2015) reported a 10% reduction in sulfur content of heavy crude oil (with initial S content of 3%) in batch BDS at 30°C and 250 rpm, within 7 days, using resting cells of *Rhodococcus* sp. strain SA11 collected in the exponential phase of DBT cultures. Moreover, total sulfur and sulfur speciation analyze of the hexane-soluble fraction of the heavy crude oil (i.e., the de-asphaltened oil fraction or the maltenes) revealed 18% total sulfur reduction including a wide range of thiophene derivatives such as DBT, benzothiophene, and their alkylated derivatives. Adlakha et al. (2016) reported a viscosity reduction of 31.28% with 76% BDS of heavy oils in a batch BDS of 1/3 oil/water ratio using growing cells of *Gordonia* sp. IITR100 at 30°C and 250 rpm, within 7 days. Martínez et al. (2016) reported the enhancement of oil BDS by engineered synthetic bacterial consortia.

It has been estimated that a BDS catalyst must have a sulfur removal activity within the range of 1–3 mmol dibenzothiophene/g dry cell weight/h in real petroleum fractions. This means the activity of the current biocatalysts is needed to be improved by 500-fold to attain a commercially viable process (Kilbane, 2006). Alves et al. (2015) accomplished a feasibility study for comparing the cost of two BDS process designs; upstream and downstream conventional HDS. The BDS costs and emission estimations were made considering the BDS of DBT as a model for sulfur compounds while HDS estimations were made based on crude oil HDS. The BDS-downstream-HDS configuration was found to be the best alternative to be applied in oil refinery based on the lower energy consumption, operational costs, besides the obtaining of almost sulfur-free fuels, with a concomitant much decrease in greenhouse gas emissions and carbon dioxide.

Certain technical issues hamper the applicability of BDS, especially the two-phase nature of the process since an aqueous phase is needed to maintain a viable biocatalyst in the BDS process, and significant quantities of water must be added to the fuel, which adds to the cost of oil separation after biotreatment. The values found in the literature generally indicate low oil-to-water volumetric ratios in the desulfurization of real fractions, ranging from 1 to 25% of oil in water, while the highest oil-to-water ratio was 50% and 80% as reported by Yu et al. (1998); and Monot et al. (2002), respectively. Most of the work on BDS was performed on middle distillates fraction. The treatment of oils using free cells has encountered some limitations such as high cost of the biocatalyst and low volumetric ratio between the organic phase and the aqueous one. Also, separation of oil product from oil-water-biocatalyst emulsion is troublesome and the free cells are difficult to recycle (Choi et al., 2003; Konishi et al., 2005). In view of the industrial application of BDS, cell immobilization is one of the most promising approaches (Mukhopadhyaya et al., 2006; Huang et al., 2012).

Adegunlola et al. (2010) investigated the ability of the immobilized spores of *Aspergillus flavus* to remove sulfur from crude oil. The spores of *A. flavus* were immobilized by mixing 80 g of the harvested spores in 1% (w/v) sterile alginate solution and then gelled into beads by dropping the suspension into a cold 15 g/L $CaCl_2$ solution. When 50 g of immobilized spores of *A. flavus* were added to crude oil for one, two, three, and seven days, the amount of sulfur removed was 27.2%, 45.2%, 90.4% and 91.7%, respectively. When 50 g of immobilized spores of *A. flavus* were introduced into the crude oil for 7 days at 35°C, 40°C and 45°C, the amount of sulfur removed was 63.2%, 55.3% and 10.5%, respectively. Finally, when 10 g, 50 g, and 100 g of immobilized spores of *A. flavus* were added to crude oil for 7 days, the amount of sulfur removed was 49.6%, 94.7% and 53.9%, respectively.

From a cost perspective, BDS has several advantages; (i) operation at low temperature and pressure, (ii) 70 to 80% lower carbon dioxide emissions and energy consumption, (iii) at adequate BDS level, the expected capital cost needed for an industrial scale BDS is two-third that of HDS (Pacheco et al., 1999). Atlas et al. (2001) estimated the HDS process cost to decrease the sulfur content from 500 to 200 ppm to be approximately 0.26 cent/L, and desulfurization cost would increase by a factor of four if the sulfur content is further lowered to 50 ppm, and (v) flexible nature.

Studies on BDS of heavy crude oil and the application of large-scale BDS technology are still limited and only at the pilot scale. By the year 2000, The US Department of Energy at the year 2000, has funded a project by $900,000 to be conducted by EBC for using a new gene-shuffling technology to develop a biocatalyst capable of desulfurizing both gasoline and diesel fuel (Grisham, 2000). The BDS rate of the available microorganisms are still low, thus, three bioreactors are essentially needed to reach low sulfur concentration. Monticello (2000) proposed a pilot process with three bioreactors to achieve very low sulfur concentration (Figure 2.8), as a start for the industrial BDS scale. Later, US-EBC has established a pilot with a working capacity of 5 barrels of oil per day (BPD) (Boniek et al., 2015). In that process, the biocatalyst and fossil fuel are mixed in a BDS bioreactor. Then the treated oil passes through a series of filters to a container for disposal of sewage. Finally, it is neutralized by a basic aqueous solution and sodium sulfate is removed in a wastewater treatment station (Boniek et al., 2015).

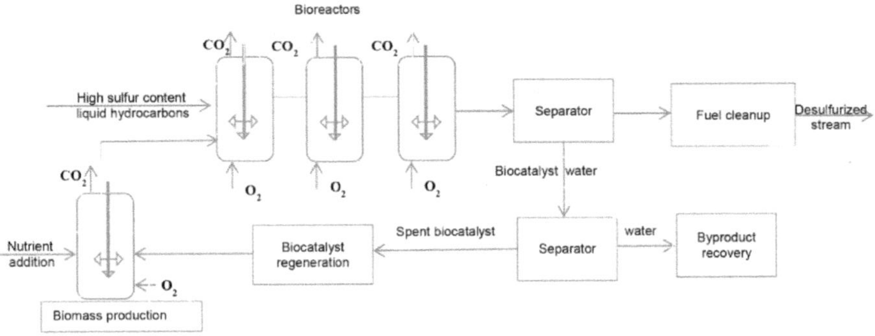

FIGURE 2.8 Biodesulfurization pilot plant (adapted from Monticello, 2000).

2.4.2 BIODENITROGENATION (BDN)

Nitrogen, like sulfur, is considered as a petroleum contaminant. Although of the low nitrogen content in crude oil, which usually ranges between 0.1 to 0.9%, however, it can reach up to approximately 2%, and nitrogen content is known to increase with the increment of the asphaltenes content (Speight, 2012, 2014). Same as sulfur, there is also a correlation between the nitrogen content and API gravity of crude oils, besides, the higher the nitrogen content, the higher the carbon residue is (Ancheyta and Speight, 2007). Nevertheless, of its low content, however, it has a significant impact on the refinery, as nitrogenous compounds poison the cracking catalysts, and contribute to gum formation in some refineries such as the domestic fuel oil. The increasing demand for petroleum feedstock for catalytic cracking and production of different petroleum refinery products has highlighted the harmful effects of nitrogen compounds, which are concentrated largely in the higher boiling fractions. Éigenson and Ivchenko (1977) reported that the higher the sulfur content in petroleum fractions, the higher its nitrogen content. Moreover, nitrogen-containing compounds coexist with sulfur-containing compounds in fossil fuels (Yi et al., 2014). Carbazole, as a model compound of the non-basic nitrogenous polyaromatic HCs negatively impacts the refining processes via two routes. (1) Carbazole is transformed into basic derivatives, during the cracking process, which consequently adsorb onto the cracking catalysts' active sites. (2) It directly inhibits the activity of the HDS catalysts. Thus, it is very recommendable to remove the nitrogenous compounds and carbazole to intensify the extent of catalytic cracking and the consequent gasoline yield. Benedik et al. (1998) estimated a major economic improvement in the low-margin high-volume refining processes upon a 90% reduction in nitrogen content with a consequent increase in gasoline yield by approximately 20%. Basic nitrogenous compounds are reported to be inhibitory to catalysts than the non-basic ones. However, the latter can be possibly converted to the inhibitory basic compounds during the refining/catalytic cracking process. Moreover, the associated metals (e.g., nickel, and vanadium) to the nitrogenous compounds in crude oil are known to be also strong catalytic inhibitors (Hegedus and McCabe, 1981; Mogollon et al., 1998; Zakaria et al., 2016).

Compared to the available rich literature published about the selective BDS of petroleum and its fractions without affecting the fuel calorific value, however, there is a little-reported about the specific biodenitrogenation (BDN). Although, it has been assessed that the petroleum BDN is more advantageous than the deep denitrogenation via the conventional costly and non-selective hydroprocessing (HDP) methods, besides its contribution to decrease the NOx emissions however, the economics of BDN process is impacted by the amount of HC lost from the fuel during the nitrogen removal (Vazquez-Duhalt et al., 2002).

Generally, the presently confirmed carbazole BDN pathway is similar to the well-known Kodama DBT degradation pathway (Figure 2.9), which results in the loss of the fuel value. This would consequently make the BDN of fuel streams economically unfeasible. Moreover, most of the carbazole-degrading microorganisms produce 2'-aminobiphenyl-2, 3-diol as an initial step in its BDN pathway. However, recovering of carbazole nitrogen as anthranilic acid or 2'-aminobiphenyl-2, 3-diol, the less inhibitory to refining catalysts, would solve part of that problem since the entire carbon content of the fuel is preserved. This can be performed by mutant or recombinant strains. Other pathways would liberate nitrogen from carbazole in the form of ammonia (Rhee et al., 1997). Microorganisms such as *Pseudomonas ayucida*, *Aneurinibacillus* sp., *Pseudomonas stutzeri*, *Yokenella* sp., and *Pseudomonas nitoreducens* were studied for BDN of fossil fuels (Kilbane et al., 2003). Degradation of carbazole by the thermophilic *Anoxybacillus rupiensis*, that can tolerate up to 80°C with maximum activity at temperature range 55–65°C, was reported, which would be advantageous for application in real petroleum processing (Fadhil et al., 2014). However, so far, no micro-organism that selectively extract nitrogen from carbazole in a microbial pathway equivalent to the 4S pathway has been isolated.

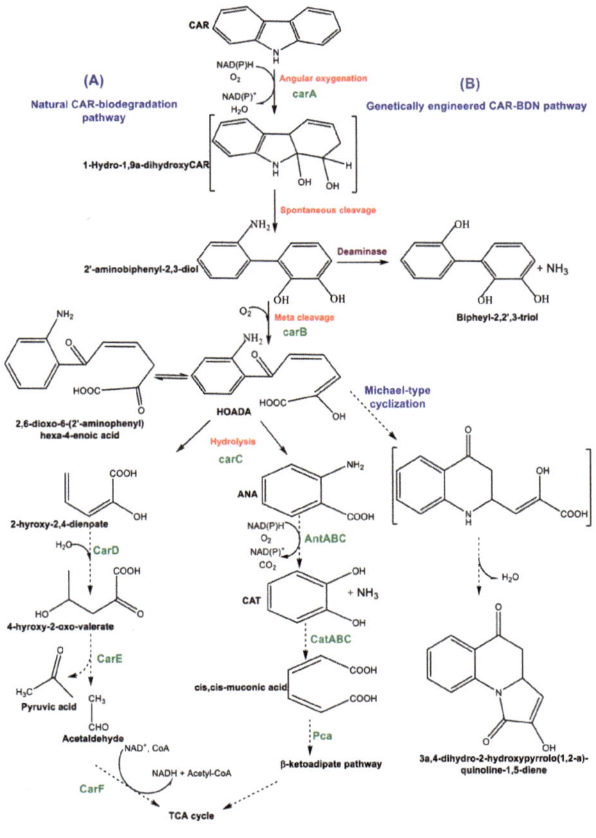

FIGURE 2.9 Pathways for biodenitrogenation (BDN) of carbazole.

Atlas and Aislabie (1992) reported the selective removal of nitrogen compounds in raw shale oil by *Pseudomonas* and *Acinetobacter* sp. where amines, nitriles, quinolines, and pyridines were converted to the water-soluble hydroxylated derivatives. Kilbane et al. (2000) reported the selective BDN of quinoline without affecting its carbon skeleton by *Pseudomonas ayucida* IGTN9m (ATCC N° PTA-806) where approximately 5% nitrogen was removed by this isolate from petroleum without affecting its energetic value. Since IGTN9m converts quinoline to 8-hydroxycoumarin without further degradation (Figure 2.10), thus no carbon is lost. Resting cells of IGTN9m selectively transformed 68% of quinoline from shale oil within 16 h. The low BDN efficiency in oil is attributed to the low abundance of quinoline relative to other NPAHs in oil and the narrow substrate range of the quinoline-degrading enzymes. Sugaya et al. (2001) reported the optimum quinoline BDN conditions using *Comamonas* sp. TKV3-2-1 for application in petroleum feed. The degradation rate reached 1.6 mmol/g cell/h at 83% (v/v) petroleum/aqueous medium with a cell density of 28.5 g/L. Moreover, Sugaya et al. (2001) suggested the application of BDN of crude oil during its storage (Figure 2.11) as it would overcome the problems of extra treatment period and cost. Sona et al. (2010) reported the BDN of crude oil (5% v/v oil/water ratio) using resting cells of *B. lichiformis* CRC-75. Anaerobic BDN was also reported to modify the properties of heavy crude oil and enhance oil recovery (Fallon et al., 2010).

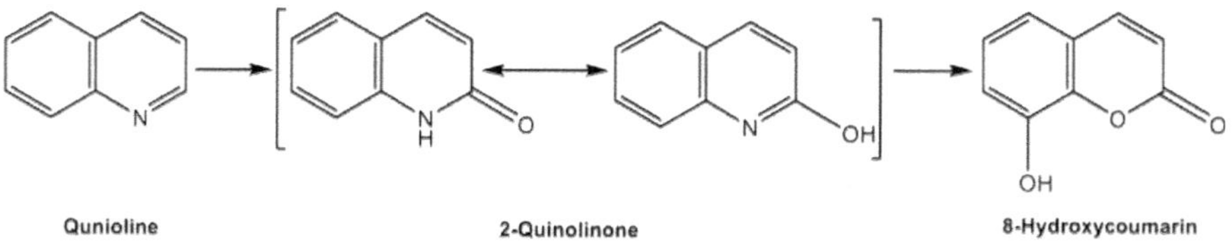

FIGURE 2.10 Pathway for quinoline biodenitrogenation by *Pseudomonas ayucida* IGTN9m.

From the practical point of view, BDN, and BDS should be integrated where nitrogen and sulfur would be removed through specific enzymatic attack on the C–N and C–S bonds without C–C bond cleavage to preserve the fuel value of the biotreated products. Moreover, a dual bioprocess for both selective BDS and BDN that overcomes the significant technical hurdles such as tolerance against solvents, high concentration of nitrogenous compounds, and high oil to water ratio, would make microbial refining processes and upgrading of petroleum and its fractions feasible on a large scale (Kilbane, 2006). *Gordonia* sp. strain F.5.25.8 can utilize DBT through the 4S pathway and carbazole as a sole source of S and N, respectively. F.5.25.8 is the first reported strain that can simultaneously metabolize DBT and carbazole (Santos et al., 2006). It could tolerate up to 42°C, which adds to its advantages for industrial application of BDS/biodenitrogenation as complementary to HDT process. Moreover, strain F.5.25.8 has a different genetic organization of the BDS (*dsz*) and BDN (*car*) gene clusters relative to *Rhodococcus erythropolis* IGTS8 and *Pseudomonas* sp. IGTN9m, respectively. Kayser and Kilbane (2004) inserted genes encoding amidases downstream of the artificial *carA* operon to accomplish the cleavage of the final C-N bond and produce biphenyl-2, 2′, 3-triol (Figure 2.9) that is reintroduced to the fuel, preserving its fuel content. The authors reported a decrease of the carbazole content in a petroleum sample by approximately 95% in 2:10 petroleum/aqueous medium system within 16 h using the genetically engineered bacterium.

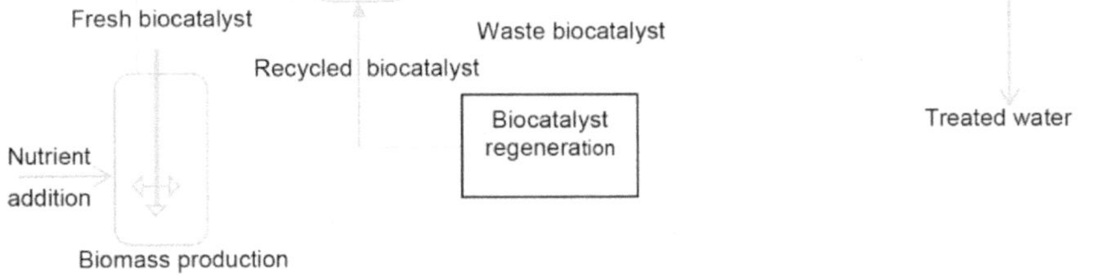

FIGURE 2.11 A process for biodenitrogenation of quinoline in crude oil under storage (as proposed by Sugaya et al., 2001).

Although the biochemistry and genetics of quinoline and carbazole metabolism as models for basic and non-basic NPAHs are available and the selective removal of nitrogen from crude oil was reported, the commercialization of a BDN process in petroleum industry has not been practically considered. This might be due to the complexity of NPAHs present in petroleum and its distillates, which is difficult to be metabolized by one microbial strain (Zakaria et al., 2015). Moreover, the selective removal of nitrogen from carbazole is still under investigation. Not only this, but also BDN process suffers from the main limitations of BDS; low nitrogen removal activity and the need for an aqueous phase. Moreover, denitrogenation is an additional step that does not produce an upgraded fuel; therefore, reaction velocities should be even higher than that of BDS.

2.4.3 BIODEMETALLIZATION

Metals such as vanadium, copper, nickel, and iron are found in petroleum as petroporphyrins and other complexes, which present primarily in the asphaltenes fraction besides other metal salts of zinc, titanium, calcium, and magnesium (Ali and Abbas, 2006; Speight, 2014). Thus, as in the case of sulfur and nitrogen contents, the higher the petroleum asphaltenes content is, the higher will be the heavy metal content (Duyck et al., 2007; Speight, 2014). The establishment of the refining process is significantly affected by the accumulation of metallic components in the higher molecular weight and polar fractions of crude oils (Panariti et al., 2000; Duyck et al., 2008; Speight, 2014, 2017). The inorganic water-soluble forms of metallic components are reported to be easily removed via a crude desalting process, where they will be concentrated in the aqueous phase. Nevertheless, the metalloporphyrins are entrenched in the tremendously complicated asphaltenes' structure; consequently, metal removal from petroporphyrins and complexes should be deeply investigated (Speight and El-Gendy, 2017).

The vanadium and nickel concentrations are reported to vary with the petroleum origin from as low as 0.1 ppm to as high as 1,200 ppm and from traces to approximately 150 ppm, respectively (Hessley, 1990; Ali and Abbas, 2006; Duyck et al., 2007). The V/Ni ratio is considered as a fingerprint and is constant in crude oils of common source rocks and varies with the rocks' geological age, for example, oils from Triassic or older age have V/Ni value greater than unity (Ball et al., 1960). Although, V/Ni ratio is known to be used for tracing the source effects (Al-Shahristani and Al-Atyia, 1972; El-Gayar et al., 2002), however, the reported literature concerning the possible biodegradation of asphaltenes and resins would impact the V/Ni ratio of these fractions (Duyck et al., 2007). Examples for petroleum feedstock with high content of organometallics are; Boscán crude oil, Cerro Negro crude oil, Maya crude oil, Wilmington crude oil, and Prudhoe Bay crude oil (Fish et al., 1984).

Heavy metals poison the refining catalysts, and they accumulate in the residual fraction, usually exposed to catalytic cracking, consequently, deposit onto the cracking catalysts decreasing their selectivity and activity. Catalyst deactivation in the cracking, hydrogenation, and HDS processes would occur due to the pore-clogging, metal deposition, as well as the deformation and destruction of the process reactors (Salehizadeh et al., 2007), which consequently led to the periodical demands for the shut-off of the catalytic process for the replacement or the clean-up of the catalyst at a vast expense (Callejas et al., 2001; Speight, 2014, 2017). It has been estimated that the annual direct material cost for the replacement of the exhausted catalysts for the United States petroleum industry is more than $1 billion (Paul and Smith, 2009). In the fluid catalytic cracking (FCC) process, sulfur in the oil feed can be converted to the readily handled H_2S, while the nonvolatile metals accumulate in the FCC unit, causing another potential serious problem as they are deposited onto the catalyst together with the coke. Besides, the presence of nickel and vanadium, which have dehydrogenation activity would promote the catalytic dehydrogenation reactions during the cracking sequence with a consequent increase in the amounts of coke and light gases at the expense of the produced gasoline yield (Elliot, 1996; Ali and Abbas, 2006). Also, the heavy metals can be liberated into the environment (in the form of ash) during fuel combustion (Xu et al., 1997).

The typical demetallization process in petroleum industry is the solvent deasphalting process where the lower molecular weight constituents are physically separated from the asphaltene constituents by mixing the heavy oil/residue with a low boiling HC liquid such as propane and/or butane (Yamada et al. 1979; Speight, 1981, 2014, 2017; Farag et al., 1989; Ancheyta and Speight, 2007). Other options for demetallization include a low-temperature cooking process in which the metals are collected in the initially-formed coke (Speight, 2014, 2017). There are also processes for the selective removal of metals from the organic moieties with a minimal conversion of the oil such as (i) metal removal using a solvent, (ii) oxidative demetallization of petroleum asphaltenes and residua, and (iii) a hydro-demetallization process (Gould, 1980; Kashima Oil Co. Ltd. Japan, 1983; Adarme et al., 1990; Bartholdy and Hannerup, 1990; Savastano, 1991; Piskorz et al., 1996; Hernandez et al., 1998).

Fedorak et al. (1993) reported the destruction of porphyrin ring systems using a chlorination reaction with *Caldariomyces fumago* chloroperoxidase leading to the biocatalytic demetallization of petroporphyrins and asphaltenes with a reduction in nickel octaethyl porphyrin and vanadyloctaethyl porphyrin of approximately 93 and 53%, respectively. However, chlorination produces chlorinated compounds, which have a negative impact on the environment. Biocatalytic removal of Ni and V from petroporphyrins by chloroperoxidase has been also reported by Mogollon et al. (1998).

Xu et al. (1997, 1998) reported the biodemetallization of petroporphyrins and crude oil through oxidation of prophyrinic rings using hemoproteins (cytochrome C reductases) from *Bacillus megaterium* and *Catharanthus roseuse* in the presence of the cofactor NADPH. In contrast to peroxidases, oxygenases from *Escherichia coli*, animal cells (such as liver or kidney cells), plant cells (such as from mung beans or *Arabidopsis thaliana*) or yeast cells (such as *Candida tropiculis*) can degrade porphyrin molecules without subjecting the HC to chlorine or peroxide to remove nickel, vanadium, cobalt, copper, iron, magnesium, and zinc. The metals can be readily removed by extraction, distillation, ion exchange and/or column chromatography. The main advantages of this process are the possibility to operate in a batch, semi-continuous or continuous mode alone or in combination with one or more biorefining processes (such as BDS) in a sealed or open vessel and in the presence or absence of light (Lateef et al., 2019).

The removal of vanadyl porphyrins (Figure 2.12) from crude oils was reported to be of great importance. The isolated fungus *Aspergillus* sp. MS-100 from a HC polluted soil in Isfahan refinery has been reported for its capability to; utilize vanadium oxide octaethyl porphyrin (VOOEP) as a sole carbon source and biodemetallization of crude oil (Salehizadeh et al., 2007). Horse myoglobin and plant peroxidase were chosen to synthesize the agents for removal of porphyrins from un-cracked fuels (Paul and Smith, 2009). Kilbane (2005) reported that, since most of the metals in petroleum are associated with organonitrogen compounds, it would be possible to simultaneously perform the BDN and biodemetallization processes to reduce the nitrogen and metals content in one process.

2.5 CONCLUSIONS AND FUTURE RESEARCH DIRECTIONS

With the continuously increasing demand for clean and ultra-clean transportation fuels and petrochemical raw materials and the depletion of high-quality petroleum reserves, the crude oil refining should experience new substantial innovation. As feedstocks of the refineries change, there must be an accompanying change in refining knowhow. This means the evolution of the so-called conventional processes of refining heavy feedstocks using the coking technologies to more innovative techniques (including the hydrogen-based and microbial-based processes), which produce ultimate amounts of high-quality liquid fuels from the feedstock and maintain emissions within the international environmental standards (Speight, 2011, 2017).

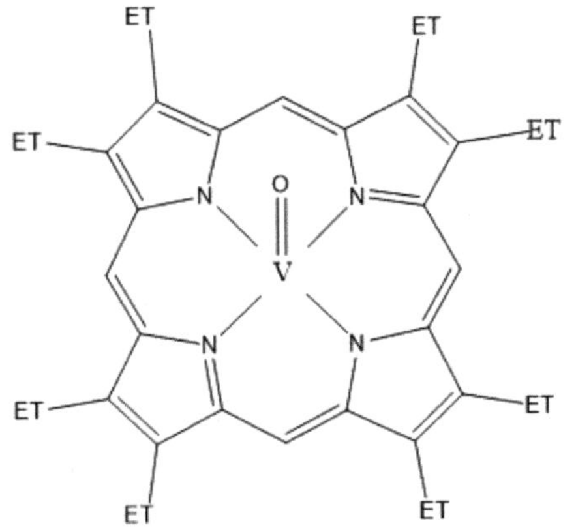

FIGURE 2.12 An example of a vanadylporphyrin.

Besides the main advanced development in the applications of petroleum biotechnology in the bioremediation of polluted environments, there are also promoted efforts to produce petroleum products via some biotransformation processes, for example; BDS, BDN, biodemetallization, and depolymerization of asphaltene constituents. (Speight and Arjoon, 2012). Among these applications, the most advanced area is BDS, for which pilot plants have been established (El-Gendy and Speight, 2015; El-Gendy and Nassar, 2018).

The well-established expensive conventional technologies in petroleum refineries cannot achieve the target of ultra-low sulfur transportation fuels in an economic manner. Alternatively, BDS as a complementary tool for HDS is recommendable to achieve the worldwide new regulations for clean fuels with lower GHG emissions. In fact, significant progress has been made toward the commercialization of crude oil BDS including (i) the isolation and characterization of microbes that are suitable for selective BDS processes, (ii) improved biocatalyst performance, lifetime, and HC tolerance (iii) development of analytical methodology, which has led to breakthroughs in the characterization of the desulfurized fuels, (iv) development of a process concept for crude oil BDS, and (v) construction and testing of a prototype bench unit.

The main hurdles to the commercialization of crude oil BDS include; (i) the specificity of the biocatalyst, (ii) the reaction rate, (iii) the reusability and lifetime of the biocatalyst, and (v) the stability of the biocatalyst under process conditions. In addition, mass transfer and separations obstacles must be overcome for crude oils with increased viscosity and density. There is continuous research to modify the biocatalyst effectiveness and to screen other organisms for additional desulfurization capabilities. Moderate thermophiles and thermotolerant species have also been reported, which are beneficial in real field operation, especially downstream the HDS process. Further, the enzyme-based catalysis also gained considerable interest because they offer several advantages over the whole-cell biocatalysts. For example, an enzyme may function in very low water content environments and, in contrast with whole-cells biocatalysts, an enzyme may even be less prone to thermal and organic solvent denaturation if the environment is hydrophobic. From the design perspective, it is usually more straightforward to modulate the kinetic and stability properties of a single protein than the properties of an enzymatic cascade, intrinsically constrained by cell metabolism.

Multiple strategies, such as immobilization, solvent engineering, and protein engineering, may be combined to enhance the characteristics of enzymes. Heavy oils require a temperature on the order of 60 to 100°C to be free-flowing fluids, and there is a need to develop microbial species and enzymes that can exist at and be active in that temperature range.

Microbial immobilization and application of nanobiotechnology are also recommendable to enhance BDS and BDN rate as it solves the problem of the cost of biocatalyst preparation, high ratio of water-to-oil, enhances the stability of the biocatalyst, facilitates its recovery and reusability, and decreases the risk of contamination.

A biorefinery in which crude oil feedstocks are converted into refined products using microbes and enzymes as biocatalysts would enable microbial biotransformation of the low-sulfur, low-nitrogen, and low-metals components of the feedstock. The biorefinery can be coupled with a gasification refinery, which would have, as the centerpiece, a gasification technology to produce synthesis gas (from the biotransformation-resistant portions of the feedstock). The latter can be used to produce liquid fuels via Fischer-Tropsch synthesis technology (Speight, 2011), thereby increasing overall refinery efficiency (conversion).

KEYWORDS

- **biodemetallization**
- **biodenitrogenation**
- **biodesulfurization**
- **biorefining**
- **biotransformation**
- **microbial enhanced oil recovery**

REFERENCES

Abdel-Waly, A. A., (1999). Laboratory study on activating indigenous microorganisms to enhance oil recovery. *J. Can. Petrol. Technol., 38*(2), 55–61.

Adarme, R., Sughrue, E. L., Johnson, M. M., Kidd, D. R., Phillips, M. D., & Shaw, J. E., (1990). Demetallization of asphaltenes: Thermal and catalytic effects with small pore catalysts. *J. Am. Chem. Soc., 35*, 614–618.

Adegunlola, G. A., Oloke, J. K., Majolagbe, O. N., Adebayo, E. A., Oyeyiola, A. G., & Adegunlola, F. O., (2010). Microbial desulphurization of crude oil using immobilized spores of *Aspergillus flavus*. *Adv. Environ. Biol., 4*(2), 155–158.

Adlakha, J., Singh, P., Ram, S. K., Kumar, M., Singh, M. P., Singh, D., Sahai, V., & Srivastava, P., (2016). Optimization of conditions for deep desulfurization of heavy crude oil and hydrodesulfurized diesel by *Gordonia* sp. IITR100. *Fuel, 184*, 761–769.

Agarwal, P., & Sharma, D. K., (2010). Comparative studies on the bio-desulfurization of crude oil with other desulfurization techniques and deep desulfurization through integrated processes. *Energy Fuels, 24*(1), 518–524.

Akhtar, N., Ghauri, M. A., & Akhtar, K., (2016). Dibenzothiophene desulfurization capability and evolutionary divergence of newly isolated bacteria. *Arch. Microbiol., 198*(6), 509–519.

Al-Bahry, S. N., Al-Wahaibi, Y. M., Al-Hinai, B., Joshi, S. J., Elshafie, A. E., Al-Bemani, A. S., & Al-Sabahi, J., (2016). Potential in heavy oil biodegradation via enrichment of spore forming bacterial consortia. *J. Pet. Explor. Prod. Technol., 6*, 787–799.

Ali, H. R., El-Gendy, N. Sh., Moustafa, Y. M., Mohamed, I., Roushdy, M. I., & Hashem, A. I., (2012). Degradation of asphaltenic fraction by locally isolated halotolerant bacterial strains. *ISRN Soil Sci.* ID 435485. https://doi.org/10.5402/2012/435485.

Ali, H. R., Ismail, D. A., & El-Gendy, N. Sh., (2014). The biotreatment of oil polluted seawater by biosurfactant producer halotolerant *Pseudomonas aeruginosa* Asph2. *Energy Source A, 36*(13), 1429–1436. https://doi.org/10.1080/15567036.2014.880092.

Ali, M. A., & Abbas, S., (2006). A review of methods for the demetallization of residual fuel oils. *Fuel Process. Technol., 87*, 573–584.

Al-Shahristani, H., & Al-Atyia, M. J., (1972). Vertical migration of oil in Iraqi oil fields: Evidence based on vanadium and nickel concentrations. *Geochim. Cosmochim. Acta, 36*, 929–938.

Al-Sulaimani, H., Al-Wahaibi, Y., Al-Bahry, S., Elshafie, A., Al-Bemani, A., Joshi, S., & Zargari, S., (2011a). Optimization and partial characterization of biosurfactants produced by *Bacillus* species and their potential for ex-situ enhanced oil recovery. *Society of Petroleum Engineers SPE J., 16*(3), 672–682.

Al-Sulaimani, H., Joshi, S., Al-Wahaibi, Y., Al-Bahry, S., Elshafie, A., & Bemani, A. A., (2011b). Microbial biotechnology for enhancing oil recovery: Current developments and future prospects. *Biotechnol. Bioinf. Bioeng., 1*(2), 147–158.

Alves, L., Paixão, S. M., Pacheco, R., Ferreira, A. F., & Silva, C. M., (2015). Biodesulphurization of fossil fuels: Energy, emissions and cost analysis. *RSC Adv., 5*, 34047–34057.

Ancheyta, J., & Speight, J. G., (2007). *Hydroprocessing of Heavy Oils and Residua.* CRC Press, Taylor & Francis Group, Boca Raton, Florida.

Ancheyta, J., (2016). *Deactivation of Heavy Oil Hydroprocessing Catalysts: Fundamentals and* Modeling. Wiley, Hoboken.

Armstrong, S. M., Sankey, B. M., & Voodoo, G., (1995). Conversion of dibenzothiophene to biphenyl by sulfate reducing bacteria isolated from oil field production facilities. *Biotechnol. Lett., 17*, 1133–1136.

Atlantic Richfield Company, (1986). *Biodesulfurization of Carbonaceous Materials.* United States Patent No. 4632906.

Atlas, R. M., & Aislabie, J., (1992). *Process for Biotechnological Upgrading of Shale Oil.* US Patent No. 5,143,827.

Atlas, R. M., Boron, D. J., Deever, W. R., Johnson, A. R., McFarland, B. L., & Meyer, J. A., (2001). *Method for Removing Organic Sulfur from Heterocyclic Sulfur Containing Organic Compounds*. US Patent H1,986.

Atlas, R. M., Boron, D. J., Deever, W. R., Johnson, A. R., McFarland, B. L., & Meyer, J. A., (1998). *Biodesulfurization of Gasoline: A Technology Roadmap* (p. 1e82). Coordinating Research Council, Atlanta Georgia. CRC Project no. E-7c.

Ayala, M., & LeBorgne, S., (2010). Microorganisms utilizing sulfur containing hydrocarbons. In: Timmis, K. N., (ed.), *Handbook of Hydrocarbon and Lipid Microbiology.* Springer, Berlin, (2010).

Ayala, M., Vazquez-Duhalt, R., Morales, M., & Le Borgne, S., (2016). Application of microorganisms to the processing and upgrading of crude oil and fractions. In: Lee, S. Y., (ed.), *Consequences of Microbial Interactions with Hydrocarbons, Oils, and Lipids: Production of Fuels and Chemicals, Handbook of Hydrocarbon and Lipid Microbiology.* Springer International Publishing, AG.

Ayala, M., Verdin, J., & Vazquez-Duhalt, R., (2007). The prospects for peroxidase-based biorefining of petroleum fuels. *Biocat. Biotransform., 25*, 114–129.

Bachman, R. T., Johnson, A. C., & Edyvean, R. G. J., (2014). Biotechnology in the petroleum industry: An overview. *Int. Biodeter. Biodegr., 86*, 225–237.

Ball, J. S., Wenger, W. J., Hyden, H. J., Horr, C. A., & Myers, A. T., (1960). Metal content of twenty-four petroleums. *J. Chem. Eng. Data., 5*, 553–557.

Bartholdy, J., & Hannerup, P. N., (1990). Hydrodemetallation in reside hydroprocessing. *J. Am. Chem. Soc., 35*, 619–625.

Benedik, M. J., Gibbs, P. R., Riddle, R. R., & Willson, R. C., (1998). Microbial denitrogenation of fossil fuels. *Trends Biotechnol., 16*, 390–395.

Bertrand, J. C., Rambeloarisoa, E., Rontani, J. F., Giusti, G., & Mattei, G., (1983). Microbial Degradation of crude oil in sea water in continuous culture. *Biotechnol. Lett., 5*, 567–572.

BGR, (2015). Energy study. *Reserves, Resources and Availability of Energy Resources.* Federal Institute of Geoscience and Natural Resources, Hannover, Germany.

Bhatia, S., & Sharma, D. K., (2010). Biodesulfurization of dibenzothiophene, its alkylated derivatives, and crude oil by a newly isolated strain of *Pantoea agglomerans* D23W3. *Biochem. Eng. J., 50*(3), 104–109.

Boniek, D., Figueiredo, D., Santos, A. F. B., & Stoianoff, M. A. R., (2015). Biodesulfurization: A mini review about the immediate search for the future technology. *Clean Technol. Environ. Policy, 17*(1), 29–37.

Bryant, R. S., & Lindsey, R. P., (1996). World-wide applications of microbial technology for improving oil recovery. In: *Proceedings of the SPE Symposium on Improved Oil Recovery* (pp. 27–134). Society of Petroleum Engineers, Richardson, Texas.

Bryant, R. S., Bailey, S. A., Step, A. K., Evans, D. B., & Parli, J. A., (1998). Biotechnology for heavy oil recovery. *SPE Paper No. 36767, Proceedings. SPE/DOE Improved Oil Recovery Symposium,* 1–7.

Bryant, R. S., Donaldson, E. C., Yen, T. F., & Chilingarian, G. V., (1989). Microbial enhanced oil recovery. In: Donaldson, E. C., Chilingarian, G. V., & Yen, T. F., (eds.), *Enhanced Oil Recovery II: Processes and Operations* (pp. 423–450). Elsevier, Amsterdam, Netherlands.

Bustos-Jaimes, I., Amador, G., Castorena, G., & Le Borgne, S., (2003). Genotypic characterization of sulfur-oxidative desulfurizing bacterial strains isolated from Mexican refineries. *Oil Gas Sci. Technol., 58*(4), 521–526.

Chen, H., Cai, Y., Zhang, W., & Li, W., (2009). Methoxylation pathway in biodesulfurization of model organosulfur compounds with *Mycobacterium* sp. *Bioresour. Technol., 100,* 2085–2087.

Choi, O., Choi, K., Ryu, H., & Chang, Y., (2003). Enhancement of phase separation by the addition of de-emulsifiers to three phase (diesel oil/biocatalyst/aqueous phase) emulsion in diesel biodesulfurization. *Biotechnol. Lett., 25,* 73–77.

Daryasafar, A., Jamialahmadi, M., Moghaddam, M. B., & Moslemi, B., (2016). Using biosurfactant producing bacteria isolated from an Iranian oil field for application in microbial enhanced oil recovery. *J. Pet. Technol., 38*(8), 739–746.

Demirbas, A., Bafail, A., & Nizami, A. S., (2016). Heavy oil upgrading: Unlocking the future fuel supply. *Petrol. Sci. Technol., 34,* 303–308.

Duarte, G. F., Rosado, A. S., Seldin, L., De Araujo, W., & Van, E. J. D., (2001). Analysis of bacterial community structure in sulfurous-oil-containing soils and detection of species carrying dibenzothiophene desulfurization (dsz) genes. *Appl. Environ. Microbiol., 67,* 1052–1062.

Duyck, C., Miekeley, N., Fonseca, T. C. O., Szatmari, P., & Santos, N. E. V. S., (2008). Trace element distributions in biodegraded crude oils and fractions from the Potiguar basin. *Brazil. J. Braz. Chem. Soc., 19*(5), 978–986.

Duyck, C., Miekeley, N., Porto Da, S. C. L., Aucélio, R. Q., Campos, R. C., Grinberg, P., & Brandão, G. P., (2007). Trace element distributions in biodegraded crude oils and fractions from the Potiguar basin, Brazil. *Spectrochim. Acta Part B., 62,* 939–951.

Eckart, V., Hieke, W., Bauch, J., & Gentzsch, H., (1981). Microbial desulfurization of petroleum and heavy petroleum fractions. 1. Studies on microbial aerobic desulfurization of Romashkino crude oil. *Chemical Abstracts, 94,* 142230q.

Eckart, V., Hieke, W., Bauch, J., & Gentzsch, H., (1982). Microbial desulfurization of petroleum and heavy petroleum fractions. 3. Change in the chemical com position of fuel-D-Oil by microbial aerobic desulfurization. *Chemical Abstracts, 97,* 147259c.

Eckart, V., Kohler, M., & Hieke, W., (1986). Microbial desulfurization of petroleum and heavy petroleum fractions. 5. Anaerobic desulfurization of Romashkino petroleum. *Central Microbiology, 141*, 291–300.

Éigenson, A. S., & Ivchenko, E. G., (1977). Distribution of sulfur and nitrogen in fractions from crude oil and residues. *Chem. Technol. Fuels Oils, 13*, 542–544.

El Mahdi, A. M., Abdul Aziz, H., Abu Amr, S. S., El-Gendy, N. Sh., & Nassar, H. N. (2016). Isolation and characterization of *Pseudomonas* sp. NAF1 and its application in biodegradation of crude oil. *Environ. Earth Sci. 75*, 380. https://doi.org/10.1007/s12665-016-5296-z.

El-Gayar, M., Mostafa, M. S., Abdelfattah, A. E., & Barakat, A. O., (2002). Application of geochemical parameters for classification of crude oils from Egypt into source-related types. *Fuel Process. Technol., 79*, 13–28.

El-Gendy, N. Sh., (2001). *Biodesulfurization of Organosulfur Compounds in Crude Oil and its Fractions*. MSc Thesis, Department of Chemistry, Cairo University, Cairo, Egypt.

El-Gendy, N. Sh., (2004). *Biodesulfurization Potentials of Crude Oil by Bacteria Isolated from Hydrocarbon Polluted Environments in Egypt*. Ph.D. Thesis, Department of Chemistry, Cairo University, Cairo, Egypt.

El-Gendy, N. Sh., (2015). Biodesulfurization of petroleum and its fraction. In: Pant, K. K., Sinha, S., Bajpai, S., & Govil, J. N., (eds.), *Advances in Petroleum Engineering, Chemical Technology Series, Advances in Petroleum Engineering II: Petrochemical* (Vol. 4, pp. 655–680). Studium Press LLC, Houston, USA.

El-Gendy, N. Sh., & Nassar, H. N. (2018). "Biodesulfurization in Petroleum Refining" The publisher; John Wiley & Sons, Inc., 111 River Street, Hoboken, NJ 07030, USA and Scrivener Publishing LLC, 100 Cummings Center, Suite 541J, Beverly, MA 01915, USA. https://doi.org/10.1002/9781119224075.

El-Gendy, N. Sh., & Speight, J. G., (2015). *Handbook of Refinery Desulfurization*. CRC Press, Taylor & Francis, Boca Raton, Florida. https://doi.org/10.1201/b19102.

El-Gendy, N. Sh., Ali, H. R., El-Nady, M. M., Deriase, S. F., Moustafa, Y. M., & Roushdy, M. I. (2014a). Effect of different bioremediation techniques on petroleum biomarkers and asphaltene fraction in oil polluted sea water. *Desalin. Water. Treat. 52*, 7484–7494. https://doi.org/10.1080/19443994.2013.831784.

El-Gendy, N. Sh., Farahat, L. A., Moustafa, Y. M., Shaker, N., & El-Temtamy, S. A., (2006). Biodesulfurization of crude and diesel oil by *Candida parapsilosis* NSh45 isolated from Egyptian hydrocarbon polluted sea water. *Biosci. Biotechnol. Res. Asia, 3*(1a), 5–16.

El-Gendy, N. Sh., Nassar, H. N., & Abu, A. S. S., (2014b). Factorial design and response surface optimization for enhancing a biodesulfurization process. *Petrol. Sci. Technol., 32*(14), 1669–1679. https://doi.org/10.1080/10916466.2014.892988.

Elliot, J. D., (1996). *Delayed coker design and Operation: Recent Trends and Innovations*. Foster Wheeler USA Corporation.

Esso Research and Engineering Company, (1961). *Bacteriological Desulfurization of Petroleum*. U.S. patent no. 2975103.

Fadhil, A. M. A., Al-Jailawi, M. H., & Mahdi, M. S., (2014). Isolation and characterization of a new thermophilic, carbazole degrading bacterium (*Anoxybacillus rupiensis*) strain Ir3 (JQ912241). *Int. J. Adv. Res., 2*, 795–805.

Fallon, R. D., Hnatow, L. L., Jackson, S. C., & Keeler, S. J., (2010). *Method for Identification of Novel Anaerobic Denitrifying Bacteria Utilizing Petroleum Components as Sole Carbon Source*. US patent 7740063 B2.

Farag, A. S., El-Din, O. I. S, Youssef, M. H., Hassan, S. I., & Farmawy, S., (1989). Solvent demetallization of heavy oil residue. *Hung. J. Ind. Chem., 17*(3), 289–294.

Fedorak, P. M., Semple, K. M., Vazquez-Duhalt, R., & Westlake, D. W. S., (1993). Chloroperoxidase-mediated modifications of petroporphyrins and asphaltenes. *Enzyme Microb. Technol., 15*, 429–437.

Finnerty, W. R., & Robinson, M., (1986). Microbial desulfurization of fossil fuels. A review. *Biotechnol. Bioeng. Symp., 16*, 205–221.

Finnerty, W. R., & Singer, M. E., (1983). Microbial enhancement of oil recovery. *Biotechnology, 1*, 47–54.

Finnerty, W. R., (1993). *Symposium on: Bioremediation and Bioprocessing Presented Before the Division of Petroleum Chemistry* (pp. 282–285). Inc. 205th National Meeting, American Chemical Society, Denver, Co.

Fish, R. H., Komlenic, J. J., & Wines, B. K., (1984). Characterization and comparison of vanadyl and nickel compounds in heavy crude petroleums and asphaltenes by reverse-phase and size-exclusion liquid chromatography/graphite furnace atomic absorption spectrometry. *Anal. Chem., 56*, 2452–2460.

Gao, C. H., & Zekri, A., (2011). Applications of microbial-enhanced oil recovery technology in the past decade. *Energy Source. A, 33*(10), 972–989.

García-Arellano, H., Buenrostro-Gonzalez, E., & Vazquez-Duhalt, R., (2004). Biocatalytic transformation of petroporphyrins by chemical modified cytochrome C. *Biotechnol. Bioeng., 85*(7), 790–798.

Gassara, F., Suri, N., Stanislav, P., & Voordouw, G., (2015). Microbially enhanced oil recovery by sequential injection of light hydrocarbon and nitrate in low- and high-pressure bioreactors. *Environ. Sci. Technol., 49*(20), 12594–12601.

Gould, K. A., (1980). Oxidative demetallization of petroleum asphaltenes and residua. *Fuel, 59*(10), 733– 736.

Gray, K. A., Pogrebinsky, O. S., Mrachko, G. T., Xi, L., Monticello, D. J., & Squires, C. H., (1996). Molecular mechanisms of biocatalytic desulfurization of fossil fuels. *Nat. Biotechnol., 14*(13), 1705.

Greenwood, P. F., Wibrow, S., George, S. J., & Tibbett, M., (2008). Sequential hydrocarbon biodegradation in a soil from arid coastal Australia, treated with oil under laboratory-controlled conditions. *Org. Geochem., 39*, 1336–1346.

Grisham, J., (2000). Biocatalytic success. *Nat. Biotechnol., 18*(7), 701.

Gunam, I. B. W., Yaku, Y., Hirano, M., Yamamura, K., Tomita, F., Sone, T., & Asano, K., (2006). Biodesulfurization of alkylated forms of dibenzothiophene and benzothiophene by *Sphingomonas subarctica* T7b. *J. Biosci. Bioeng., 101*, 322–327.

Gunam, I. B. W., Yamamura, K., Sujaya, N., Antara, N. S., Aryanta, W. R., Tanaka, M., Tomita, F., et al., (2013). Biodesulfurization of dibenzothiophene and its derivatives using resting and immobilized cells of *Sphingomonas subarctica* T7b. *J. Microbiol. Biotechnol., 23*(4), 473–482.

Gupta, R. K., & Gera, P., (2015). Process for the upgradation of petroleum residue: Review. *Int. J. Adv. Technol. Eng. Sci., 3*(2), 643–656.

Head, I. M., Jones, D. M., & Röling, W. F. M., (2006). Marine microorganisms make a meal of oil. *Nat. Rev. Microbiol., 4*, 173–182.

Hegedus, L. L., & McCabe, R. W., (1981). Catalyst poisoning. *Catal. Rev. Sci. Eng., 23*, 377–476.

Hernandez, A., Mellado, R., & Martinez, J., (1998). Metal accumulation and vanadium induced multidrug resistance by environmental isolates of *Escherichia hermani* and *Enterobacter cloacae*. *Appl. Environ. Microbiol., 64*, 4377–4320.

Hernández-López, E. L., Ayala, M., & Vazquez-Duhalt, R., (2015). Microbial and enzymatic biotransformations of asphaltenes. *Petrol. Sci. Technol., 33*(9), 1017–1029.

Hernández-López, E. L., Perezgasga, L., Huerta-Saquero, A., Mouriño-Pérez, R., & Vazquez-Duhalt, R., (2016). Biotransformation of petroleum asphaltenes and high molecular weight polycyclic aromatic hydrocarbons by *Neosartorya fischeri*. *Environ. Sci. Pollut. Res., 23*, 10773–10784.

Hessley, R. A., (1990). In: Speight, J. G., (ed.), *Fuel Science and Technology Handbook*. Marcel Dekker, Inc., New York, NY, USA.

Hitzmann, D. O., (1962). *Treatment of Hydrocarbon*. US Pat. 3,069,325.

Huang, T., Qiang, L., Zelong, W., Daojang, Y., & Jiamin, X., (2012). Simultaneous removal of thiophene and dibenzothiophene by immobilized *Pseudomonas delafieldii* R-8 cells. *Chin. J. Chem. Eng., 20*, 47–51.

Hutzler, M. J., Sitzer, S., Holtberg, P. D., Conti, J., Kendell, J. M., & Kydes, A. S., (2003). *Annual Energy Outlook 2003 with Projection to 2025*. The United States Energy Information Administration, US Department of Energy, Washington, DC, 20585, USA.

Ibrahim, M. K., Salama, M. F., Abdel-Moghny, T., & El-Shahawy, M. R., (2004). A study on biodesulfurization of an Egyptian crude oil. II- valuation of microbial desulfurization capacity of some local bacterial isolates. *Egypt. J. Biotechnol., 17*, 329–335.

Jack, T. R., (1991). Microbial enhancement of oil recovery. *Curr. Opin. Biotechnol., 2*(3), 444–449.

Jahromi, H., Fazaelipoor, M. H., Ayatollahi, S., & Niazi, A., (2014). Asphaltenes biodegradation under shaking and static conditions. *Fuel, 117*, 230–235.

Jiang, X., Yang, S., & Li, W., (2014). Biodesulfurization of model compounds and de-asphalted bunker oil by mixed culture. *Appl. Biochem. Biotechnol., 172*, 62–72.

Jinfeng, L., Lijun, M., Bozhong, M., Rulin, L., Fangtian, N., & Jiaxi, Z., (2005). The field pilot of microbial enhanced oil recovery in a high temperature petroleum reservoir. *Petrol. Sci. Eng., 48*, 265–271.

Kareem, S. A., Aribike, D. S., Susu, A. A., & Nwachukwu, S. C. U. (2016). Anaerobic Biodesulfurization of Kerosene Part II: Investigating Its Kinetics. Chem. *Eng. Process Tech. 2*, 1029.

Kareem, S. A., Susu, A. A., Aribike, D. S., & Nwachukwu, S. C. U. (2014). Investigating the kinetics of biodesulfurization of diesel. Assumption University, *J. Technol. 18*, 27–35.

Karimi, M., Mahmoodi, M., Niazi, A., Al-Wahaibi, Y., & Ayatollahi, S., (2012). Investigating wettability alteration during MEOR process, a micro/macro-scale analysis. *Colloids Surf. B: Biointerfaces., 95*, 129–136.

Kashima Oil Co., Ltd. Japan, (1983). *Removal of Heavy Metals from Heavy Petroleum Oil*. JP Patent No. 58096681. Assigned to Kashima Oil Co., Ltd. Japan.

Kaufman, E. N., Harkins, J. B., & Borole, A. P., (1998). Comparison of batch stirred and electrospray reactors for biodesulfurization of dibenzothiophene in crude oil and hydrocarbon feedstocks. *Appl. Biochem. Biotechnol., 73*, 127–144.

Kayser, K. J., & Kilbane, J. J., (2004). *Method for Metabolizing Carbazole in Petroleum*. US Patent 6943006.

Kilbane, J. J., & Jackoswki, F., (1992). Biodesulfurization of water soluble coal-derived material by *Rhodococcus rhodochrous* IGTS8. *Biotechnol. Bioeng., 40*(9), 1107–1114.

Kilbane, J. J., (2005). Biotechnological upgrading of petroleum. In: Ollivier, B., & Magot, M., (eds.), *Petroleum Microbiology* (pp. 239–256). ASM Press, Washington, USA.
Kilbane, J. J., (2006). Microbial biocatalyst developments to upgrade fossil fuels. *Curr. Opin. Biotechnol., 17*, 305–314.
Kilbane, J. J., Ranganathan, R., Cleveland, L., Kayser, K. J., Ribiero, C., & Linhares, M. M., (2000). Selective removal of nitrogen from quinoline and petroleum by *Pseudomonas ayucida* IGTN9m. *Appl. Environ. Microbiol., 66*(2), 688–693.
Kilbane, J. J., Ribeiro, C. M. S., & Linhares, M. M., (2003). *Bacterial Cleavage of Only Organic C-N Bonds of Carbonaceous Materials to Reduce Nitrogen Content.* US patent 6541240 B1.
Kim, B. Y., Kim, H. Y., Kim, T. S., & Park, D. H., (1995). Selectivity of desulfurization activity of Desulfovibrio desulfuricans M6 on different petroleum products. *Fuel Process. Technol., 43*, 87–94.
Kim, H. Y., Kim, T. S., & Kim, B. H., (1990a). Degradation of organic compounds and the reduction of dibenzothiophene to biphenyl and hydrogen sulfide by *Desulfovibrio desulfuricans* M6. *Biotechnol. Lett., 12*, 761–764.
Kim, J. S., & Crowley, D. E., (2007). Microbial diversity in natural asphalts of the Rancho La Brea tar pits. *Appl. Environ. Microbiol., 73*, 4579–4591.
Kim, T. S., Kim, H. Y., & Kim, B. H., (1990b). Petroleum desulfurization by *Desulfovibrio desulfuricans* M6 using electrochemically supplied reducing equivalent. *Biotechnol. Lett., 12*, 757–760.
Kleshchev, S. M., (1989). *Method of Removing Organic Sulfur Compounds from Petroleum Products*. RU-Patent: 1,505,960.
Kohler, M., Genz, I. L., Schicht, B., & Eckart, V., (1984). Microbial desulfurization of petroleum and heavy petroleum fractions. 4. Anaerobic degradation of organic sulfur compounds. *Zentralbl. Mikrbiol., 139*, 239–247.
Konishi, J., Ishii, Y., Okumura, K., & Suzuki, M., (2000). *High Temperature Desulfurization by Microorganisms*. US Patent 6,130,081.
Konishi, M., Kishimoto, M., Omasa, I., Katakura, Y., Shioya, S., & Ohtake, H., (2005). Effect of sulfur sources on specific desulfurization activity of *Rhodococcus erythropolis* KA251 in exponential fed-batch culture. *J. Biosci. Bioeng., 24*, 49–54.
Lacotte, D. J., Mille, G., Acquaviva, M., & Bertrand, J. C., (1996). Arabian light 150 asphaltene biotransformation with n-alkanes as co-substrate. *Chemosphere, 32*, 1755–1761.
Lateef, S. A., Ajumobi, O. O., & Onaiz, S. A., (2019). Enzymatic desulfurization of crude oil and its fractions: A mini review on the recent progresses and challenges. *Arab. J. Sci. Eng., 44*, 5181–5193.
Lavania, M., Cheema, S., Sarma, P. M., Mandal, A. K., & Lal, B., (2012). Biodegradation of asphalt by *Garciaella petrolearia* TERIG02 for viscosity reduction of heavy oil. *Biodegradation, 23*, 15–24.
Lazar, I., Petrisor, I., & Yen, T., (2007). Microbial enhanced oil recovery (MEOR). *Petrol. Sci. Technol., 25*(11), 1353–1366.
Le Borgne, S., & Quintero, R., (2003). Biotechnological processes for refining of petroleum. *Fuel Process. Technol., 81*, 155–169.
Leòn, V., & Kumar, M., (2005). Biological upgrading of heavy crude oil. *Biotechnol. Bioprocess Eng., 10*, 471–481.
Li, W., & Jiang, X., (2013). Enhancement of bunker oil biodesulfurization by adding surfactant. *World J. Microbiol. Biotechnol., 29*(1), 103–108.

Li, Y. G., Ma, J., Zhang, Q. Q., Wang, C. S., & Chen, Q., (2007). Sulfur-selective desulfurization of dibenzothiophene and diesel oil by newly isolated *Rhodococcus erythropolis* NCC-1. *Chin. J. Org. Chem., 25*, 400–405.

Liu, Q., Lin, J., Wang, W., Huang, H., & Li, S., (2015). Production of surfactin isoforms by *Bacillus subtilis* BS-37 and its applicability to enhanced oil recovery under laboratory conditions. *Biochem. Eng., 93*, 31–37.

Liu, T., Cao, G., & Ba, Y., (2012). Activation of the indigenous microorganism and field test in Zhan 3 block [J]. *Chem. Eng. Oil Gas, 41*(4), 411–478.

Malik, K. A., (1978). Microbial removal of organic Sulfur from crude oil and the environment: Some new perspectives. *Process Biochem., 13*, 10–13.

Martínez, I., Mohamed, M. E., Rozas, D., García, J. L., & Díaz, E., (2016). Engineering synthetic bacterial consortia for enhanced desulfurization and revalorization of oil sulfur compounds. *Metab. Eng., 35*, 46–54.

McFarland, B. L., (1999). Biodesulfurization. *Curr. Opin. Microbiol., 2*, 257–264.

McInerney, M. J., Duncan, K. E., Youssef, N., Fincher, T., Maudgalya, S. K., Folmsbee, M. J., Knapp, R., et al., (2005). *Development of Microorganisms with Improved Transport and Biosurfactant Activity for Enhanced Oil Recovery.* Report to the Department of Energy, DE-FE-02NT15321, Washington, DC.

Mogollón, L., Rodríguez, R., Larrota, W., Ortiz, C., & Torres, R., (1998). Biocatalytic removal of nickel and vanadium from Petro porphyrins and asphaltenes. *Appl. Biochem. Biotechnol., 70–72*, 765–777.

Mohamed, M. E., Al-Yacoub, Z. H., & Vedakumar, J. V., (2015). Biocatalytic desulfurization of thiophenic compounds and crude oil by newly isolated bacteria. *Front. Microbiol., 6*(112). doi: 10.3389/fmicb.2015.0 0112.

Mohebali, G., & Ball, A. S., (2016). Biodesulfurization of diesel fuels - past, present and future perspectives. *Int. Biodeter. Biodegrad., 110*, 163–180.

Monot, F., Abbad-Andaloussi, S., & Warzywoda, M., (2002). *Biological Culture Containing Rhodococcus Erythropolis and/or Rhodococcus Rhodnii and Process for Desulfurization of Petroleum Fraction.* United States Patent 6,337,204.

Monticello, D. J., & Finnerty, W. R., (1985). Microbial desulfurization of fossil fuels. *Ann. Rev. Microbiol., 39*, 371–389.

Monticello, D. J., (2000). Biodesulfurization and upgrading of petroleum distillates. *Curr. Opin. Biotechnol., 11*, 540–546.

Monticello, D. J., Haney, III., & William, M., (1996). *Biocatalytic Process for Reduction of Petroleum Viscosity*. US Pat. 5,529,930.

Montiel, C., Quintero, R., & Aburto, J., (2009). Petroleum biotechnology: Technology trends for the future. *Afr. J. Biotechnol., 8*(12), 2653–2666.

Morales, M., & Le Borgne, S., (2014). Protocols for the isolation and preliminary characterization of bacteria for biodesulfurization and biodenitrogenation of petroleum-derived fuels. In: McGenity, T. J., Timmis, K. N., & Nogales, B., (eds.), *Hydrocarbon and Lipid Microbiology Protocols*. Springer, Berlin.

Moustafa, Y. M. M., El-Gendy, N. Sh., Farahat, L. A., Abo-State, M. A., & El-Temtamy, S. A., (2006). Biodesulfurization of Ras Badran crude oil and its constituents with special emphasis on its asphaltene fraction. *Egypt. J. Petrol., 15*(1), 21–30.

Mukhopadhyaya, M., Chowdhury, R., & Bhattacharya, P., (2006). Biodesulfurization of hydrodesulfurized diesel in a trickle bed reactor-experiments and modeling. *J. Sci. Ind. Res., 65*, 432–436.

Nassar, H. N., Abu Amr, S. A., El-Gendy, N. Sh. (2021b). Biodesulfurization of refractory sulfur compounds in petro-diesel by a novel hydrocarbon tolerable strain *Paenibacillus glucanolyticus* HN4. *Environ Sci Pollut Res Int, 28*, 8102–8116. https://doi.org/10.1007/s11356-020-11090-7 (accessed on 23 December 2021).

Nassar, H. N., Ali, H. R., & El-Gendy, N. Sh. (2021a). Waste prosperity: Mandarin (*Citrus reticulata*) peels inspired SPION for enhancing diesel oil biodesulfurization efficiency by *Rhodococcus erythropolis* HN2. Fuel. 294: 120534. https://doi.org/10.1016/j.fuel.2021.120534 (accessed on 23 December 2021).

Nassar, H. N., Ismail, A. R., El-Salamony, R. A., Aboelazayem, O., Abu Amr, S. A., El-Gendy, N. Sh. (2021c). Animal bone affluence for environmental reclamation: Biodiesel, petro-diesel biodesulfurization and xenobiotic polluted water photo-treatment. *Biofuel. Bioprod. Biorefin. 5*, 770–792. https://doi.org/10.1002/bbb.2194 (accessed on 24 December 2021).

Nezammahalleh, H., (2015). Biodesulfurization of light crude oil using *Bacillus subtilis* Wb600. *J. Biol. Today's World., 4*(3), 67–73.

Ohshiro, T., Suzuki, K., & Izumi, Y., (1996). Regulation of dibenzothiophene degrading enzyme activity of *Rhodococcus erythropolis* D-1. *Ferm. Bioeng., 82*, 121–124.

Pacheco, M. A., Lange, E. A., Pienkos, P. T., Yu, L. Q., Rouse, M. P., Lin, Q., & Linquist, L. K., (1999). Recent advances in desulfurization of diesel fuel. *NPRA Annual Meeting, NPRA AM-99–27* (pp. 1–26). San Antonio.

Panariti, N., Del, B. N. A., Del, P. G., & Marchionna, M., (2000). Petroleum residue upgrading with dispersed catalysts. Part 1. Catalysts activity and selectivity. *Appl. Catal. A: Gen., 204*(2), 203– 213.

Paul, J. A. K., & Smith, M. L., (2009). *Method for Purification of Uncatalyzed Natural Fuels from Metal Ions by Means of at Least One Hemeprotein and Use of the at Least on Hemeprotein.* United States Patent 8,475,652.

Peng, P., Morales-Izquierdo, A., Hogg, A., & Strausz, O. P., (1997). Molecular structure of Athabasca asphaltene: Sulfide, ether, and ester linkages. *Energy Fuels, 11*, 1171–1187.

Pifferi, P., Lanzarini, G., & Malleuzzi, D., (1990). *The Desulfurization of Heavy Oils and Residua.* EU Patent 401922.

Piskorz, J., Radlein, D., Majerski, P., & Scott, D., (1996). *Hydrotreating of Heavy Hydrocarbon Oils in Supercritical Fluids.* US Patent No. 5,496,464. Assigned to Natural Resources Canada.

Rhee, S. K., Lee, K. S., Chung, J. C., & Lee, S. T., (1997). Degradation of pyridine by *Nocardioides* sp. strain OS4 isolated from the oxic zone of a spent shale column. *Can. J. Microbiol., 43*(2), 205–209.

Rontani, J. F., Bosser-Joulak, F., Rambeloarisoa, E., Bertrand, J. C., & Faure, G. R., (1985). Analytical study of asphalt crude oil and asphaltenes biodegradation. *Chemosphere, 14*, 1413–1422.

Salama, M. F., Ibrahim, M. K., & El-Shahawy, M. R., (2004). A study on biodesulfurization of an Egyptian crude oil. I-evaluation of biodesulfurization capacity of *Rhodococcus rhodochrous* ATCC 53968. *Egypt. J. Biotechnol., 17*, 321–328.

Salehizadeh, H., Mousavi, M., Hatamipour, S., & Kermanshahi, K., (2007). Microbial demetallization of crude oil using *Aspergillus* sp.: Vanadium oxide octaethyl porphyrin (VOOEP) as a model of metallic petroporphyrins. *Iran. J. Biotechnol., 5*(4), 226–231.

Santos, S. C., Alviano, D. S., Alviano, C. S., Padula, M., Leitao, A. C., Martins, O. B., Ribeiro, C. M., et al., (2006). Characterization of *Gordonia* sp. strain F.5.25.8 capable of dibenzothiophene desulfurization and carbazole utilization. *Appl. Microbiol. Biotechnol., 71*, 355–362.

Savastano, C. A., (1991). Solvent extraction approach to petroleum demetallation. *Fuel Sci. Technol. Int., 9*(7), 833–871.

Setti, L., Lanzarini, G., & Pifferi, P. G., (1995). Dibenzothiophene biodegradation by a *Pseudomonas* sp. in model solutions. *Process Biochem., 30*(8), 721–728.

Setti, L., Lanzarini, G., & Pifferi, P. G., (1997). Whole cell biocatalysis for an oil desulfurization process. *Fuel Process. Technol., 52*, 145–153.

Setti, L., Rossi, M., Lanzarini, G., & Pifferi, P. G., (1993). The effect of n-alkanes in the degradation of dibenzothiophene and of organic sulfur compounds in heavy oil by a *Pseudomonas* sp. *Biotechnol. Lett., 14*(6), 515–520.

She, Y. H., Zhang, F., Xia, J. J., Kong, S. Q., Wang, Z. L., Shu, F. C., & Hu, J. M., (2011). Investigation of biosurfactant-producing indigenous microorganisms that enhance residue oil recovery in an oil reservoir after polymer flooding. *Appl. Biochem. Biotechnol., 163*, 223–234.

Simpson, D. R., Natraj, N. R., Mclnerney, M. J., & Duncan, K. E., (2011). Biosurfactant producing Bacillus are present in produced brines from Oklahoma oil reservoirs with a wide range of salinities. *Appl. Biochem. Biotechnol., 91*, 1083–1093.

Sona, K., Maryam, M., & Esmat, F., (2010). Quinoline biodegradation by *Bacillus licheniformis* strain CRC-75. *Iran. J. Chem. Chem. Eng., 29*(2), 151–158.

Souayeh, M., Al-Wahaibi, Y., Al-Bahry, S., Elshafie, A., Al-Bemani, A., Joshi, S., Al Hashmi, A., & Al-Mandhari, M., (2014). Optimization of a low-concentration *Bacillus subtilis* strain biosurfactant toward microbial enhanced oil recovery. *Energy Fuels, 28*(9), 5606–5611.

Speight, J. G., & Arjoon, K. K., (2012). *Bioremediation of Petroleum and Petroleum Products.* Scrivener Publishing, Beverly, Massachusetts.

Speight, J. G., & El-Gendy, N. Sh., (2017). *Introduction to Petroleum Biotechnology*. Gulf Professional Publishing, Cambridge, MA, USA. https://doi.org/10.1016/C2015-0-02007-X.

Speight, J. G., & Islam, M. R., (2016). *Peak Energy - Myth or Reality*. Scrivener Publishing, Beverly, Massachusetts.

Speight, J. G., (1981). *The Desulfurization of Heavy Oils and Residua*. Marcel Dekker, Inc., New York, USA.

Speight, J. G., (2011). *The Refinery of the Future*. Gulf Professional Publishing, Elsevier, Oxford, United Kingdom.

Speight, J. G., (2012). *Crude Oil Assay Database*. Knovel, Elsevier, New York. http://www.knovel.com/web/portal/browse/display?_EXT_KNOVEL_DISPLAY_bookid=5485&VerticalID=0 (accessed on 25 October 2021).

Speight, J. G., (2013a). *Heavy Oil Production Processes*. Gulf Professional Publishing, Elsevier, Oxford, United Kingdom.

Speight, J. G., (2013b). *Heavy and Extra Heavy Oil Upgrading Technologies*. Gulf Professional Publishing, Elsevier, Oxford, United Kingdom.

Speight, J. G., (2014). *The Chemistry and Technology of Petroleum* (5th edn.). CRC Press, Taylor & Francis Group, Boca Raton, Florida.

Speight, J. G., (2015). *Handbook of Petroleum Product Analysis* (2nd edn.). John Wiley & Sons Inc., Hoboken, New Jersey.

Speight, J. G., (2017). *Handbook of Petroleum Refining*. CRC Press, Taylor & Francis Group, Boca Raton, Florida.

Stanislaus, A., Marafi, A., & Rana, M. S., (2010). Recent advances in the science and technology of ultra-low sulfur diesel (ULSD) production. *Catal. Today, 153*, 1–6.

Stringfellow, W. T., (2001). *Biological Upgrading of Heavy Oil for Viscosity Reduction.* Chevron Research and Technology Company, Richmond CA Texaco Exploration and Production Technology Department, Houston, TX.

Sugaya, K., Nakayama, O., Hinata, N., Kamekura, K., Ito, A., Yamagiwa, K., & Ohkawa, A., (2001). Biodegradation of quinoline in crude oil. *J. Chem. Technol. Biotechnol., 76*, 603–611.

Sun, S., Zhang, Z., Luo, Y., Zhong, W., Xiao, M., Yi, W., et al., (2011). Exopolysaccharide production by a genetically engineered *Enterobacter* cloacae strain for microbial enhanced oil recovery. *Bioresour. Technol., 102*, 6153–6158.

Tavassoli, T., Mousavi, S. M., Shojaosadati, S. A., & Salehizadeh, H., (2012). Asphaltene biodegradation using microorganisms isolated from oil samples. *Fuel, 93*, 142–148.

Thouand, G., Bauda, P., Oudot, J., Kirsch, G., Sutton, C., & Vidalie, J. F., (1999). Laboratory evaluation of crude oil biodegradation with commercial or natural microbial inocula. *Can. J. Microbiol., 45*, 106–115.

Tilstra, L., Eng, G., Olson, G. J., & Wang, F. W., (1992). Reduction of sulfur from polysulphidic model compounds by the hyperthermophilic *Archaebacterium Pyrococcus furiosus. Fuel, 71*, 779–783.

Torkamani, S., Shayegan, J., Yaghmaei, S., & Alemzadeh, I., (2008a). Study of a newly isolated thermophilic bacterium capable of kuhemond heavy crude oil and dibenzothiophene biodesulfurization following 4S pathway at 60°C. *J. Chem. Technol. Biotechnol., 83*(12), 1689–1693.

Torkamani, S., Shayegan, J., Yaghmaei, S., & Alemzadeh, I., (2008b). Study of the first isolated fungus capable of heavy crude oil biodesulfurization. *Ind. Eng. Chem. Res., 47*(19), 7476–7482.

Torkamani, S., Shayegan, J., Yaghmaei, S., & Alemzadeh, I., (2009). *Annual Report of 'Heavy Crude oil Biodesulfurization Project' Initiated by Petroleum Engineering Development Company (PEDEC)*. A subsidiary of National Iranian Oil Company.

Uraizee, F. A., Venosa, A. D., & Suidan, M. T., (1998). A model for diffusion-controlled bioavailability of crude oil components. *Biodegradation, 8*, 287–296.

Vaz, D. A., Gudina, E. J., Alameda, E. J., Teixeira, J. A., & Rodrigues, L. R., (2012). Performance of a biosurfactant produced by a *Bacillus subtilis* strain isolated from crude oil samples as compared to commercial chemical surfactants. *Colloids Surf B Biointerfaces, 89*, 167–174.

Vazquez-Duhalt, R., Torres, E., Valderrama, B., & Le Borgne, S., (2002). Will biochemical catalysis impact the petroleum refining industry? *Energy Fuels, 16*, 1239–1250.

Xu, G. W., Mitchell, K. W., & Moticello, D. J., (1997). *Process for Demetallizing a Fossil Fuel*. United States Patent 5,624,844.

Xu, G. W., Mitchell, K. W., & Moticello, D. J., (1998). *Fuel Product Produced by Demetallizing a Fossil Fuel with an Enzyme*. United States Patent 5,726,056.

Xu, P., Yu, B., Li, F. L., Cai, X. F., & Ma, C. Q., (2007). Microbial degradation of sulfur, nitrogen and oxygen heterocycles. *Trends Microbiol., 14*, 398–405.

Yamada, K. O., Morimoto, M., & Tani, Y., (2001). Degradation of dibenzothiophene by sulfate-reducing bacteria cultured in the presence of only nitrogen gas. *J. Biosci. Bioeng., 91*, 91–93.

Yamada, Y., Matsumoto, S., Kakiyama, H., & Honda, H., (1979). *Assigned to Agency of Industrial Sciences and Technology,* Japan. Japanese patent No. JP patent 54110206.

Yang, C., & Lou, Z., (1997). Screening standard and geological bases for microbial enhanced oil recovery. *Petrol. Explor. Develop., 24*(2), 69–72.

Yi, N., Xue, G., HogShuai, G., XiangPing, Z., & SuoJiang, Z., (2014). Simultaneous desulfurization and denitrogenation of liquid fuels using two functionalized group ionic liquids. *Sci. China, 57*(12), 1766–1773.

Yu, B., Xu, P., Shi, Q., & Ma, C., (2006). Deep desulfurization of diesel oil and crude oils by a newly isolated *Rhodococcus erythropolis* strain. *Appl. Environ. Microbiol., 72*, 54–78.

Yu, L., Meyer, T. A., & Folsom, B. R., (1998). *Oil/Water/Biocatalyst Three Phase Separation Process*. US Patent No. 5,772,901.

Zakaria, B. S., Nassar, H. N., Abu Amr, S. S., & El-Gendy, N. Sh. (2015). Applying factorial design and response surface methodology to enhance microbial denitrogenation by Tween 80 and yeast extract. *Pet. Sci. Technol. 33,* 880–892. https://doi.org/10.1080/10916466.2015.1021013 (accessed on 23 December 2021).

Zakaria, B. S., Nassar, H. N., El-Gendy, N. Sh., El-Temtamy, S. A., & Sherif, S. M. (2016). Denitrogenation of carbazole by a novel strain *Bacillus clausii* BS1 isolated from Egyptian coke. *Energy Sources A: Recovery Util. Environ. Eff. 38,* 1840–1851. https://doi.org/10.1080/15567036.2015.1004389 (accessed on 23 December 2021).

Zhang, F., She, Y., Banat, I. M., Chai, L., Yi, S., Yu, G., & Hou, D., (2014). Potential microorganisms for prevention of paraffin precipitation in a hypersaline oil reservoir. *Energy Fuels, 28*(2), 1191–1197.

Zhou, Z. Y., & Zhang, K., (2004). Development situation and prospect of oil fields in China. *Petrol. Explor. Develop., 31*, 84–87.

Zobell, C. E., (1953). *Process of Removing Sulfur from Petroleum Hydrocarbons and Apparatus*. US Patent 2,641,564.

Zou, C., Wang, M., Xing, Y., Lan, G., Ge, T., Yan, X., & Gu, T., (2014). Characterization and optimization of biosurfactants produced by *Acinetobacter baylyi* ZJ2 isolated from crude oil contaminated soil sample toward microbial enhanced oil recovery applications. *Biochem. Eng. J., 90*, 49–58.

Zrafi-Nouira, I., Guermazi, S., Chouari, R., Safi, N. M. D., Pelletier, E., Backhrouf, A., Saidane-Moshabi, D., & Sghir, A., (2009). Molecular diversity analysis and bacterial population dynamics of an adapted seawater microbiota during the degradation of *Tunisian zarzantine* oil. *Biodegradation, 20*, 467–486.

CHAPTER 3

Novel Insights into Bioremediation of Petroleum-Polluted Environments and Bacterial Catabolic Pathways

LAURA RODRÍGUEZ-CASTRO, ROBERTO E. DURÁN, CONSTANZA C. MACAYA, FLAVIA DOROCHESI, LISETTE HERNÁNDEZ, FELIPE SALAZAR-TAPIA, VANESSA AYALA-ESPINOZA, PATRICIO SANTIS-CORTÉS, XIMENA BÁEZ-MATUS, and MICHAEL SEEGER

Molecular Microbiology and Environmental Biotechnology Laboratory, Department of Chemistry and Center of Biotechnology Daniel Alkalay Lowitt, Universidad Técnica Federico Santa María, Valparaiso, Chile, E-mail: michael.seeger@gmail.com (M. Seeger)

3.1 INTRODUCTION

Oil-derived hydrocarbons (HCs) are major environmental pollutants due to their persistence and high toxicity for biological systems (Prince, 2010; Logeshwaran et al., 2018). The clean-up of HC-polluted sites is a critical challenge to restore the affected ecosystems (Fuentes et al., 2014). The most frequently used bioremediation strategies in HC-polluted environments are biostimulation and bioaugmentation. Biostimulation promotes native microbial metabolism by adding nutrients or biosurfactants, while bioaugmentation consists of the inoculation of autochthonous or allochthonous microorganisms (Fuentes et al., 2014; Macaya et al., 2019a). For the optimal development of a biostimulation strategy, the nutritional requirements and the physicochemical properties of each impacted matrix should be studied (Hazen, 2010; Prince, 2010). Bioaugmentation is employed mainly in contaminated environments that possess a low number of HC-degrading microorganisms (Brusseau and Maier, 2004; Fuentes et al., 2014, 2016). Bacterial strains from soil and marine environments belonging to the genera

Achromobacter, *Acinetobacter*, *Alcaligenes*, *Alcanivorax*, *Arthrobacter*, *Bacillus*, *Brevibacterium*, *Burkholderia*, *Cycloclasticus*, *Marinobacter*, *Nocardia*, *Oleispira*, *Paraburkholderia*, *Pseudomonas*, *Rhodococcus*, *Roseobacter*, and *Sphingomonas* with the capability to degrade HCs have been described (Logeshwaran et al., 2018; Unimke et al., 2018; Macaya et al., 2019a). In aliphatic and aromatic HC catabolism, key enzymes are required to activate the contaminants for further degradation. Different alkane hydroxylases are required to oxidize short-, medium-, and long-chain alkanes (Rojo, 2009; Gibu et al., 2019). Bacterial aromatic degradation involves two steps: the activation of the aromatic ring, and the subsequent cleavage reaction. Aerobic catabolic pathways initially modify the aromatic ring by mono- or dioxygenases, incorporating molecular oxygen to the HC (Seeger et al., 2001; Wang et al., 2017). Aerobic degradation of aromatic compounds is catalyzed by bacterial multicomponent monooxygenases (BMMs) and ring-hydroxylating dioxygenases (RHDs). The aromatic HCs are funneled through several peripheral catabolic pathways into diverse central pathways (Fuentes et al., 2014; El-Naas et al., 2014). Bacterial polycyclic aromatic hydrocarbon (PAH) degradation is mainly achieved by PAH-RHDs, which possess a low substrate specificity, enabling the oxidation of a wide spectrum of aromatic compounds (Notomista et al., 2003; Parales et al., 2008; Wang et al., 2017). In addition, hybrid aerobic pathways are involved in the bacterial degradation of aromatic compounds such as toluene and ethylbenzene (Ismail and Gescher, 2012; Grishin and Cygler, 2015). Anaerobic degradation of HCs has been less characterized than the aerobic catabolic routes. Anoxic activation of the aromatic ring or the aliphatic chain occurs mainly by addition of fumarate, carboxylation, or hydroxylation. In anaerobic pathways, HCs are commonly transformed into acetyl-CoA and CO_2. In the absence of oxygen, HC degradation in bacteria is linked to the use of nitrate, iron, sulfate, and other compounds as electron acceptors (Kniemeyer et al., 2003; Fuchs et al., 2011; Meckenstock et al., 2016; Rabus et al., 2016; Boll and Estelmann, 2020). The microbial communities during petroleum pollution events and bioremediation have been studied to characterize their dynamics and understand the roles of main microbial members involved in HC biodegradation (Allen et al., 2007; Fuentes et al., 2016; Mahjoubi et al., 2018; Franzetti et al., 2020). The characterization of novel hydrocarbonoclastic microorganisms and the elucidation of their aliphatic and aromatic HC catabolic pathways are required to understand the complexities of the degradation of HCs and to establish novel bioremediation processes for specific polluted sites towards a more sustainable development. This chapter discusses the main bioremediation strategies for petroleum-polluted sites, and describes the key bacterial enzymes and catabolic pathways for the degradation of aliphatic and aromatic HCs.

3.2 PETROLEUM HYDROCARBONS: COMPOSITION AND PERSISTENCE

The term "petroleum" derives from the Latin language —*petra,* which means rock, and *oleum*, which means oil—, and is used to refer to a complex mixture of HC compounds. The term 'oil' refers to the naturally occurring liquid petroleum, also called crude oil, to differentiate it from their gaseous components, natural gas, or from solid petroleum, such as asphalt, pitch, or bitumen. Petroleum is present as accumulations on rock pores and, in some cases, as solid tar deposits or surface leakages (Speight, 2014; Alsadi, 2018). In a global context, petroleum is crucial for industrial development and manufacturing processes, constituting a critical resource for the entire world. Plastics, solvents, fertilizers, pesticides, dyes, and textiles, among many other products are obtained from different oil fractions (Jafarinejad, 2016; Alsadi, 2018).

The chemical composition of petroleum mostly consists of two basic elements, hydrogen (H) and carbon (C). According to the number of carbon and hydrogen atoms, as well as the type of covalent carbon-carbon bonds, HC molecules could have different forms, sizes, and physicochemical properties. HCs can possess more than 60 carbon atoms, where a higher molecule size correlates with a higher boiling point (Fuentes et al., 2014). Petroleum is mainly composed of HCs, but also contains a minor presence of nitrogen, oxygen, and sulfur compounds, along with heavy metals and metalloids. The geological environment where the oil has been generated, migrated, and/or accumulated, determines the physicochemical properties of the crude oil mixtures. These properties are density, viscosity, surface tension, flash point, refraction index, caloric value, and electrical conductivity (Alsadi, 2018).

According to their chemical structures, petroleum compounds are classified into four fractions: aliphatics, aromatics, resins, and asphaltenes. Aliphatic HCs, which constitute the most abundant fraction, are linear or branched molecules with either unsaturated or saturated carbon-carbon covalent bonds (*n*-alkanes, isoalkanes, cycloalkanes, terpenes, and steranes). Aromatic HCs possess one or more benzene rings in their structure. Monocyclic aromatic hydrocarbons (MAHs) have only one aromatic ring.

Benzene, toluene, ethylbenzene, and the three isomers of xylene (BTEX) are the most volatile and water-soluble aromatic HCs, which explains their broad environmental distribution. Benzene stands out in this group as a model compound and was classified by the International Agency for Research on Cancer as a member of the Group 1 of carcinogenic and toxic agents to humans and environmental health. Due to its toxicity, the maximum levels of benzene in air are regulated in the European Union. PAHs have two or more benzene rings. Low-molecular-weight PAHs, such as naphthalene, anthracene, phenanthrene and fluorene, have two or three benzene rings. High-molecular weight PAHs, such as pyrene and fluoranthene, have four or more benzene rings. Resins and asphaltenes constitute less than 25% of the total composition of crude oils, containing polar functional groups formed with nitrogen, sulfur, oxygen, and trace metal (e.g., nickel, vanadium, and iron), alongside aromatic compounds containing long alkyl chains. Asphaltenes are high molecular weight compounds colloidally dispersed in saturated and aromatic fractions (Macaya et al., 2019a).

3.2.1 ENVIRONMENTAL IMPACT OF PETROLEUM HYDROCARBONS

Pollution and wastes are associated with petroleum industrial activities, such as exploration, development, HC processing (refineries and petrochemical plants), storage, transport, and distribution. In every step of the transformation of petroleum into usable energy, air emissions, wastewater, and solid waste can be released into the environment, with the potential to cause negative effects on the impacted area and harm living organisms (Wang et al., 2011a).

Despite the environmental threats associated with the petrochemical industry, the economy has promoted the constant increase of the distribution and storage networks of crude oil and processed fuels. In 2020, the International Energy Agency (IEA) estimated that the total usable storage capacity of oil in land and sea would reach 99.9 million barrels in the world. In April 2020, oil storage reached 63% of its capacity due to the state of coronavirus pandemic and the concomitant strong decrease in human mobility (Saefong, 2020). Oil production is concentrated in the Middle East and Russia, but specific countries of America and Africa are also strongly involved. The countries with the largest reservoirs are Saudi Arabia, Canada, Iran, Iraq,

Kuwait, United Arab Emirates, Venezuela, Russia, Libya, Nigeria, Kazakhstan, United States, China, Qatar, Algeria, Brazil, and Mexico (Jafarinejad, 2016).

Most of the petroleum transport activities run through pipelines (land transport) and tankers (marine transport). Any failure of these processes may cause negative impacts to the environment in all its dimensions (e.g., air, water bodies, sediments, and soils), due to the high content of volatile and recalcitrant compounds. Although construction technologies, methods, and materials, as well as navigation systems and climate prediction have been improved, the cases in which a release has caused significant political, economic, social, and environmental impacts are not unusual. In fact, between 1970 and 2019 there have been more than 10,000 oil spills into the sea from tankers —three of which occurred during 2019, equivalent to 1,000 tons of crude oil or derivatives— which translates into 5.86 million tons of oil spilled in the last 50 years, according to the International Tanker Owners Pollution Federation (ITOPF, 2019).

In 2020 up to August, two major oil spills from tankers have occurred: (i) Russia (May 29, 2020), where a pond released between 20,000 and 21,000 tons of diesel in the vicinity of the Ambarnaya River, caused by the melting of permafrost, which weakened one of the ponds supports (Nechepurenko, 2020); and (ii) Republic of Mauritius (July 25, 2020), as a result of the stranding of the Japanese ship MV Wakashio in Pointe d'Esny. In the last spill, 1,000 to 1,200 tons of fuel oil leaked into the sea from the vessel that was carrying 4,000 tons (Singh-Khadka, 2020). In the first case, besides the damages caused to the environment near to the river, the main concern was the arrival of fuel to Lake Pyasino and Pyasino River, which leads to the Kara Sea, part of the Arctic Ocean. Russia's Federal Supervisory Natural Resources Management Service (Rosprirodnadzor) values the damage caused by the spill at 2 billion US $ (Ilyushina, 2020). In the second case, although the episode was not a massive spill, the environment of Mauritius that is rich in coral, seagrasses, and mangroves was highly impacted by this pollution event (Lewis, 2020).

Oil-derived HCs are major environmental pollutants due to their persistence and high toxicity to biological systems (Logeshwaran et al., 2018). During petroleum processing, heavy metals and metalloids accumulate in the surrounding areas, increasing the environmental risks of oil-related activities. Even at low levels, some of these heavy metals are highly toxic and may destabilize ecosystems and bioaccumulate in macroorganisms, thus severely affecting their health (Pathak and Mandalia, 2012; Jafarinejad, 2016).

The clean-up of petroleum HC-polluted sites is a critical challenge to restore the impacted ecosystems for a sustainable development. The main bioremediation strategies of petroleum-polluted sites, and the key bacterial enzymes and the metabolic pathways for the degradation of aliphatic and aromatic HCs will be discussed.

3.3 BIOREMEDIATION: STRATEGIES AND MICROBIAL COMMUNITIES

Conventional physicochemical treatments for the clean-up of oil spills include extraction with pumps, barriers, skimmers, and adsorbent materials, which rarely achieve complete removal of HC contaminants (Ossai et al., 2020). Another method is the application of chemical dispersants, which breaks the oil into small droplets using emulsifiers and solvents (Grote et al., 2016). However, this treatment can only be applied on oil spills in the sea when the coastline depth is more than 20 meters due to its toxicity to marine organisms. For the clean-up of polluted soils, biological treatments are part of the dominant technologies (20–40% of the applied treatments) in European countries such as the Netherlands, Belgium, Italy, and Slovakia (van Liedekerke et al., 2014). Bioremediation is a technology of increasing application due to its attractive properties: low-cost, bio-based, eco-friendly, with reduced impact on global warming (Ron, 2000; Navia and Seeger, 2006; Fuentes et al., 2014; Macaya et al., 2019a). This technology takes advantage of microbial catabolic potential to accelerate natural degradative processes in contaminated environments. Bioremediation improves natural microbial degradation in the impacted area by modifying environmental factors such as nutrients and electron acceptors (e.g., oxygen), and by activating or adding microorganisms to reduce, transform, and mineralize the pollutants (Ron, 2000; Navia and Seeger, 2006; Hernández et al., 2012; Fuentes et al., 2014; Macaya et al., 2019a).

The most frequently applied bioremediation strategies in HC-polluted environments are biostimulation and bioaugmentation (Navia and Seeger, 2006; Fuentes et al., 2014, 2016; Macaya et al., 2019a). Biostimulation promotes microbial metabolism by adding nutrients (e.g., nitrogen and phosphorus), or biosurfactants/bioemulsifiers. Bioaugmentation consists of the inoculation of autochthonous or allochthonous microorganisms to increase the degradation rate. The bioremediation strategy should consider the physicochemical properties of the polluted matrix (Macaya et al., 2019a).

For water, the most used bioremediation strategies are biostimulation, bioaugmentation, and bioventing (air injection to improve aerobic metabolism). For soil, biostimulation, bioaugmentation, and landfarming (wetting and turning the contaminated soil to improve aeration and homogenization) are the most commonly employed processes (Atlas, 2005; Macaya et al., 2019a). In aquatic environments affected by oil spills, oxygen, and moisture requirements are not an issue, but in soil, these factors may reduce the biodegradation rate due to diffusional limitation and matrix heterogeneity (Atlas, 1991).

Another critical factor of the remediation strategy is the selection of *on-site* or *off-site* bioremediation processes. *On-site* bioremediation is performed at the polluted site and could be applied *in situ* (without extracting the polluted soil or water) o *ex-situ* (e.g., depositing the excavated soil in biopiles or extracting water to a treatment plant in the polluted site). In contrast, *off-site* bioremediation requires transport of the polluted matrix to the final treatment location, increasing air pollution during transport and the total remediation cost (Chandra et al., 2013; Atlas, 2005). Table 3.1 shows an overview of the widely employed bioremediation strategies.

Contamination events disturb the diversity, structure, and abundance of bacterial, fungal, and archaeal communities that inhabit the polluted environment. Oil spills on soils and aquatic environments are highly toxic for microbial communities. However, specific microorganisms are tolerant to HCs and capable of degrading the pollutants. Bacteria are the predominant microorganisms that degrade HCs. Marine obligate hydrocarbonoclastic bacteria such as *Alcanivorax*, *Marinobacter*, and *Oleispira* have been widely characterized. In pristine ecosystems, these bacterial genera are present at low or undetectable levels, representing less than 0.1% of the total community. After an oil contamination event, the abundance of hydrocarbonoclastic bacteria increases and may represent more than 60% of the microbial community (Fuentes et al., 2016; Mahjoubi et al., 2018). Fuentes et al. (2016) reported a short-lapse dynamics of soil microbial communities in response to diesel pollution (30 g kg^{-1} dry soil) and different bioremediation treatments. Bacterial community analyses showed an operational taxonomic unit (OTU) related to the *Alkanindiges* genus, which strongly increased from ~ 0.1% in the initial microbiota to >60% after six weeks (Fuentes et al., 2016). The characterization of the microbial communities in pristine fields, wellhead, and storage tanks of an oil production facility indicated that Proteobacteria (Pseudomonadota) predominate in the virgin field and the wellhead, while the storage tank showed a higher presence of Firmicutes (Bacillota). The

TABLE 3.1 Estimated Costs, Advantages, and Disadvantages of Remediation Strategies for Hydrocarbon-Polluted Matrices

Matrix	Strategy	Treating Site	Cost[a]	Units	Advantages	Disadvantages
Soil and sediment	Biological					
	Biostimulation	*In situ*	30–100	[US $/m^3]	Use of native microorganisms.	Long treatment time.
	Bioaugmentation	*In situ*	30–100	[US $/m^3]	High efficiency, specific pollutants degradation.	Dependency on environmental factors and site characteristics.
	Bioventing	*In situ*	79–970	[US $/m^3]	Readily available equipment.	Requires high soil permeability.
	Biopiles	*Ex-situ*	130–260	[US $/m^3]	Low environmental impact and simple operation.	Requires excavation of contaminated soils.
	Composting	*Ex-situ*	629–756	[US $/m^3]	Low maintenance.	Demands availability of large physical space.
	Landfarming	*Ex-situ*	<100	[US $/m^3]	Simple and cost-effective.	Demands availability of large physical space.
	Physicochemical					
	Thermal treatment	*In situ*	38–81	[US $/m^3]	Low atmospheric pollutants emission.	High energy costs, low removal efficiency of some compounds.
	Soil vapor extraction	*In situ*	405–1,485	[US $/m^3]	Minimal physical disturbance to the matrix.	High cost, requires air treatment, low removal efficiency of some contaminants.
	Other Treatments					
	Off-site disposal	*Ex-situ*	270–460	[US $/ton]	Low treatment time.	No removal of pollutants, pollution by soil transport, high environmental impact.

TABLE 3.1 *(Continued)*

Matrix	Strategy	Treating Site	Cost[a]	Units	Advantages	Disadvantages
Ground and surface water	Biological					
	Biostimulation (oxygen sparging)	*In situ*	10–20	[US $/m^3]	Improves the aerobic degradation by native microorganisms.	May affect other natural on-site remediation processes (anaerobes).
	Biostimulation (calcium peroxide)	*In situ*	3–5[b]	[US $/m^3]	Improves the aerobic degradation by native microorganisms.	May affect other natural on-site remediation processes (anaerobes).
	Bioaugmentation (pump and treatment)	*In situ*	10–40	[US $/m^3]	High efficiency, specific pollutant degradation.	High energy cost (pumping), frequent maintenance.
	Physicochemical					
	Air sparging	*In situ*	24–84	[US $/m^3]	High availability of technology.	No removal of pollutants, requires homogenous environmental conditions.
Air emission/ off-gas	Biological					
	Biofiltration	*In situ*	5–10	[US $/kg of contaminant]	Low operation cost.	High temperature dependence.
	Physicochemical					
	Carbon adsorption	*In situ*	35–59	[US $/m^3 per min]	Simple operation.	Fast saturation for highly contaminated air volumes.

[a]Data from the US Federal Remediation Technologies Roundtable (2020). Available at: http://www.frtr.gov/matrix2/top_page.html.
[b]Operation and maintenance costs are not included.

predominant genera were *Ochrobactrum* and *Acinetobacter* in the virgin field, *Lactococcus*, and *Pseudomonas* in wellhead, and *Prauseria* and *Bacillus* in the storage tank (Liu et al., 2019b). Microbiological analysis in sediments showed that high proportions of cultivable HC-degrading microorganisms correlate with an increase in the underground HC concentration (Allen et al., 2007).

Interestingly, HC-degrading bacteria have also been found in unconventional environments. A microbial community from an aircraft tank that grows on fuel-HCs has been reported (Itah et al., 2009). Atmospheric HCs could be a determinant of microbial community structure in the phyllosphere of plants in urban areas (Gandolfi et al., 2017). An increase in atmospheric PAHs in urban areas led to high HC levels on tree leaves, raising the abundance of PAH-degrading microorganisms that degrade these pollutants on the phyllosphere (Franzetti et al., 2020).

3.3.1 BIOSTIMULATION

Biostimulation is the addition of nutrients, electron acceptors or surfactants in a contaminated matrix to stimulate the growth rate and the catabolic activity of endogenous microorganisms (Macaya et al., 2019a; Hoang et al., 2020). The addition of these compounds in a HC-polluted ecosystem promotes microbial metabolism and enhances HC degradation by the native community (Fuentes et al., 2014; Varjani et al., 2020).

Several naturally occurring pollutants are usually present in the environment. Thus, in a large number of environments, native microorganisms are exposed to pollutants for long periods of time, developing adaptive capabilities to survive and proliferate. Natural environments that lack pollutant-degrading microorganisms are scarce. Even in pristine environments, the presence of bacteria that possess genes to catabolize recalcitrant organic compounds has been reported (Fredrickson et al., 1988). Nevertheless, the long-term exposure to pollutants determines the microbial community composition (Nikolopoulou and Kalogerakis, 2010). Therefore, a higher proportion of pollutant degraders is observed in environments with chronic contamination than in non-polluted sites.

During an oil spill event, a nutritional imbalance in the impacted environment is generated due to the increasing carbon content compared to other elements. The addition of nutrients allows to compensate this nutritional imbalance, mainly in the ratio of carbon (C), nitrogen (N), and phosphorus (P) content. The nutrient requirements will be site-specific according to the HC components and the baseline value of nutrients in a specific polluted environment, although an optimal ratio of C:N:P ~ 30:5:1 is required for a successful bioremediation (Ossai et al., 2020). However, the *in-situ* application of nutrients is challenged by the heterogeneous distribution of the contaminants.

Organic amendments are composed of organic moieties derived from biomass and/or organisms (e.g., compost, wood chips, biochar, animal manure, and straw). These substances are rich in organic matter as well as macro- and microelements (Wang et al., 2017). These organic compounds are frequently used in treatments of polluted soil. Inorganic amendments commonly contain minerals associated with soil fertility (e.g., gypsum, coal combustion byproducts, fly ash, paraffinized urea, dioctyl-pyrophosphate). Slow-release mixtures that include inorganic nutrients coated in solid forms or paraffinized are used in the treatment of large water bodies in order to avoid the diffusion of the components into the surrounding water, focusing their presence and action in the impacted area (Prince, 2010; Macaya et al., 2019a).

Another critical parameter to bioremediate HC-polluted systems, especially in aquatic environments, is the bioavailability of the pollutants. Non-polar tails of long-chain HCs have low solubility in water, limiting microbial colonization. Chemical detergents or biosurfactants are applied to reduce surface tension and enhance the microbial accessibility to HCs. The operation effectiveness depends on the physicochemical properties of the HC pollutants and the water body. Specific microorganisms such as *Pseudomonas fluorescens*, *Bacillus subtilis*, *Acinetobacter radioresistens* and *Candida bombicola* may be applied for the production of biosurfactants and bioemulsifiers such as sophorolipids, rhamnolipids, and alasan (Macaya et al., 2019a, b).

For the optimal development of a biostimulation strategy, the nutritional requirements of each impacted matrix should be determined by evaluating the matrices on a case-by-case basis. Environmental impacts of the biostimulation process in open waters should be analyzed since the migration of nutrients to the surrounding water may generate eutrophication. In soil, the application of inorganic nutrients may change the biogeochemical characteristics of the site, precipitating metals and changing the pH, redox potential, and conductivity (Hazen, 2010; Prince, 2010).

3.3.2 BIOAUGMENTATION

Bioaugmentation consists in the addition of microbial cultures, previously selected based on their specific degradation capability, adaptive response, genetic stability, and compatibility with native microbial communities, to improve pollutant removal rate and to reduce treatment times in polluted matrices (Venosa et al., 2002; Fuentes et al., 2014; Orellana et al., 2018). Pollutant-degrading microorganisms may be isolated by selective enrichment (using the target compounds as the sole carbon or nitrogen sources) from the polluted site (autochthonous microorganisms) to be subjected to the clean-up or isolated from similar polluted sites (allochthonous microorganisms) (Fuentes et al., 2016; Méndez et al., 2017, 2018; Speight, 2017; Macaya et al., 2019a). Bioaugmentation is employed mainly in contaminated environments that possess a low number of native microorganisms capable of degrading petroleum-derived pollutants, as well as in environments with native microbial communities with low degradation efficiency due to the toxicity of the pollutants (Brusseau and Maier, 2004; Fuentes et al., 2014).

Microorganisms may be added to a polluted environment as a single strain or a consortium (Fuentes et al., 2016; Macaya et al., 2019a). Microbial consortia have advantages compared to individual strains due to the broader spectrum of petroleum-derived compounds to be degraded, the potential synergic interactions of different strains, and the higher probability of their environmental adaptation (Fuentes et al., 2016; Speight, 2017). For example, the interactions between fermentative and HC-degrading bacteria, which produce volatile fatty acids that are used by syntrophic organisms such as *Smithella* and *Syntrophus* strains to produce acetate, formate, and hydrogen. These intermediates are metabolized to methane and carbon dioxide by acetoclastic and hydrogenotrophic methanogenic microorganisms, funneling degradation under anaerobic conditions into an association called "degradation of syntrophic methanogenic HCs" (Gray et al., 2011; Tan et al., 2013; Gieg et al., 2014).

Other approaches of bioaugmentation include genetically modified microorganisms (GMOs) or catabolic genes encoded in a vector that could be transferred by conjugation to native microorganisms (Saavedra et al., 2010; Rojas et al., 2011; Seeger et al., 2011, 2014; Bravo, 2017; Bravo et al., 2020). The degradation capabilities of these GMOs may be superior to native microorganisms (Saavedra et al., 2010; Rojas et al., 2011; Seeger et al., 2011, 2014). Nevertheless, their use in field-scale may be not successful, due to their low adaptation to an ecological niche, or their potential adsorption on solid surfaces, limiting their distribution in the contaminated ground. A critical issue is the public concern and regulatory norms of releasing GMOs into the environment (Brusseau and Maier, 2004).

Full-scale applications of bioaugmentation strategies are challenged by various complexities such as: (1) limitation of microorganisms to reach the pollutants; (2) concentration of contaminants that does not support the metabolism of the inoculated cells; (3) inoculated microorganisms may use substrates other than the contaminants; (4) the occurrence of leaching or predation (the growth rate of the degrading organism may be slower than the removal rate); and (5) microbial inhibitory compounds present in the environment (Bon and Verstraete, 2010).

3.3.3 BIOELECTROCHEMICAL SYSTEMS (BES)

The microbial conversion of organic matter into electrical current was reported for the first time more than a century ago (Potter, 1911; Cohen, 1931). Microbial fuel cell (MFC) is a system in which microorganisms catalyze the conversion of chemical energy into electrical energy, where whole cells and enzymes are used as biocatalysts to achieve oxidation and/or reduction reactions (Berk and Canfield, 1964; van Hees, 1965; Rabaey et al., 2010). In bioelectrochemical systems (BES), insoluble electrons from microbial respiration are accepted by the anode (Rabaey et al., 2010). This approach can be used to stimulate microbial activity in environments where electron acceptors are deficient or where acceptors have been depleted due to the presence of high amounts of carbon sources, for example, after an oil pollution event (Daghio et al., 2016, 2017).

BES may be used as a clean-up strategy when traditional bioremediation is not successful due to the absence of suitable electron acceptors (Espinoza-Tofalos et al., 2020). In addition, microbial metabolism in a BES can be stimulated with the application of an overpotential, thus increasing the rate of pollutant degradation. In BES, the degradation of HCs may be concomitant with the production of electrical energy (Morris and Jin, 2007). BES with pure microbial cultures have been applied for the degradation of BTEX compounds (Zhang et al., 2010; Sun et al., 2014; Lai et al., 2016; Espinoza-Tofalos et al., 2018). Microbial communities have also been used for the degradation of aromatic compounds (Lin et al., 2014; Wu et al., 2014; Daghio et al., 2016, 2018). However, the main challenge in this field is to scale-up the BES technology for commercial remediation applications (Espinoza-Tofalos et al., 2020).

3.3.4 SUCCESSFUL BIOREMEDIATION TRIALS

Bioremediation processes of HC-polluted soils at laboratory scale have been performed in different countries. Successful bioaugmentation of a crude oil-polluted soil (30,000 ppm) in India with a bacterial consortium (*Stenotrophomonas maltophilia*, *Pseudomonas aeruginosa* and *Ochrobactrum* sp. strains) resulted in > 80% removal after 75 days (Varjani et al., 2015). Bioaugmentation of soil polluted with anthracene, phenanthrene, and pyrene (1,000 ppm) in Brazil by a consortium of five bacteria (*Gordonia polyisoprenivorans*, *Mycobacterium fortuitum*, *Bacillus cereus*, *Microbacterium* sp. and a *Microbacteriaceae* strain) and the fungus *Fusarium oxysporum*, achieved a high degradation (96%) of these aromatic pollutants after 70 days (Jacques et al., 2008). Biostimulation of soils highly polluted with total petroleum hydrocarbons (TPH) (>380,000 ppm) in South Africa, by mixing the soil with wood chips in a ratio of 1:1 and later adding sewage sludge in a ratio of 1:4, removed 99% TPH after 19 months (Atagana, 2010). In contrast, only 17% of TPH were removed by natural attenuation. Biostimulation of a soil polluted with TPH (41,000 ppm) in Argentina, by mixing soil with apple filter cake in a ratio of 7:3 in a rotating barrel, achieved high degradation of the alkanes in gasoline range organics (GRO) (>92%) and diesel range organics (DRO) (>86%) after 90 days (Medaura and Ercoli, 2008). A combination of biostimulation and bioaugmentation was used for the clean-up of weathered HC-polluted soils (100,000 ppm) in Mexico, applying sugarcane bagasse and sugarcane bagasse-associated microorganisms, removed 60% TPH after 16 days (Pérez-Armendáriz et al., 2004).

Diverse successful cases of full-scale bioremediation projects have been described. In Poland, the bioremediation of contaminated soils (30,000 ppm TPH) impacted with acidic petroleum sludge by biostimulation using nutrients and a surfactant in biopiles was studied (Hazen et al., 2003). The main pollutants were BTEX and benzo[a]pyrene. After 20 months, more than 81% of the petroleum HCs were removed. In India, a consortium of four native bacterial strains was applied in different soils polluted with oil refinery wastes, achieving the removal of 48,914 tons of pollutants distributed in 44 separate cases (Mandal et al., 2012). The contaminated soils were treated in biopiles, and the consortia were inoculated by spraying a solution supplemented with nitrogen, phosphorus, and potassium. The initial pollutant concentration ranged from 84 to 531 g of TPH/kg of soil, reaching a final concentration of less than 10 g/kg soil after bioremediation in periods up to 12 months.

3.4 AEROBIC BACTERIAL DEGRADATION OF PETROLEUM HYDROCARBONS (HCS)

Diverse microorganisms possess the capability to degrade petroleum HCs (Fuentes et al., 2014, 2016; Macaya et al., 2019a). HC-degrading bacterial strains of the genera *Achromobacter*, *Acinetobacter*, *Arthrobacter*, *Bacillus*, *Brevibacterium*, *Burkholderia*, *Nocardia*, *Paraburkholderia*, *Pseudomonas*, *Rhodococcus*, *Rhodotorula*, *Sphingomonas*, *Sporobolomyces*, and *Vibrio* have been isolated from various environments. Fungal isolates belonging to *Candida*, *Hansenula*, *Penicillium*, *Pichia*, *Saccharomyces*, and *Yarrowia* genera have also been described as degraders of crude oil components. In aquatic environments, hydrocarbonoclastic bacteria from *Alcaligenes*, *Alcanivorax*, *Cycloclasticus*, *Dietzia*, *Marinobacter*, and *Roseobacter* taxa have been reported. HC-degrading bacteria from soil niches, mainly belonging to the *Achromobacter*, *Acinetobacter*, *Micrococcus*, *Paraburkholderia*, *Pseudomonas*, and *Rhodococcus* genera have been extensively characterized for HC catabolism. This group includes some widely known HC-degrading model bacteria such as *Acinetobacter venetianus* RAG-1, *Pseudomonas putida* mt-2, and *Rhodococcus jostii* RHA1. Fungal strains from the *Hansenula*, *Pichia*, *Saccharomyces*, and *Yarrowia* genera degrade medium-chain alkanes (Logeshwaran et al., 2018; Unimke et al., 2018; Macaya et al., 2019a).

3.4.1 ALKANE CATABOLISM

Alkanes are saturated HCs that can be linear (*n*-alkanes), branched (*iso*-alkanes) or cyclic (*cyclo*-alkanes) (Scoma et al., 2015). Alkanes are the most abundant family of HCs with an estimated abundance of 20–50% in crude oil and are commonly found in oil-contaminated environments (Rojo, 2009). Aerobic and anaerobic microorganisms utilize alkanes as sole carbon and energy sources (Sierra-Garcia and Oliveira, 2013). Aerobic degradation by bacteria usually occurs through the sequential oxidation of one or both terminal methyl groups of the molecule (Figure 3.1). This process is usually initiated by an alkane hydroxylase (alkane monooxygenase), which converts alkanes to alkanols. Alkanols are oxidized by sequential reactions of alcohol and aldehyde dehydrogenases (ALDH) into the corresponding fatty acids, which are assimilated through the β-oxidation pathway. Alkane hydroxylases are key enzymes in the degradation, since they activate the alkanes, allowing their subsequent degradation (Rojo, 2009; Gibu et al., 2019). Most of the studies that describe the mechanisms associated with bacterial degradation of alkanes have focused on the enzymes that catalyze the initial oxidation step. Depending on the length of the substrate chain, different alkane hydroxylases are required to oxidize alkanes (Shuai et al., 2019). These enzymatic systems are classified into three categories:

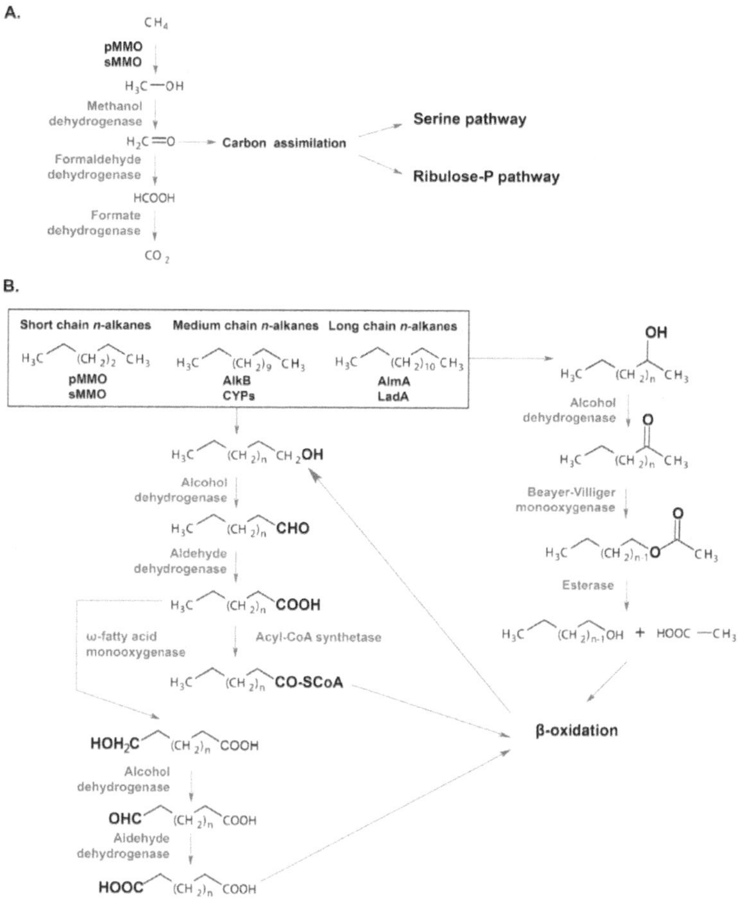

FIGURE 3.1 An overview of aerobic *n*-alkanes degradation pathways in bacteria. Degradation pathways of methane (A) and short- (> C_2), medium- (> C_5) and long-chain (> C_{17}) *n*-alkanes (B) are represented. Only alkane monooxygenases are specific for each alkane group. The remaining catabolic enzymes are shared by all these catabolic pathways, independently of their chain length. For each molecule, bold lettering indicates the functional group modified by the involved enzymes. [*Abbreviations:* pMMO: particulate methane monooxygenase; sMMO: soluble methane monooxygenase; AlkB: alkane monooxygenase; CYP: cytochrome P450; LadA/AlmA: long-chain alkane monooxygenase].

1. **Short-Chain Alkanes (C_1–C_4):** Aerobic bacterial metabolism of short-chain alkanes has received less attention compared to the catabolism of medium- and long-chain alkanes (Cappelletti et al., 2015). Methane-utilizing bacteria or methanotrophs oxidize methane to methanol by methane monooxygenase-like enzymes (MMO). *Thauera butanivorans* is a model Gram negative bacterium for the degradation of gaseous *n*-alkanes. The insertion of one oxygen from dioxygen (O_2) into the short-chain alkane is catalyzed by an iron- or copper-containing monooxygenase (MO), which may be a particulate MMO (pMMO) or a soluble MMO (sMMO) (Hamamura et al., 2001). pMMOs are present in all methanotrophs, while sMMOs were described in some methanotrophs and are expressed only under low copper availability (Kotani et al., 2006). Crystal structures of pMMOs from methanotrophs reveal a trimeric structure ($\alpha_3\beta_3\gamma_3$), consisting of three copies of PmoB, PmoA, and PmoC subunits (Culpepper and Rosenzweig, 2012). The sMMO system is composed of three components: a hydroxylase containing three polypeptides arranged as a $\alpha_2\beta_2\gamma_2$ dimer (MMOH; comprising the MmoX, MmoY, and MmoZ subunits), a reductase (MmoC) that transfers electrons from NADH to MMOH (responsible for O_2 activation), and a regulatory protein (MmoB) that is required for its activity (Zhang et al., 2006). *Rhodococcus* sp. BCP1 is able to grow on gaseous and short-chain *n*-alkanes (up to *n*-heptane). Strain BCP1 possesses both *prmABCD* and *smoABCD* gene clusters, which encode a di-iron monooxygenase that is involved in gaseous *n*-alkanes oxidation (Cappelletti et al., 2015). *Nocardioides* sp. CF8 is able to grow on alkanes ranging from C_2 to C_{16} and possesses two methane monooxygenases (pMMO and sMMO). In contrast to other methanotrophs, strain CF8 simultaneously synthesizes both monooxygenases (Hamamura and Arp, 2000; Hamamura et al., 2001).
2. **Medium-Chain Alkanes (C_5–C_{16}):** Aerobic medium-chain *n*-alkanes degradation by bacteria has been widely studied. Medium-chain *n*-alkanes are oxidized generally by an integral membrane non-heme di-iron AlkB-type monooxygenase, but also a cytochrome P450 (CYP) oxidizes these substrates (Liu et al., 2011; Laczi et al., 2015). Alkane degradation pathway in *Pseudomonas putida* GPo1 has been well characterized (van Beilen et al., 2001). The first step proceeds by hydroxylating the alkane substrate at the terminal position by AlkB. The two soluble electron transfer proteins rubredoxin (AlkG) and

rubredoxin reductase (AlkT) deliver electrons to AlkB to carry out the hydroxylation step. AlkT transfers electrons from NADH via a FAD cofactor to AlkG and eventually to AlkB. The alkanol is converted by an alcohol dehydrogenase to an aldehyde which is oxidized by an ALDH into a fatty acid that enters the β-oxidation pathway (Rojo, 2009; Gibu et al., 2019). In *P. putida* GPo1 the degradation genes are encoded in the *alkBFGHJKL* and *alkST* operons of the plasmid OCT. The *alkBFGHJKL* operon encodes enzymes involved in the degradation of alkanes. The *alkST* operon encodes the transcriptional regulator AlkS and rubredoxin reductase AlkT. The transcriptional regulator AlkS is activated by short-chain alkanes (C_5–C_{11}) and induces the expression of the *alkBFGHJKL* operon.

More than 60 *alkB* genes (*alkB* or *alkM*) have been reported in Gram-negative bacteria (including α-, β-, and γ-proteobacteria) and high C+G content Gram-positive bacteria (e.g., *Rhodococcus erythropolis* NRRL B-15631, *Mycobacterium tuberculosis* H37Rv, *Gordonia* sp. SoCg, *Nocardioides* sp. CF8, *Dietzia* sp. DQ12-45-1b) (Smits et al., 1999, 2002; Hara et al., 2004; Piccolo et al., 2011; Wang et al., 2011b; Zampolli et al., 2014). Diverse bacteria contain more than one *alkB* gene. For example, *R. erythropolis* strains Q15 and NRRLB16531 possess four alkane monooxygenase genes (*alkB1*, *alkB2*, *alkB3* and *alkB4*). The *alkB1* and *alkB2* genes are parts of the *alk* gene cluster containing two rubredoxins (*rubA1*, *rubA2*; *rubA3*, *rubA4*) and a rubredoxin reductase *rubB* gene (homologous to *alkG* and *alkT* genes in *P. putida*), whereas the *alkB3* and *alkB4* genes are distributed in the genome. The heterologous expression of the *alkB1* and *alkB2* genes of *R. erythropolis* strains Q15 and NRRLB16531 in *P. fluorescens* KOB21 allows KOB21 to grow on C_{10}–C_{16} and C_{18} *n*-alkanes (Whyte et al., 2002). The marine *Alcanivorax borkumensis* strain SK2 contains the alkane monooxygenase-encoding *alkB1* and *alkB2* genes; the disruption of the *alkB1* gene affects the degradation of *n*-hexane (Hara et al., 2004). The functional differences of the multiple *alkB* genes should be further studied.

The AlkB alkane hydroxylases possess high sequence diversity and are classified in two groups. The type-1 integral membrane alkane hydroxylases (AH1), which are mainly present in *Pseudomonas* and other γ-proteobacteria, oxidize short- and medium-chain alkanes (van Beilen and Funhoff, 2007). For example, AlkB of *P. putida* GPo1 and AlkB1 of *A. borkumensis* SK2 oxidize C_5–C_{13} alkanes (Hara et al.,

2004; Johnson and Hyman, 2006). Most AlkB proteins such as AlkB2 of *R. erythropolis* Q15 and AlkBa of *Rhodococcus ruber* DSM43338 belong to the type-2 integral membrane alkane hydroxylase (AH2) group that oxidizes long-chain alkanes (van Beilen and Funhoff, 2007; Amouric et al., 2010). A novel gene encoding an AlkB fused with the rubredoxin domain at the C-terminal, AlkW, was reported in *Dietzia* strains. AlkW is responsible for the degradation of C_{14}–C_{36} alkanes (Nie et al., 2014).

Cytochromes P450 (CYPs) are heme-thiolate enzymes ubiquitously distributed among the six kingdoms of life and well known for their role in primary and secondary metabolism (Ogburn et al., 2010; Ortiz De Montellano, 2010). CYPs belong to a superfamily with more than 1,000 members and are subclassified into different families (Maier et al., 2001). Most of the P450 members are monooxygenases, catalyzing the incorporation of an oxygen atom into the substrate, initiating the oxidation of medium-chain *n*-alkanes (C_5–C_{10}) and aromatic HCs (Bell and Wong, 2007; Hannemann et al., 2007; Rojo, 2009; Urlacher and Girhard, 2012; Khmelevtsova et al., 2017). These oxidation reactions can lead to hydroxylation, demethylation, dealkylation, and sulfoxidation (Girvan and Munro, 2016). NAD(P)H is the main coenzyme for the delivery of two electrons required for the ferric heme-iron reduction of CYP. In prokaryotes, the major redox partner systems are ferredoxin and ferredoxin reductase (McLean et al., 2015). There are several genes encoding ferredoxin and ferredoxin reductase in bacteria and fungi (Chun et al., 2007; Kelly and Kelly, 2013). In the soil actinomycetes *Steptomyces coelicolor* A3(2) and *Mycobacterium smegmatis* and the fungus *Aspergillus oryzae* more than sixty different CYPs have been described (Lamb et al., 2002; Kelly et al., 2005; Nazmul et al., 2010). The cytochromes CYP249 of *Rhodococcus* and *Gordonia* strains are involved in the degradation of petroleum additives such as methyl *tert*-butyl ether, ethyl *tert*-butyl ether and *tert*-amyl butyl ether (Malandain et al., 2010). Interestingly, CYPs have also an important role in the degradation of aromatic HCs.

3. **Long-Chain Alkanes (>C_{17}):** Most alkanes are easily degraded but long-chain alkanes degradation is a challenge for HC-degrading microorganisms. *n*-Alkanes with 18 or more carbons atoms are found usually in a solid-state at room temperature, causing clogging in oil pipes (Wang and Shao, 2012). The expression of alkane hydroxylases

is a critical issue (van Beilen et al., 2001). In the long-chain alkane degrader *Acinetobacter* sp. M-1, the alkane hydroxylases *alkMa* and *alkMb* genes are differentially expressed in response to $>C_{22}$, and C_{16}–C_{22} *n*-alkanes, respectively (Tani et al., 2001). The *alkB1* and *alkB2* genes of *P. aeruginosa* strains PA01 and PR1 were preferentially expressed at the stationary and early exponential phase during growth on C_{10} and C_{22}–C_{24} alkanes, respectively (Marín et al., 2003). Different transcriptional regulators have been described in *Acinetobacter*, *Alcanivorax*, and *Pseudomonas* species. The transcriptional repressor AlmR regulates the expression of the *almA* gene in marine bacteria (Wang and Shao, 2014). Nevertheless, bacterial sensing, signal transduction and absorption mechanisms involved in alkane metabolism are still largely unknown. *Acinetobacter calcoaceticus* S30, *Alcaligenes odorans* P20, *A. borkumensis* SK2, *Alkanivorax dieselolei* B-5 and *Alcanivorax hongdengensis* A-11-3 degrade up to C_{33} long-chain *n*-alkanes (Lal and Khanna, 1996; Kasai et al., 2001, 2002; Syutsubo et al., 2001; Liu and Shao, 2005; Wu et al., 2008). *P. aeruginosa* WatG degraded up to C_{40} *n*-alkanes (147,5 ng L^{-1}) after 14 days at 30°C (Hasanuzzaman et al., 2007). *Acinetobacter* strains degrade up to C_{44} *n*-alkanes, however, only in the presence of chemical emulsifiers (Sakai et al., 1994).

To degrade the high diversity of alkanes, microorganisms require to activate enzymes and transcriptional regulators. Diverse monooxygenase systems for *n*-alkane oxidation have been presented above. For long-chain alkanes activation, the long-chain alkane monooxygenase LadA encoded in pLW1071 plasmid of the thermophilic *Geobacillus thermodenitrificans* strain NG80-2 that was isolated from a deep oil reservoir in China was characterized (Feng et al., 2007). Strain NG80-2 converts C_{15}–C_{36} long-chain alkanes by a terminal oxidation pathway, generating primary alcohols in the absence of rubredoxin and rubredoxin reductase. Alcohol dehydrogenases (ADH1 and ADH2), ALDH and tricarboxylic cycle (TCA) enzymes are similar to reported enzymes (Tourova et al., 2016; Feng et al., 2007). According to functional classification, LadA is a flavin-dependent monooxygenase, belonging to the bacterial luciferase SsuD subfamily (Li et al., 2008). Three *ladA* genes (*ladAαB23*, *ladAβB23* and *ladBB23*) are located within a gene cluster that contains the gene encoding a NADPH-dependent flavin mononucleotide (FMN) reductase in *Geobacillus thermoleovorans* B23 (Boonmak et al., 2014). Similar gene clusters have been described in other *Geobacillus* strains (Tourova et al., 2016). Genes encoding flavin-binding alkane monooxygenases *almA* were reported in *Acinetobacter* sp. DSM 17874 and *Acinetobacter baylyi* ADP1 (Vaneechoutte et al., 2006; Throne-Holst et al., 2007; Wentzel et al., 2007). An *almA* gene mutant of *Acinetobacter* sp. DSM 17874 is able to grow only on up to C_{24} alkanes, while wild type cells grow on up to C_{36} alkanes (Throne-Holst et al., 2007). The *almA* genes are also present in *Acinetobacter venetianus* RAG-1, *Acinetobacter* sp. MP-1, *Marinobacter aquaeolei* VT8, *Oceanobacter* sp. RED65, *Ralstonia* spp., *Mycobacterium* spp., *Photorhabdus* spp., *Psychrobacter* spp. and *Nocardia farcinica* IFM10152 (Reisfeld et al., 1972; Sakai et al., 1994; Vaneechoutte et al., 2006; Throne-Holst et al., 2007).

3.4.2 AROMATIC HYDROCARBON CATABOLISM

Aromatic HCs are ubiquitous crude oil components and are characterized by a high stability and recalcitrance to physicochemical and biological degradation. The stability of the apolar C-H bond and the delocalization of π orbitals of aromatic HCs implicate an energetic and mechanistic challenge for microbial metabolism. Bacterial degradation of aromatics involves two steps: the activation of the aromatic ring and the subsequent cleavage reaction (Seeger et al., 1995; Rabus et al., 2016). Aerobic catabolic pathways initially modify the aromatic ring by an oxygenation reaction (e.g., mono- or di-oxygenation), incorporating molecular oxygen into the HC substrate (Seeger et al., 1995; McKay et al., 1997; Seeger et al., 1999, 2001; Wang et al., 2017). Successive oxidation reactions catalyzed by hydroxylases and monooxygenases, or a single dioxygenation step by non-heme Rieske dioxygenases, allow the conversion of the aromatic substrates to dihydroxylated intermediates (Seeger et al., 1997, 2003; Varjani et al., 2017; Unimke et al., 2018). These compounds are further metabolized by intra- (*ortho* cleavage) or extra-diol (*meta* cleavage) ring-cleaving dioxygenases that open the aromatic ring (Seeger et al., 1997; Wang et al., 2017). Degradation pathways of aromatic HCs commonly converge into the central metabolites catechol, protocatechuate, and gentisate (Chain et al., 2006). Occasionally, dead-end metabolites with higher toxicity than the original contaminant are produced during microbial degradation (Cámara et al., 2004; Pieper and Seeger, 2008).

Aerobic enzymatic systems, such as di-iron-containing hydroxylases/monooxygenases, non-heme Rieske, intra- and extra-diol dioxygenases are based on oxygen activation and hydroxyl radical incorporation into target aromatic molecules. Bacterial multicomponent monooxygenases (BMMs) involved in aromatic HC degradation are composed of six polypeptides, where five components are conserved (α, β, γ, N, and R subunits). The hydroxylase complex is composed of α, β and γ subunits, while the N and R components mediate the electron transfer to the di-iron center (Notomista et al., 2003; Parales et al., 2008). Iron-dependent non-heme Rieske dioxygenases are multi-component enzymatic systems that may have three components: a terminal oxygenase, a ferredoxin and a reductase; or two components: an oxygenase and a bifunctional ferredoxin/reductase subunit (Wang et al., 2017). The oxygenase component catalyzes the activation of oxygen and the subsequent *cis*-hydroxylation of the aromatic ring, whereas the ferredoxin and reductase subunits carry out the electron transfer to the catalytic center (Wang et al., 2017). Commonly, the oxygenase component is composed of a large α-subunit, which possesses the catalytic site consisting of a [2Fe-2S] iron-sulfur cluster with a mononuclear iron, and a small β-subunit. Intradiol and extradiol dioxygenases also possess a non-heme iron center, responsible for oxygen activation (Wang et al., 2017). In general, a low substrate specificity seems to be a general feature of hydroxylases and RHDs responsible for the degradation of aromatic HCs, although some substrates have a higher affinity to each enzymatic system, affecting their turnover rates (Seeger et al., 1999, 2001; Suenaga et al., 2006; Fuentes et al., 2014).

1. **Monoaromatic Hydrocarbons:** Aerobic degradation of BTEX compounds in bacteria by BMMs and RHDs have been extensively described (Figure 3.2). BTEX HCs are funneled through several peripheral catabolic pathways into the central metabolites catechol, (di)methylcatechol, ethylcatechol, and protocatechuate (El-Naas et al., 2014). Group 1 BMMs, primarily phenol hydroxylase (PH), toluene/benzene 2-monooxygenase (TB2MO) and toluene 2-monooxygenase (T2MO), and group 2 BMMs such as benzene monooxygenase (BMO), toluene monooxygenase (TMO), toluene/*o*-xylene monooxygenase (ToMO) and alkene monooxygenase (AMO), have been described in Proteobacteria (Pseudomonadota) and Actinobacteria (Actinomycedota) (Notomista et al., 2003). These bacterial enzymes transform monocyclic aromatic HCs by successive hydroxylation reactions.

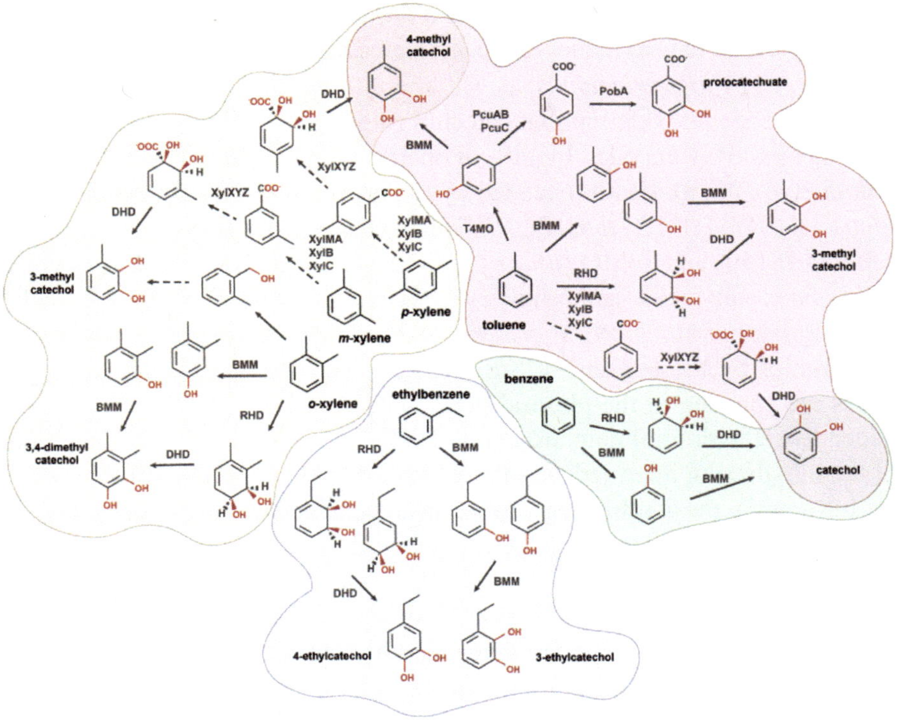

FIGURE 3.2 Bacterial aerobic pathways of benzene, toluene, ethylbenzene, and xylene degradation. The solid arrows indicate one enzymatic reaction, while dashed arrows illustrate multiple steps. [*Abbreviations:* BMM: bacterial multicomponent monooxygenase; RHD: ring-hydroxylating dioxygenase; DHD: dihydrodiol dehydrogenase; T4MO: toluene 4-monooxygenase; XylMA: xylene monooxygenase; XylB: benzyl alcohol dehydrogenase; XylC: benzaldehyde dehydrogenase; XylXYZ: benzoate 1,2-dioxygenase; PcuAB: *p*-cresol methyl hydroxylase; PcuC: *p*-hydroxybenzaldehyde dehydrogenase; PobA: *p*-hydroxybenzoate hydroxylase].

Aerobic catabolism of benzene, the most toxic and hazardous member of the BTEX group, can proceed by the enzymatic action of di-iron-containing PHs, which oxidize benzene into phenol, and phenol into catechol. PH is involved in benzene degradation in *Pseudomonas* sp. M1, *Arhodomonas* sp. strain Rozel and *Alcaligenes aquatilis* QD168 (Santos and Sá-Correia, 2007; Dalvi et al., 2012; Durán et al., 2019). The most studied phenol degradation pathway via PH has been described in *Pseudomonas* sp. CF600, whose enzymes are encoded in the *dmp* operon located in plasmid pVI150 (Powlowski and Shingler, 1994). The upper pathway *dmpKLMONP* gene cluster in strain CF600 encodes a multicomponent PH involved in the hydroxylation of phenol, while the lower pathway *dmpQBCDEFGHI* operon encodes a catechol 2,3-dioxygenase and the enzymes for the hydrolytic and the 4-oxalocrotonate branches of the *meta*-cleavage pathway (Powlowski and Shingler, 1994). *Pseudomonas* sp. M1 and *A. aquatilis* QD168 also possess a PH for benzene oxidation, but catechol ring-fission follows via the *ortho*-cleavage pathway encoded by the *cat* operon (Santos and Sá-Correia, 2007; Durán et al., 2019). BMO (encoded by the operon *bmoABCD* from *P. aeruginosa* JI104), TBMO (*tbmABCDEF* gene cluster from *Burkholderia cenocepacia* JS150) and ToMO (*touABCDEF* genes from *Pseudomonas stutzeri* OX1) also catalyze the same enzymatic peripheral reaction (Notomista et al., 2003; Parales et al., 2008). *P. putida* ML2 oxidizes benzene to *cis*-benzene dihydrodiol by the benzene 1,2-dioxygenase (BDO) BedC1C2BA, whose genes are encoded in the plasmid pHMT112 (Tan et al., 1993). The NAD-dependent dihydrodiol dehydrogenase BedD catalyzes the re-aromatization of the dihydrodiol intermediate into catechol (Bilal and Iqbal, 2019).

In bacteria, the aerobic degradation of toluene occurs via at least six catabolic pathways (Parales et al., 2008; Fuentes et al., 2014). The first pathway, identified in *P. putida* strains F1 and UV4, includes toluene oxidation into toluene *cis*-dihydrodiol by toluene 2,3-dioxygenase (TDO) TodC1C2BA, and a subsequent dehydrogenation reaction by toluene *cis*-dihydrodiol dehydrogenase TodD to produce 3-methylcatechol. BDO and TDO can perform a similar dioxygenation reaction with benzene or toluene, although TDO has a higher specificity towards alkylbenzenes (Bagnéris et al., 2005). A second pathway was described in *Burkholderia vietnamiensis* G4, where a T2MO (*tomA0A1A2A3A4A5* gene cluster) catalyzes a first mono-hydroxylation of toluene to *o*-cresol and a second hydroxylation to 3-methylcatechol (Parales et al., 2008). In the third catabolic pathway reported in *Ralstonia pickettii* PKO1, toluene 3-monooxygenase (T3MO, *tbuA1UBVA2*) hydroxylates toluene in *meta*-position leading to *m*-cresol, followed by a second hydroxylation through a phenol/cresol hydroxylase (*tbuD*) with the concomitant formation of 3-methylcatechol. In *Pseudomonas mendocina* KR1, toluene oxidation occurs in *para*-position by toluene 4-monooxygenase (T4MO, *tmoABCDE* gene cluster) to produce *p*-cresol. The methyl radical is then oxidized to yield 4-hydroxybenzoate and a second hydroxylation leads to protocatechuate. Another T4MO from *Burkholderia* sp. JS150 produces *p*-cresol from toluene, followed by another hydroxylation to form 4-methylcatechol. The fifth route is mediated by the non-specific ToMO (*touFEDCBA* genes) described in *P. stutzeri* OX1. Toluene is hydroxylated by ToMO in *ortho*, *meta* or *para* position, yielding *o*-cresol, *m*-cresol or *p*-cresol, respectively. Cresols are then converted by ToMO and a multicomponent phenol/cresol hydroxylase (*touKLMNOP* genes) into 3- or 4-methylcatechol.

A sixth mechanism for toluene degradation has been characterized in the TOL pathway carried by the pWW0 plasmid of *P. putida* mt-2, which is one of the most studied pathways. This pathway involves a two-component monooxygenase, three dehydrogenases, and a RHD (*xylMABCXYZL* genes) to transform toluene via benzoate into catechol (Fuentes et al., 2014).

Xylene-degrading bacteria are divided into two groups: those capable of degrading both *m*- and *p*-xylene, and those able to degrade *o*-xylene. Both degradation pathways are hardly present in a single bacterium (Di Canito et al., 2018). Some toluene-degrading enzymes are also capable of catabolizing xylene. T4MO hydroxylates xylene to produce methylbenzyl alcohol (Lee et al., 2019). ToMO hydroxylates *o*-xylene in positions 3 and 4, leading to 3,4-dimethylcatechol (Di Canito et al., 2018). Two *o*-xylene degradation pathways have been described in *Rhodococcus* strains. The first pathway starts with an oxidation of the methyl group to form 2-methylbenzyl alcohol, followed by a second oxidation to yield 3-methylcatechol. A second pathway produces 3,4-dimethylcatechol after two oxidations (Di Canito et al., 2018). *Rhodococcus* sp. DK17 oxidizes *o*-xylene by *o*-xylene 3,4-dioxygenase AkbA1A2A3A4 and produces 3,4-dimethylcatechol, which is further metabolized via the *meta*-cleavage pathway. *O*-xylene 3,4-RHD can also convert toluene and ethylbenzene to their respective *cis*-dihydrodiols (Kim et al., 2013). The TOL pathway of the pWW0 plasmid from *P. putida* mt-2 has been extensively studied due to its role in *m*-xylene and *p*-xylene catabolism. As another group 2 BMMs, the methyl radical is oxidized by a two-component xylene monooxygenase that continues with two successive dehydrogenations (*xylUWCMABN* gene cluster) forming the corresponding (alkyl)benzoate. Enzymes of the lower pathway TOL operon (*xylXYZLTEGFJQKIH*) catalyze dioxygenation of the (alkyl)benzoate to produce a (methyl)catechol, which is *meta*-cleaved (Fuentes et al., 2014). The *m*/*p*-xylene-degrading strain *Halioxenophilus aromaticivorans* KU68F^{T} possesses the upper pathway *xylCMABN* operon (Iwaki et al., 2018).

Broad-substrate RHDs involved in ethylbenzene dihydroxylation and degradation in bacteria have been described. *O*-xylene dioxygenase from strain DK17 metabolizes ethylbenzene, forming 3-ethylcatechol as the major central metabolite probably via *cis*-dihydrodiol 3,4-ethylbenzene (Kim et al., 2004). Ethylbenzene dioxygenases from *R. jostii* RHA1 are encoded by the *etbA4*, *etbA1A2*, and *ebdA1A2A3* gene clusters that are induced by ethylbenzene. EtbA/EbdA RHDs expressed in *R. erythropolis* IAM1399 are able to metabolize ethylbenzene, benzene, toluene, biphenyl, 4-chlorobiphenyl, polychlorobiphenyls, naphthalene, phenanthrene, dibenzofuran, and dibenzo-*p*-dioxin (Iwasaki et al., 2007). *Sphingomonas* sp. PNB AhdA1bA2b

RHD oxidizes ethylbenzene to dihydrodiol, PAHs (*viz.* naphthalene, phenanthrene, anthracene, biphenyl, acenaphthene, benzo[a]pyrene, benz[a]anthracene) and other aromatic compounds (*viz.* propylbenzene, *p*-cumene, and *p*-cymene) (Khara et al., 2014). Some broad-range BMMs are capable to metabolize ethylbenzene as a non-preferred substrate. ToMO and T3MO degrade ethylbenzene, while the latter yields 3- and 4-ethylphenol with 10% of its normal enzymatic activity (Parales et al., 2008).

2. **Polycyclic Aromatic Hydrocarbons (PAHs):** Bacterial PAH degradation is achieved mainly by PAH-RHDs, which possess a low substrate specificity that permits the oxidation of a wide spectrum of aromatic compounds (Figure 3.3). For example, NidAB, and NidA3B3 RHDs of *Mycobacterium vanbaalenii* PYR-1 are involved in oxygenation of BTEX (toluene, *m*-xylene), low-molecular-weight (naphthalene, anthracene, phenanthrene, fluoranthene) and high-molecular-weight (pyrene, benz[a]anthracene, and benzo[a]pyrene) PHAs (Fuentes et al., 2014). Multicomponent enzymatic systems involved in PAH degradation oxidize one ring first (Varjani et al., 2017).

Aerobic naphthalene degradation proceeds via the gentisate or the catechol central pathways. A series of six successive reactions transform naphthalene to salicylate. Naphthalene dioxygenase (NagAaAbAcAd), *cis*-naphthalene dihydrodiol dehydrogenase (NagB), 1,2-dihydronaphthalene dioxygenase (NagC), 2-hydroxychromene-2-carboxylate isomerase (NagD), *trans-o*-hydroxybenzylidenepyruvic hydratase-aldolase (NagE) and salicylaldehyde dehydrogenase (NagF) are the enzymes of the naphthalene upper pathways. Salicylate is transformed by salicylate 5-hydroxylase (NagH) or salicylate 1-hydroxylase (NagG) to gentisate or catechol, respectively (Peng et al., 2008; Lee et al., 2019).

In *Martelella* sp. AD-3, two pathways are involved in aerobic anthracene degradation. A first mechanism involves dioxygenation at C-1 and C-2 positions of the aromatic ring, followed by subsequent reactions that transform anthracene into salicylate. Similar to naphthalene degradation, salicylate may undergo 5- or 1-hydroxylation by NagH or NagG to produce catechol or gentisate, respectively. In the second pathway, the first oxidation occurs at C-9, 10, forming *cis*-9, 10-dihydrodiol, which is converted non-enzymatically to the dead-end product 9,10-anthraquinone (Zeinali et al., 2008; Cui et al., 2014).

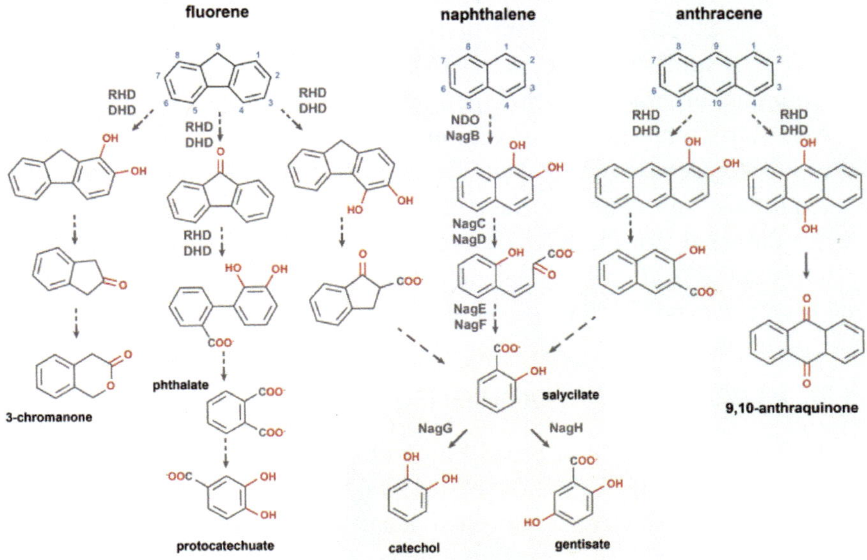

FIGURE 3.3 Biodegradation of polycyclic aromatic hydrocarbons (PAHs) under aerobic conditions. Central intermediates involved in PAHs bacterial catabolic pathways usually converge into salicylate, followed by a hydroxylation at position C_1 or C_5, yielding catechol or gentisate, respectively. The solid arrows indicate one enzymatic reaction, while dashed arrows illustrate multiple steps. [*Abbreviations:* RHD: ring-hydroxylating dioxygenase; DHD: dihydrodiol dehydrogenase; NDO: naphthalene dioxygenase; NagB: naphthalene dihydrodiol dehydrogenase; NagC: 1,2-dihydronaphthalene dioxygenase; NagD: 2-hydroxychromene-2-carboxylate isomerase; NagE: *trans-o*-hydroxybenzylidenepyruvate hydratase-aldolase; NagF: salicylaldehyde dehydrogenase; NagG: salicylate 1-hydroxylase; NagH: salicylate 5-hydroxylase].

Three catabolic pathways were reported for aerobic fluorene degradation in bacteria. The first proceeds via dioxygenation at C-1,2 of the aromatic ring and the subsequent formation of 3-chromanone after several reactions. A second pathway starts with a 3,4-dioxygenation, leading to the formation of salicylate, which is further metabolized via the naphthalene pathway. A third pathway, described in *Brevibacterium* sp. DPO1361 and *Pseudomonas* sp. F274, involves an initial monooxygenation at C-9, converting fluorene to 9-fluoroenol, which is dehydrogenated to 9-fluorenone. An angular carbon dioxygenation transforms 9-fluorenone into 1,10-dihydro-11,10-dihydroxy-fluorene-9-one. Subsequently, this intermediate is transformed to phthalate and eventually to protocatechuate (Mrozik et al., 2003; Seo et al., 2009).

Cytochrome P450 monooxygenase (CYP108J1), ferredoxin (*fdx*) and ferredoxin reductase (*hcaD*) are involved in the degradation of PAHs by *Rhodococcus* sp. P14 (Luo et al., 2016). The CYP108J1 gene is induced by phenanthrene, anthracene, benz[a]anthracene, and pyrene. The degradation of PAHs starts with the hydroxylation of the aromatic ring by cytochrome P450 monooxygenases (CYPs) and RHDs through the addition of one or two oxygen atoms. Subsequently, the pathway proceeds to generate metabolites that are funneled into the Krebs cycle (Luo et al., 2016; Peng, et al., 2020). Cyclohexane, diesel or naphthalene induce the expression of the CYP153 gene in *R. erythropolis* (Sazykin et al., 2019).

3.4.3 HYBRID AEROBIC CATABOLIC PATHWAYS

Another strategy used by a broad range of bacteria to overcome aromatic stability involves the formation of aryl-CoA epoxide metabolites during the degradation of benzoate and phenylacetate (Ismail and Gescher, 2012; Grishin and Cygler, 2015). An initial activation of the aromatic compound by the thioesterification through the attachment of a CoA-moiety yields a CoA thioester intermediate, which is further oxidized to an epoxide and finally cleaved by an enoyl-CoA hydratase. These pathways minimize the requirements of molecular oxygen, which is only required for the epoxide formation by a monooxygenase complex. These hybrid routes are useful in microaerobic environments where aerobic catabolism could not be achieved (Ismail and Gescher, 2012). Although hybrid pathways have been only characterized for the degradation of phenylacetate (*paa* genes) and benzoate (*badA* and *box* genes), ethylbenzene and toluene catabolism via these central pathways has been proposed.

Pseudomonas sp. Y2 and *Pseudomonas fluorescens* CA-4 degrade ethylbenzene, probably oxidizing the side chain to produce 2-phenylethanol or 1-phenylethanol (Utkin et al., 1991; Corkery et al., 1994). 2-Phenylethanol in strain CA-4 undergoes two consecutive oxidations by alcohol and ALDHs to produce phenylacetate, whereas meta-cleavage of the aromatic ring was also observed (Corkery et al., 1994). However, ethylbenzene is degraded mainly by oxidation of the aromatic ring in *P. fluorescens* CA-4 and *P. putida* 39/D by ethylbenzene dioxygenase, suggesting that the side chain oxidation is a minor route (Gibson et al., 1973; Corkery et al., 1998). Toluene degradation by *Burkholderia* sp. K24 induces benzoate-CoA ligase (BadA) and the enzymes encoded by the *boxABCD* operon and the *boxB* gene (Lee et al., 2016). The regulation of the hybrid catabolic pathway at transcriptional and protein synthesis levels suggests that toluene may be degraded via benzoyl-CoA/epoxybenzoyl-CoA. Further analyses are required to confirm the involvement of the hybrid aerobic pathways in the bacterial degradation of ethylbenzene and toluene.

3.5 ANAEROBIC BACTERIAL DEGRADATION OF PETROLEUM HYDROCARBONS

3.5.1 ALKANE CATABOLISM

Microbial degradation of HCs is limited under anaerobic conditions, due to the absence of molecular oxygen as the final electron acceptor (Abbasian et al., 2015). Under anoxic conditions, several facultative and obligate anaerobes have the ability to degrade *n*-alkanes with six or more carbons, particularly hexadecane (C_{16}) (Rojo, 2009; Bian et al., 2015). However, the bacterial growth and HC degradation are much slower than under oxic conditions. Anaerobic HC catabolism occurs in deep and anoxic environments, such as natural oil and gas leakages/seeps on land or sea sediments, and also has been observed below the surface of areas exposed to oxygen but where this element has been completely consumed (Abbasian et al., 2015; Musat, 2015). In the absence of oxygen, microorganisms utilize nitrate, nitrite, nitrous oxide, sulfate, thiosulfate, carbonate, H_2 and oxidized metals (Fe^{3+}, Mn^{4+}, U^{6+}) as electron acceptors (Pilloni et al., 2011; Abbasian et al., 2015). Based on the selection of the electron acceptors available that provide the highest potential energy for HCs degradation, stratification of the microbial communities ocurrs, starting with nitrate- and nitrite-reducing microorganisms, followed by metal-reducing bacteria, sulfate-reducing bacteria (SRB) and methanogenic microorganisms (Dojka et al., 1998; Widdel et al., 2010).

Fumarate addition is the predominant activation mechanism of *n*-alkane degradation under anoxic conditions (Ji et al., 2019), and this occurs via two main mechanisms (Rojo, 2009; Sierra-Garcia and Oliveira, 2013) (Figure 3.4). The first mechanism involves activation of the alkane substrate by the addition of fumarate at the subterminal carbon, producing an alkylsuccinate derivative, whereas in the second mechanism, fumarate is added at the terminal carbon. Activation of alkanes by nitrate-, iron-, and sulfate-reducing and methanogenic alkane-degrading bacteria has been reported (Abbasian et al., 2015). Anaerobic HC-degrading bacteria that use fumarate in Proteobacteria (Pseudomonadota) and Firmicutes (Bacilliota) phyla were identified by metagenome analyses in marine sediments at seeps (Stagars et al., 2016). Metagenomic analyzes of oil reservoirs revealed that members of the Atribacteria (Atribacterota) phylum are also potentially capable of HC degradation via fumarate addition (Mbadinga et al., 2011; Liu et al., 2019a). Sulfidogenic and denitrifying *Desulfatibacillum alkenivorans* AK-01 and *Aromatoleum* sp. HxN1 are bacterial models to investigate the activation by fumarate addition (Grundmann et al., 2008; Bian et al., 2015). *D. alkenivorans* AK-01 is able to grow on C_{13}–C_{18} *n*-alkanes, while *Aromatoleum* sp. HxN1 is able to grow on C_6–C_8 *n*-alkanes (Callaghan et al 2006; Grundmann et al., 2008). The addition of fumarate is catalyzed by strictly anaerobic glycyl radical enzymes such as alkylsuccinate synthase and (1-methylalkyl) succinate synthase (Grundmann et al., 2008; Callaghan et al 2012). Anaerobic alkane activation has been proposed based on cell fatty acids and isotope labeling analysis (Callaghan et al 2006, Grundmann et al., 2008; Zedelius et al., 2011). A carboxylate rearrangement is observed in the (methyl)alkylsuccinate, presumably as a CoA thioester, followed by a decarboxylation reaction (Widell and Grundmann, 2010; Rabus et al., 2016). The resulting fatty acid can then undergo β-oxidation. If the degradation produces propionate, fumarate could be regenerated through the methylmalonyl-CoA pathway (Widdel et al., 2006; Mbadinga et al., 2011; Callaghan et al 2012; Fuentes et al., 2014) (Figure 3.4(A)). Although the *D. alkenivorans* AK-01 genome was sequenced, only the alkylsuccinate synthase genes have been identified (Callaghan et al., 2012). Alkylsuccinate metabolites are biochemical indicators for *in situ* anaerobic microbial degradation of oil alkanes. Alkylsuccinates have been reported in oil-contaminated environments as well as in oilfield facilities (Grundmann et al., 2008; Sierra-Garcia and Oliveira, 2013).

Despite growing evidence supporting the preponderant role of anaerobic activation of *n*-alkanes by fumarate, the anaerobic activation by carboxylation, was described in the bacterium *Desulfococcus oleovorans* Hxd3 (Aeckersberg et al., 1991; Zhou et al., 2016). Strain Hdx3, isolated from an oil production plant, can grow on *n*-hexadecane under strictly anoxic and oxygen-limited conditions. After eleven months, the Hdx3 strain was able to degrade 90% of *n*-hexadecane, with the concomitant reduction of sulfate. Several experiments based on $^{13}CO_2$-labeling, proposed a novel pathway for alkanes degradation via carboxylation. The initial attack on the *n*-alkane substrate proceeds via subterminal carboxylation at the C-3 position with HCO_3^- or CO_2. The adjacent carbon atoms (C-1 and C-2) of the intermediate are removed to produce a fatty acid that is one carbon shorter than the original alkane, which contains a carboxyl group added exogenously by the carboxylation reaction (So et al., 2003). Thus, the resulting fatty acid can be oxidized to CO_2 by mineralization (Figure 3.4(B)). Alternatively, it was reported that the carboxylated fatty acid can undergo modifications, such as chain elongation, or methylation at the C-10 carbon (Rock et al., 1996).

Interestingly, strain Hdx3 harbors genes that encode an ethylbenzene dehydrogenase-like enzyme, whose function is to catalyze the hydroxylation of ethylbenzene but also *n*-alkanes under anaerobic conditions, suggesting a new activation mechanism (Rabus et al., 2016; Zhou et al., 2016). *n*-Alkane activation starts with a subterminal O_2-independent hydroxylation of the aliphatic chain by the ethylbenzene dehydrogenase-like enzyme, followed by the formation of a ketone by an oxidation reaction, to finally carboxylate at C-3 of the aliphatic chain (Figure 3.4(B)).

3.5.2 CATABOLISM OF MONOAROMATIC HYDROCARBONS

Under anaerobic conditions, aromatic compounds activation occurs by coupling fumarate or carbon dioxide, while nitrate, ferric ion, sulfate or manganese act as the final electron acceptors (Figure 3.5) (Fuentes et al., 2014; Logeshwaran et al., 2018). A general mechanism for anaerobic degradation involves the addition of fumarate to BTEX compounds catalyzed by alkyl and arylalkylsuccinate synthases. Toluene and methyl-substituted monoaromatic HCs degradation via benzylsuccinate synthase depends on activation by fumarate addition (Fuchs et al., 2011; Rabus et al., 2016), except for the *p*-cymene-degrading *Aromatoleum aromaticum* pCyN1. Benzylsuccinate synthase catalyzes the reaction between fumarate and the HC substrate, activating the aromatic ring and eventually yielding (*R*)-benzylsuccinate via a benzyl radical intermediate. (*R*)-benzylsuccinate is activated to a CoA thioester by a succinyl-CoA dependent CoA-transferase. Benzylsuccinyl-CoA is catabolized to benzoyl-CoA and succinyl-CoA. Interestingly, a phylogenetic analysis of the toluene-activating benzylsuccinate synthase (*bssA*) of the nitrate-reducing bacteria *Thauera aromatica* K172 revealed that this enzyme is grouped in a cluster with enzymes of sulfate-reducing and iron-reducing bacteria, suggesting that these strains that use nitrate, iron or sulfate as electron acceptors share this metabolic pathway (Rabus et al., 2016). Toluene degradation has been reported in BES under anoxic conditions, showing for the first time the use of toluene as the sole carbon source under nitrate-reducing conditions, removing up to 73% of the pollutant (Espinoza-Tofalos et al., 2018).

Anaerobic benzene degradation could occur by several catabolic pathways (Meckenstock et al., 2016; Rabus et al., 2016). The first pathway was reported in *A. aromaticum* EbN1, where benzene is carboxylated to benzoate

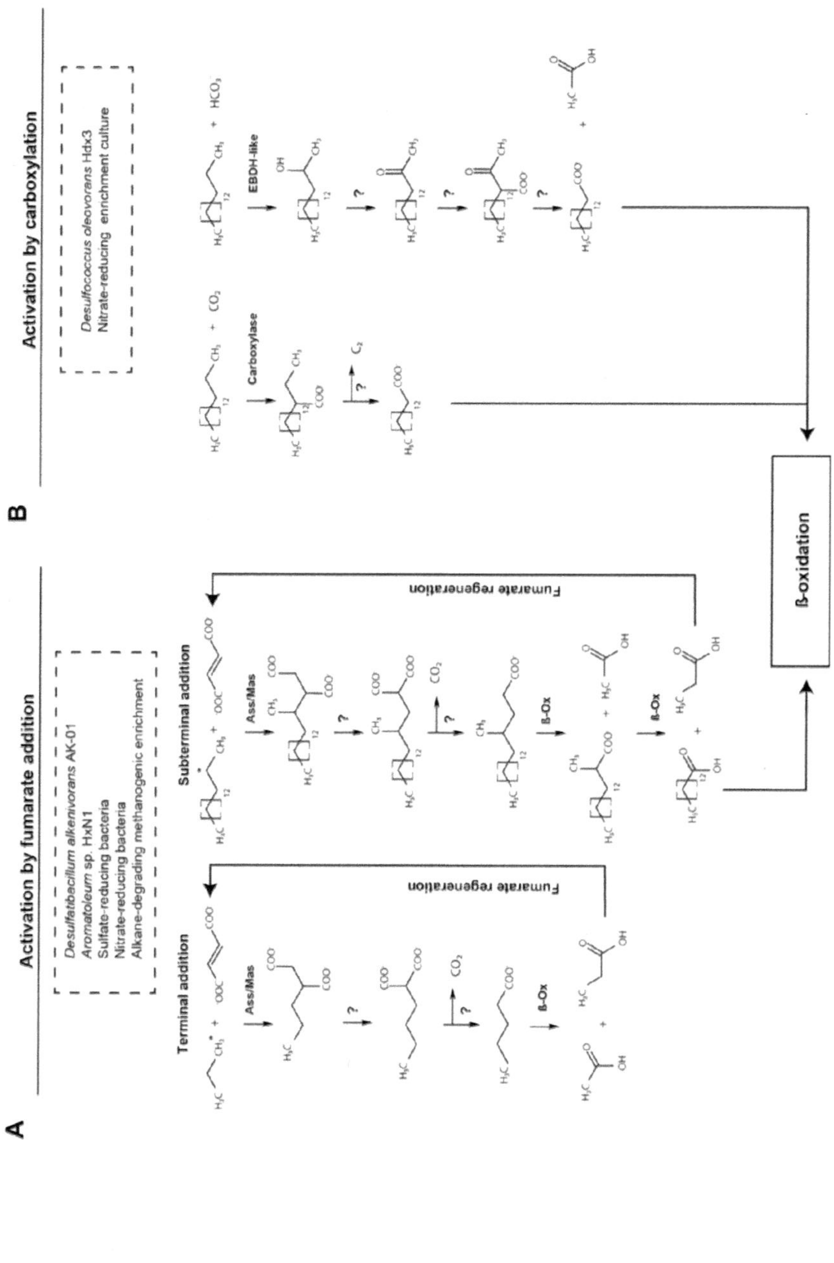

FIGURE 3.4 Anaerobic *n*-alkane degradation pathways in bacteria via fumarate addition and carboxylation. [*Abbreviations:* Ass: alkylsuccinate synthase; Mas: (1-methylalkyl) succinate synthase; EBDH-like: ethylbenzene dehydrogenase-like enzyme; β-ox: β-oxidation].

by an anaerobic benzene carboxylase, and benzoate is then activated to benzoyl-CoA. A second pathway was described in the iron-reducing bacterium *Geobacter metallireducens* and involved the hydroxylation of benzene to phenol. When Fe(III) or Mn(IV) are used as electron acceptors benzoate and/or phenol were detected during benzene degradation. The third pathway involves the methylation of benzene to toluene, and subsequent degradation via the toluene pathway. Previous studies on benzene catabolism suggested an initial methylation to produce toluene and subsequent transformation to benzoate in the presence of nitrate as the final electron acceptor.

Anaerobic ethylbenzene degradation in denitrifying bacteria starts with ethylbenzene hydroxylation by ethylbenzene dehydrogenase, leading to (*S*)-1-phenylethanol production (Fuchs et al., 2011). This intermediate is oxidized to acetophenone by a short-chain alcohol dehydrogenase. A complex carboxylase catalyzes the carboxylation of acetophenone to benzoylacetate. This compound is activated by a CoA ligase, yielding benzoylacetyl-CoA that is cleaved to benzoyl-CoA and acetyl-CoA.

Benzoyl-CoA produced during the anaerobic benzene, toluene, and ethylbenzene degradation could be degraded via an ATP-dependent pathway, described in *Rhodopseudomonas palustris*, or an ATP-independent pathway, described in *G. metallireducens* (Fuchs et al., 2011; Rabus et al., 2016). Bacterial species that use the ATP-dependent pathway such as *R. palustris*, *T. aromatica* and *Azoarcus evansii* are facultative anaerobes that obtain energy from photosynthesis or anaerobic respiration. An ATP-independent pathway must be employed by fermenting and strict anaerobic bacteria such as *G. metallireducens* to gain ATP in the process. In strict anaerobes without a respiratory chain, benzoyl-CoA is degraded into three acetyl-CoA that can be converted into three molecules of ATP. In both pathways, benzoyl-CoA is transformed to cyclohex-1,5-diene-carboxy-CoA either with class I (in the ATP-dependent pathway) or class II (in the ATP-independent pathway) benzoyl-CoA reductase. These enzymes differ biochemically: class I enzymes possess four subunits and use ferredoxin as the electron donor, while class II enzymes possess eight subunits and Fe–S clusters, tungsten, zinc, flavin, and selenocysteine as cofactors. After several different enzymatic reactions, benzoyl-CoA is converted to 3-hydroxypimelyl-CoA, and subsequently degraded to acetyl-CoA and CO_2.

Another ethylbenzene catabolic pathway was proposed in the sulfate-reducing strain EbS7 (Kniemeyer et al., 2003). Ethylbenzene is activated via fumarate to produce (1-phenylethyl)-succinate. This intermediate, after reaction with CoA, yields (1-phenylethyl)-succinyl-CoA. Some consecutive reactions lead to 4-phenylpentanoyl-CoA, however, there is no experimental evidence for the degradation intermediate. Probably, a β-oxidation step and a thiolytic cleavage occur, yielding 2-phenylpropionyl-CoA, which could be finally degraded to acetyl-CoA.

3.5.3 DEGRADATION OF POLYCYCLIC AROMATIC HYDROCARBONS

Anaerobic degradation of naphthalene and 2-methylnaphthalene under sulfate- and iron-reducing conditions has been reported (Meckenstock et al., 2016; Rabus et al., 2016; Boll and Estelmann, 2020). A CO_2 molecule is added to naphthalene by a naphthalene carboxylase, yielding 2-naphthoate, which is subsequently transformed to 2-naphthoyl-CoA by 2-naphthoate-CoA ligase. 2-Methylnaphthalene is activated via fumarate addition by the enzyme (2-naphthylmethyl)succinate synthase to (2-naphthylmethyl) succinate, which is also converted to 2-naphthoyl-CoA after five enzymatic reactions. 2-Naphthoyl-CoA is transformed to 5,6-dihydro-2-naphthoyl-CoA by a 2-naphthoyl-CoA reductase and subsequently to 5,6,7,8-tetrahydro-2-naphthoyl-CoA that is degraded to CO_2 and acetyl-CoA (Meckenstock et al., 2016; Rabus et al., 2016; Boll and Estelmann, 2020). 2-Naphthoate was also observed during growth on naphthalene with iron as the electron acceptor (Meckenstock et al., 2016).

The anaerobic degradation of PAHs with more than two rings has been scarcely studied (Boll and Estelmann, 2020). Previous reports indicate the carboxylation of phenanthrene to 2-phenanthrene carboxylic acid under sulfate-reducing conditions by bacteria of the *Desulfobacteriaceae* family. Probably after activation to phenanthroyl-CoA, a series of consecutive enzymatic reactions of dearomatization and modified β-oxidation funnel its catabolism to acetyl-CoA (Boll and Estelmann, 2020). Under methanogenic conditions, naphthalene, anthracene, phenanthrene, and methylnaphthalene are degraded with concomitant methane generation (Meckenstock et al., 2016; Rabus et al., 2016).

3.6 CONCLUSIONS AND FUTURE RESEARCH DIRECTIONS

A worldwide growing demand for non-renewable resources derived from oil increases the probabilities of pollution events during extraction, transport, storage, and distribution of petroleum and refined products. In comparison with conventional physicochemical processes, bioremediation is an eco-friendly, efficient, and low-cost technology for the decontamination of

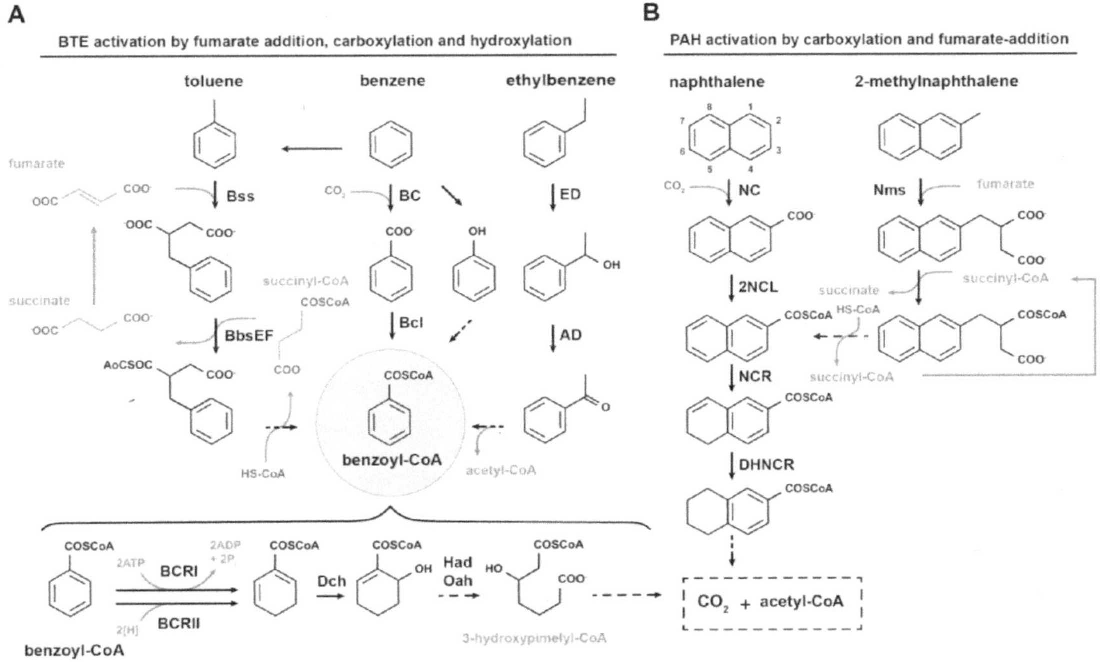

FIGURE 3.5 An overview of aromatic hydrocarbons degradation pathways under anaerobic conditions. Solid arrows indicate one enzymatic reaction, while dashed arrows illustrate multiple steps. (A) BTE degradation pathways in bacteria. Toluene is anaerobically activated via fumarate addition. Benzene can be catabolized via three bacterial pathways: methylation to yield toluene, carboxylation to produce benzoate, or hydroxylation to form phenol. Ethylbenzene could be hydroxylated or activated by fumarate addition (not shown). Regardless of the activation mechanism, BTE degradation pathways converge into benzoyl-CoA. [*Abbreviations:* Bss: benzylsuccinate synthase; BbsEF: succinyl-CoA dependent CoA-transferase; BC: benzene carboxylase; Bcl: benzoate-CoA ligase; ED: ethylbenzene dehydrogenase; AD: alcohol dehydrogenase; BCRI: benzoyl-CoA reductase class I; BCRII: benzoyl-CoA reductase class II; Dch: dienoyl-CoA hydratase; Had: hydroxyacyl-CoA dehydrogenase; Oah: oxoacyl-CoA hydrolase]. (B) Anaerobic naphthalene and methylnaphthalene activation via carboxylation and fumarate addition, respectively. [*Abbreviations:* NC: naphthalene carboxylase; Nms: (2-naphthylmethyl) succinate synthase; 2NCL: 2-naphthoate CoA-ligase; NCR: 2-naphthoyl-CoA reductase; DHNCR: 5,6-dihydro-2-naphthoyl-CoA reductase].

oil-polluted environments. Once an oil spill occurs, specific native bacteria express or acquire the capability to degrade HCs. Bioremediation exploits HC-degrading microorganisms to clean-up polluted matrices, either by supplementing the native microbes with nutrients and surfactants, or by incorporating autochthonous or allochthonous HC-degrading bacteria. Therefore, the microbial prospecting of oil-polluted, but also pristine environments is a crucial strategy for the search for novel native HC-degrading microorganisms that may be key biocatalysts for bioremediation processes. Bioprospecting applies not only to commonly impacted matrices (e.g., soil and marine sediment), but also to unconventional matrices such as the phyllosphere of urban trees. Therefore, the characterization of hydrocarbonoclastic microorganisms and the elucidation of their HC catabolic pathways are required to understand the complexities of the degradation of HCs and to establish novel bioremediation processes for specific polluted sites towards a more sustainable development.

ACKNOWLEDGMENTS

The authors gratefully acknowledge PhD ANID (L.R-C., C.C.M., F.D., L.H.) and PUCV (L.R-C., X.B-M.) fellowships, FONDECYT 1200756 (M.S., R.E.D., C.C.M., L.R-C., F.D.) (http://www.fondecyt.cl), CONICYT Programa de Investigación Asociativa (PIA) Anillo GAMBIO ACT172128 (M.S., V.A-E. R.E.D., C.C.M.) and USM (M.S., R.E.D., F.S-T., P.S-C.) (http://www.usm.cl) grants.

KEYWORDS

- **aliphatic hydrocarbon catabolism**
- **aromatic hydrocarbon catabolism**
- **bacterial hydrocarbon catabolism**
- **bioremediation**
- **hydrocarbons**
- **microbial communities**
- **oil spill**
- **petroleum**

REFERENCES

Abbasian, F., Lockington, R., Mallavarapu, M., et al., (2015). A comprehensive review of aliphatic hydrocarbon biodegradation by bacteria. *Appl. Biochem. Biotechnol., 176*(3), 670–699.

Aeckersberg, F., Bak, F., & Widdel, F., (1991). Anaerobic oxidation of saturated hydrocarbons to CO2 by a new type of sulfate-reducing bacterium. *Arch. Microbiol., 156*, 5–14.

Allen, J., Atekwana, E., Duris, J., et al., (2007). The microbial community structure in petroleum-contaminated sediments corresponds to geophysical signatures. *Appl. Environ. Microbiol., 73*(9), 2860–2870.

Alsadi, H., (2018). The petroleum substance. *Elements of Petroleum Science* (pp. 1–39). Partridge Publishing: Singapore.

Amouric, A., Quéméneur, M., Grossi, V., et al., (2010). Identification of different alkane hydroxylase systems in *Rhodococcus ruber* strain SP2B, a hexane-degrading actinomycete. *J. Appl. Microbiol., 108*(6), 1903–1916.

Atagana, H., (2010). Compost bioremediation of hydrocarbon-contaminated soil inoculated with organic manure. *Afr. J. Biotechnol., 7*(10), 1516–1525.

Atlas, R., (1991). Microbial hydrocarbon degradation-bioremediation of oil spills. *J. Chem. Tech. Biotechnol., 52*(2), 149–156.

Bagnéris, C., Cammack, R., & Mason, J., (2005). Subtle difference between benzene and toluene dioxygenases of *Pseudomonas putida*. *Appl. Environ. Microbiol., 71*(3), 1570–1580.

Barlow, L. R., & Philp, J. C., (2005). Suspicions to solutions: Characterizing contaminated land. In: Atlas, R. M., & Philip, J. C., (eds.), *Bioremediation: Applied Microbial Solutions for Real-World Environment Cleanup* (pp. 49–85). American Society for Microbiology Press, Washington, DC.

Bell, S., & Wong, L., (2007). P450 enzymes from the bacterium *Novosphingobium aromaticivorans*. *Biochem. Biophys. Res. Commun., 360*(3), 666–672.

Berk, R., & Canfield, J., (1964). Bioelectrochemical energy conversion. *Appl. Microb., 12*(1), 10–12.

Bian, X., Mbadinga, S., Liu, Y., et al., (2015). Insights into the anaerobic biodegradation pathway of *n*-alkanes in oil reservoirs by detection of signature metabolites. *Sci. Rep., 5*, 9801.

Bilal, M., & Iqbal, H., (2019). Microbial-derived biosensors for monitoring environmental contaminants: Recent advances and future outlook. *Process Saf. Environ. Prot., 124*, 8–17.

Boll, M., & Estelmann, S., (2020). Catabolic pathways and enzymes involved in the anaerobic degradation of polycyclic aromatic hydrocarbons. In: Boll, M., (ed.), *Anaerobic Utilization of Hydrocarbons, Oils, and Lipids* (pp. 135–150). Springer Nature Switzerland AG, Cham.

Boon, N., & Verstraete, W., (2010). Bioaugmentation of hydrocarbons. In: Timmis, K. N., McGenity, T., Van, D. M. J. R., & De Lorenzo, V., (eds.), *Handbook of Hydrocarbon and Lipid Microbiology* (pp. 2521–2530). Springer, Berlin.

Boonmak, C., Takahashi, Y., & Morikawa, M., (2014). Cloning and expression of three *ladA*-type alkane monooxygenase genes from an extremely thermophilic alkane-degrading bacterium *Geobacillus thermoleovorans* B23. *Extremophiles, 18*(3), 515–523.

Bravo, G., (2017). *Bioremediation of Mercury Polluted Agricultural Soils using Cupriavidus Metallidurans MSR33 in a Rotary Drum Reactor and its Effects on Microorganisms Involved in the Nitrogen Cycle*. Biotechnology PhD thesis, Universidad Federico Santa María - Pontificia Universidad Católica de Valparaíso, Valparaíso, Chile.

Bravo, G., Vega-Celedón, P., Gentina, J. C., et al., (2020). Effects of mercury II on *Cupriavidus metallidurans* strain MSR33 during mercury bioremediation under aerobic and anaerobic conditions. *Processes, 8*(8), 889.

Brusseau, M., & Maier, R., (2004). Soil and groundwater remediation. In: Artiola, J. F., Pepper, I. L., & Brusseau, M. L., (eds.), *Environmental Monitoring and Characterization* (pp. 335–356). Academic Press.

Callaghan, A., Gieg, L., Kropp, K., et al., (2006). Comparison of mechanisms of alkane metabolism under sulfate-reducing conditions among two bacterial isolates and a bacterial consortium. *Appl. Environ. Microbiol., 72*(6), 4274–4282.

Callaghan, A., Morris, B., Pereira, I., et al., (2012). The genome sequence of *Desulfatibacillum alkenivorans* AK-01: A blueprint for anaerobic alkane oxidation. *Environ. Microbiol., 14*(1), 101–113.

Cámara, B., Herrera, C., Gonzales, M., et al., (2004). From PCBs to highly toxic metabolites by the biphenyl pathway. *Env. Microbiol., 6*(8), 842–850.

Cappelletti, M., Presentato, A., Milazzo, G., et al., (2015). Growth of *Rhodococcus* sp. strain BCP1 on gaseous *n*-alkanes: New metabolic insights and transcriptional analysis of two soluble di-iron monooxygenase genes. *Front. Microbiol., 6*, 393.

Chain, P., Denef, V., Konstantinidis, K., et al., (2006). *Burkholderia xenovorans* LB400 harbors a multi-replicon, 9.7 M bp genome shaped for versatility. *Proc. Nat. Acad. Sci. USA., 113*, 15280–15287.

Chandra, S., Sharma, R., Singh, K., et al., (2013). Application of bioremediation technology in the environment contaminated with petroleum hydrocarbon. *Ann. Microbiol., 63*(2), 417–431.

Chun, Y. J., Shimada, T., Sanchez-Ponce, R., et al., (2007). Electron transport pathway for a *Streptomyces* cytochrome P450: Cytochrome P450 105D5-catalyzed fatty acid hydroxylation in *Streptomyces coelicolor* A3(2). *J. Biol. Chem., 282*(24), 17486–17500.

Cohen, B., (1931). The bacterial culture as an electrical half-cell. *J. Bacteriol., 21*(1), 18–19.

Corkery, D., & Dobson, A., (1998). Reverse transcription-PCR analysis of the regulation of ethylbenzene dioxygenase gene expression in *Pseudomonas fluorescens* CA-4. *FEMS Microbiol. Lett., 166*(2), 171–176.

Corkery, D., O'Connor, K., Buckley, C., et al., (1994). Ethylbenzene degradation by *Pseudomonas fluorescens* strain CA-4. *FEMS Microbiol. Lett., 124*(1), 23–27.

Cui, C., Ma, L., Shi, J., et al., (2014). Metabolic pathway for degradation of anthracene by halophilic *Martelella* sp. AD-3. *Int. Biodeterior. Biodegradation, 89*, 67–73.

Culpepper, M., & Rosenzweig, A., (2012). Architecture and active site of particulate methane monooxygenase. *Crit. Rev. Biochem. Mol. Biol., 47*(6), 483–492.

Daghio, M., Aulenta, F., Vaiopoulou, E., et al., (2017). Electro bioremediation of oil spills. *Water Res., 114*, 351–370.

Daghio, M., Espinoza, A., Leoni, B., et al., (2018). Bioelectrochemical BTEX removal at different voltages: Assessment of the degradation and characterization of the microbial communities. *J. Hazard. Mater., 341*, 120–127.

Daghio, M., Vaiopoulou, E., Patil, S. A., et al., (2016). Anodes stimulate anaerobic toluene degradation via sulfur cycling in marine sediments. *Appl. Environ. Microbiol., 82*(1), 297–307.

Dalvi, S., Azetsu, S., Patrauchan, M., et al., (2012). Proteogenomic elucidation of the initial steps in the benzene degradation pathway of a novel halophile, *Arhodomonas* sp. strain rozel, isolated from a hypersaline environment. *Appl. Environ. Microbiol., 78*(20), 7309–7316.

Di Canito, A., Zampolli, J., Orro, A., et al., (2018). Genome-based analysis for the identification of genes involved in o-xylene degradation in *Rhodococcus opacus R7. BMC Genomics, 19*(1), 587.

Dojka, M., Hugenholtz, P., Haack, S., et al., (1998). Microbial diversity in a hydrocarbon-and chlorinated-solvent-contaminated aquifer undergoing intrinsic bioremediation. *Appl. Environ. Microbiol., 64*(10), 3869–3877.

Durán, R., Méndez, V., Rodriguez-Castro, L., et al., (2019). Genomic and physiological traits of the marine bacterium *Alcaligenes aquatilis* QD168 isolated from Quintero Bay, Central Chile, reveal a robust adaptive response to environmental stressors. *Front. Microbiol., 10*, 528.

El-Naas, M., Acio, J., & El Telib, A., (2014). Aerobic biodegradation of BTEX: Progresses and prospects. *J. Environ. Chem. Eng., 2*(2), 1104–1122.

Espinoza-Tofalos, A., Alviz-Gazitua, P., Franzetti, A., et al., (2021). Bio-electrochemical remediation of petroleum hydrocarbons (pp. 269–285). In: Kumar, P., & Kuppam, C., (eds.), *Bioelectrochemical Systems.* Springer, Singapore.

Espinoza-Tofalos, A., Daghio, M., González, M., et al., (2018). Toluene degradation by *Cupriavidus metallidurans* CH34 in nitrate-reducing conditions and in bioelectrochemical systems. *FEMS Microbiol. Lett., 365*(12), 1–15.

Feng, L., Wang, W., Cheng, J., et al., (2007). Genome and proteome of long-chain alkane degrading *Geobacillus thermodenitrificans* NG80-2 isolated from a deep-subsurface oil reservoir. *Proc. Natl. Acad. Sci. U. S. A., 104*(13), 5602–5607.

Franzetti, A., Gandolfi, I., Bestetti, G., et al., (2020). Plant-microorganisms interaction promotes removal of air pollutants in Milan (Italy) urban area. *J. Hazard. Mater., 384*, 121021.

Fredrickson, J. K., Hicks, R. J., Li, S. W., et al., (1988). Plasmid incidence in bacteria from deep subsurface sediments. *Appl. Environ. Microbiol., 54*(12), 2916–2923.

Fuchs, G., Boll, M., & Heider, J., (2011). Microbial degradation of aromatic compounds — from one strategy to four. *Nat. Rev. Microbiol., 9*(11), 803–816.

Fuentes, S., Barra, B., Caporaso, J. G., et al., (2016). From rare to dominant: A fine-tuned soil bacterial bloom during petroleum hydrocarbon bioremediation. *Appl. Environ. Microbiol., 82*(3), 888–896.

Fuentes, S., Méndez, V., Aguila, P., et al., (2014). Bioremediation of petroleum hydrocarbons: Catabolic genes, microbial communities, and applications. *Appl. Microbiol. Biotechnol., 98*(11), 4781–4794.

Gandolfi, I., Canedoli, C., Imperato, V., et al., (2017). Diversity and hydrocarbon-degrading potential of epiphytic microbial communities on *Platanus x acerifolia* leaves in an urban area. *Environ. Pollut., 220*, 650–658.

Gibson, D. T., Gschwendt, B., Yeh, W. K., et al., (1973). Initial reactions in the oxidation of ethylbenzene by *Pseudomonas putida. Biochemistry, 12*(8), 1520–1528.

Gibu, N., Kasai, D., Ikawa, T., et al., (2019). Characterization and transcriptional regulation of *n*-alkane hydroxylase gene cluster of *Rhodococcus jostii* RHA1. *Microorganisms, 7*(11), 479.

Gieg, L. M., Fowler, S. J., & Berdugo-Clavijo, C., (2014). Syntrophic biodegradation of hydrocarbon contaminants. *Curr. Opin. Biotech., 27*, 21–29.

Girvan, H. M., & Munro, A. W., (2016). Applications of microbial cytochrome P450 enzymes in biotechnology and synthetic biology. *Curr. Op. Chem. Biol., 31*, 136–145.

Gray, N. D., Sherry, A., Grant, R. J., et al., (2011). The quantitative significance of *Syntrophaceae* and syntrophic partnerships in methanogenic degradation of crude oil alkanes. *Environ. Microbiol., 13*(11), 2957–2975.

Grishin, A. M., & Cygler, M., (2015). Structural organization of enzymes of the phenylacetate catabolic hybrid pathway. *Biology, 4*(2), 424–442.

Grote, M., Nagel, A., Nies, H., Rauterberg, J., & Wahrendorf, D. S., (2016). *The Use of Dispersants to Combat Oil Spills in Germany at Sea*. Federal Institute for Risk Assessment: Berlin.

Grundmann, O., Behrends, A., Rabus, R., et al., (2008). Genes encoding the candidate enzyme for anaerobic activation of *n*-alkanes in the denitrifying bacterium, strain HxN1. *Environ. Microbiol., 10*(2), 376–385.

Hamamura, N., & Arp, D. J., (2000). Isolation and characterization of alkane-utilizing *Nocardioides* sp. strain CF8. *FEMS Microbiol. Lett., 186*(1), 21–26.

Hamamura, N., Yeager, C. M., & Arp, D. J., (2001). Two distinct monooxygenases for alkane oxidation in *Nocardioides* sp. strain CF8. *Appl. Environ. Microbiol., 67*(11), 4992–4998.

Hannemann, F., Bichet, A., Ewen, K. M., et al., (2007). Cytochrome P450 systems-biological variations of electron transport chains. *Biochim. Biophys. Acta, Bioenerg., 1770*(3), 330–344.

Hara, A., Baik, S. H., Syutsubo, K., et al., (2004). Cloning and functional analysis of *alkB* genes in *Alcanivorax borkumensis* SK2. *Environ. Microbiol., 6*(3), 191–197.

Hasanuzzaman, M., Ueno, A., Ito, H., et al., (2007). Degradation of long-chain *n*-alkanes (C_{36} and C_{40}) by *Pseudomonas aeruginosa* strain WatG. *Int. Biodeter. Biodegr., 5*(1), 40–43.

Hazen, T. C., (2010). *In Situ*: Groundwater bioremediation. In: Timmis, K. N., McGenity, T., Van, D. M. J. R., & De Lorenzo, V., (eds.), *Handbook of Hydrocarbon and Lipid Microbiology* (pp. 2583–2596). Springer: Berlin.

Hazen, T. C., Tien, A. J., Worsztynowicz, A., et al., (2003). Biopiles for remediation of petroleum-contaminated soils: A polish case study. In: Šašek, V., Glaser, J. A., & Baveye, P., (eds.), *The Utilization of Bioremediation to Reduce Soil Contamination: Problems and Solutions* (1st edn., pp. 229–246). Nato Science Series (Series IV: Earth and Environmental Sciences). Springer.

Hernández, M., Morgante, V., Villalobos, P., et al., (2012). *S-Triazine-Herbicide-Degrading Bacteria Product for the Bioremediation and Method of Bioremediation.* U.S. Patent 8,263,385B2.

Hoang, S. A., Lamb, D., Seshadri, B., et al., (2020). Rhizoremediation as a green technology for the remediation of petroleum hydrocarbon-contaminated soils. *J. Hazard. Mater., 401*, 1–80.

Ilyushina, M., (2020). *The Russian Whistleblower Risking it All to Expose the Scale of an Arctic Oil Spill Catastrophe*. CNN News [Online]. Russia's Federal Supervisory Natural Resources Management Service (Rosprirodnadzor). https://edition.cnn.com/2020/07/10/europe/arctic-oil-spill-russia-whistleblower-intl/index.html (accessed on 25 October 2021).

Oil Tanker Spill Statistic, (2019). International Tanker Owners Pollution Federation (ITOPF).

Ismail, W., & Gescher, J., (2012). Epoxy coenzyme A thioester pathways for degradation of aromatic compounds. *Appl. Environ. Microbiol., 78*(15), 5043–5051.

Itah, A. Y., Brooks, A. A., Ogar, B. O., et al., (2009). Biodegradation of international jet A-1 aviation fuel by microorganisms isolated from aircraft tank and joint hydrant storage systems. *Bull. Environ. Contam. Toxicol., 83*(3), 318–327.

Iwaki, H., Yamamoto, T., & Hasegawa, Y., (2018). Isolation of marine xylene-utilizing bacteria and characterization of *Halioxenophilus aromaticivorans* gen. nov., sp. nov. and its xylene degradation gene cluster. *FEMS Microbiol. Lett., 365*(7), 1–8.

Iwasaki, T., Takeda, H., Miyauchi, K., et al., (2007). Characterization of two biphenyl dioxygenases for biphenyl/PCB degradation in a PCB degrader, *Rhodococcus* sp. strain RHA1. *Biosci., Biotechnol., Biochem., 71*(4), 993–1002.

Jacques, R. J., Okeke, B. C., Bento, F. M., et al., (2008). Microbial consortium bioaugmentation of a polycyclic aromatic hydrocarbons contaminated soil. *Bioresour. Technol., 99*(7), 2637–2643.

Jafarinejad, S., (2016). *Petroleum Waste Treatment and Pollution Control.* Butterworth-Heinemann: Oxford.

Ji, J. H., Liu, Y. F., Zhou, L., et al., (2019). Methanogenic degradation of long *n*-alkanes requires fumarate-dependent activation. *Appl. Environ. Microbiol., 85*(16), e00985–19.

Johnson, E. L., & Hyman, M. R., (2006). Propane and *n*-butane oxidation by *Pseudomonas putida* GPo1. *Appl. Environ. Microbiol., 72*(1), 950–952.

Jones, D. M., Head, I. M., Gray, N. D., et al., (2008). Crude-oil biodegradation via methanogenesis in subsurface petroleum reservoirs. *Nature, 451*(7175), 176–180.

Kasai, Y., Kishira, H., Sasaki, T., et al., (2002). Predominant growth of *Alcanivorax* strains in oil-contaminated and nutrient-supplemented sea water. *Environ. Microbiol., 4*(3), 141–147.

Kasai, Y., Kishira, H., Syutsubo, K., et al., (2001). Molecular detection of marine bacterial populations on beaches contaminated by the Nakhodka tanker oil-spill accident. *Environ. Microbiol., 3*(4), 246–255.

Kelly, S. L., Kelly, D. E., Jackson, C. J., et al., (2005). The diversity and importance of cytochrome P450 in microbes. In: Ortiz De, M. P. R., (ed.), *Cytochrome P450* (pp. 585–617). Springer Nature, Switzerland.

Kelly, S., & Kelly, D., (2013). Microbial cytochromes P450: Biodiversity and biotechnology. Where do cytochromes P450 come from, what do they do and what can they do for us? *Philos. Trans. R. Soc., B, 368*(1612), 20120476.

Khara, P., Roy, M., Chakraborty, J., et al., (2014). Functional characterization of diverse ring-hydroxylating oxygenases and induction of complex aromatic catabolic gene clusters in *Sphingobium* sp. PNB. *FEBS Open Bio., 4*, 290–300.

Khmelevtsova, L. E., Sazykin, I. S., Sazykina, M. A., et al., (2017). Prokaryotic cytochromes P450. *Appl. Biochem. Biotechnol., 53*(4), 401–409.

Kim, D., Chae, J., Zylstra, G., et al., (2004). Identification of a novel dioxygenase involved in metabolism of *o*-xylene, toluene, and ethylbenzene by *Rhodococcus* sp. strain DK17. *Appl. Environ. Microbiol., 70*(12), 7086–7092.

Kim, D., Yoo, M., Choi, K. Y., et al., (2013). Characterization and engineering of an *o*-xylene dioxygenase for biocatalytic applications. *Bioresour. Technol., 145*, 123–127.

Kniemeyer, O., Fischer, T., Wilkes, H., et al., (2003). Anaerobic degradation of ethylbenzene by a new type of marine sulfate-reducing bacterium. *J. Appl. Environ. Microbiol., 69*(2), 760–768.

Kotani, T., Kawashima, Y., Yurimoto, H., et al., (2006). Gene structure and regulation of alkane monooxygenases in propane-utilizing *Mycobacterium* sp. TY-6 and *Pseudonocardia* sp. TY-7. *J. Biosci. Bioeng., 102*(3), 184–192.

Laczi, K., Kis, Á., Horváth, B., et al., (2015). Metabolic responses of *Rhodococcus erythropolis* PR4 grown on diesel oil and various hydrocarbons. *Appl. Microbiol. Biotechnol., 99*(22), 9745–9759.

Lai, B., Yu, S., Bernhardt, P. V., et al., (2016). Anoxic metabolism and biochemical production in *Pseudomonas putida* F1 driven by a bioelectrochemical system. *Biotechnol. Biofuels, 9*(1), 1–13.

Lal, B., & Khanna, S., (1996). Degradation of crude oil by *Acinetobacter calcoaceticus* and *Alcaligenes odorans*. *J. Appl. Bacteriol., 81*(4), 355–362.

Lamb, D. C., Skaug, T., Song, H., et al., (2002). The cytochrome P450 complement (CYPome) of *Streptomyces coelicolor* A3(2). *J. Biol. Chem., 277*(27), 24000–24005.

Lee, S. Y., Kim, G. H., Yun, S. H., et al., (2016). Proteogenomic characterization of monocyclic aromatic hydrocarbon degradation pathways in the aniline-degrading bacterium *Burkholderia* sp. K24. *PloS One, 11*(4), 1–13.

Lee, Y., Lee, Y., & Jeon, C. O., (2019). Biodegradation of naphthalene, BTEX, and aliphatic hydrocarbons by *Paraburkholderia aromaticivorans* BN5 isolated from petroleum-contaminated soil. *Sci. Rep., 9*(1), 1–13.

Lewis, D. (2020). Cleaning up after Mauritius oil spill. Nature, 585, 172.

Li, L., Liu, X., Yang, W., et al., (2008). Crystal structure of long-chain alkane monooxygenase (LadA) in complex with coenzyme FMN: Unveiling the long-chain alkane hydroxy- lase. *J. Mol. Biol., 376*(2), 453–465.

Lin, C. W., Wu, C. H., Chiu, Y. H., et al., (2014). Effects of different mediators on electricity generation and microbial structure of a toluene powered microbial fuel cell. *Fuel, 125*, 30–35.

Liu, C., & Shao, Z., (2005). *Alcanivorax dieselolei* sp. nov., a novel alkane-degrading bacterium isolated from sea water and deep-sea sediment. *Int. J. Syst. Evol. Microbiol., 55*(3), 1181–1186.

Liu, C., Wang, W., Wu, Y., et al., (2011). Multiple alkane hydroxylase systems in a marine alkane degrader, *Alcanivorax dieselolei* B-5. *Environ. Microbiol., 13*(5), 1168–1178.

Liu, J., Wu, J., Lin, J., et al., (2019b). Changes in the microbial community diversity of oil exploitation. *Genes, 10*(8), 556.

Liu, Y. F., Qi, Z. Z., Shou, L. B., et al., (2019a). Anaerobic hydrocarbon degradation in candidate phylum '*Atribacteria*' (JS1) inferred from genomics. *ISME J., 13*(9), 2377–2390.

Logeshwaran, P., Megharaj, M., Chadalavada, S., et al., (2018). Petroleum hydrocarbons (PH) in groundwater aquifers: An overview of environmental fate, toxicity, microbial degradation and risk-based remediation approaches. *Environ. Technol. Inno., 10*, 175–193.

Lovley, D. R., Holmes, D. E., & Nevin, K. P., (2004). Dissimilatory Fe(III) and Mn(IV) reduction. *Adv. Microb. Physiol., 49*, 219–286.

Luo, A., Wu, Y. R., Xu, Y., et al., (2016). Characterization of a cytochrome P450 monooxygenase capable of high molecular weight PAHs oxidization from *Rhodococcus* sp. P14. *Process Biochem., 51*(12), 2127–2133.

Macaya, C. C., Durán, R. E., Hernández, L., et al., (2019a). *Bioremediation of Petroleum* (pp. 1–6). Reference Module in Life Sciences; Elsevier.

Macaya, C. C., Méndez, V., Durán, R. E., et al., (2019b). Complete genome sequence of hydrocarbon-degrading halotolerant *Acinetobacter radioresistens* DD78 isolated from the Aconcagua River mouth in central Chile. *Microbiol. Resour. Announce., 8*(33), 00601–00619.

Mahjoubi, M., Cappello, S., Souissi, Y., et al., (2018). Microbial bioremediation of petroleum hydrocarbon-contaminated marine environments. In: Zoveidavianpoor, M., (ed.), *Recent Insights in Petroleum Science and Engineering* (Vol. 16, p. 325). IntechOpen: London, UK.

Maier, T., Förster, H. H., Asperger, O., et al., (2001). Molecular characterization of the 56-kDa CYP153 from *Acinetobacter* sp. EB104. *Biochem. Biophys. Res. Commun., 286*(3), 652–658.

Malandain, C., Fayolle-Gulchard, F., & Vogel, T. M., (2010). Cytochromes P450-mediated degradation of fuel oxygenates by environmental isolates. *FEMS Microbiol. Ecol., 72*(2), 289–296.

Mandal, A. K., Sarma, P. M., Jeyaseelan, C. P., et al., (2012). Large scale bioremediation of petroleum hydrocarbon contaminated waste at Indian oil refineries: Case studies. *Int. J. Life Sci. Pharma Res., 2*(4), 114–128.

Marín, M. M., Yuste, L., & Rojo, F., (2003). Differential expression of the components of the two alkane hydroxylases from *Pseudomonas aeruginosa. J. Bacteriol., 185*(10), 3232–3237.

Mbadinga, S. M., Wang, L. Y., Zhou, L., et al., (2011). Microbial communities involved in anaerobic degradation of alkanes. *Int. Biodeter. Biodegr., 65*(1), 1–13.

McKay, D., Seeger, M., Zielinski, M., et al., (1997). Heterologous expression of biphenyl dioxygenase-encoding genes from a gram-positive broad-spectrum PCB degrader and characterization of chlorobiphenyl oxidation by the gene products. *J. Bacteriol., 179*, 1924–1930.

McLean, K. J., Luciakova, D., Belcher, J., et al., (2015). Biological diversity of cytochrome P450 redox partner systems. *Adv. Exp. Med. Biol., 851*, 299–317.

Meckenstock, R. U., Boll, M., Mouttaki, H., et al., (2016). Anaerobic degradation of benzene and polycyclic aromatic hydrocarbons. *J. Mol. Microbiol. Biotechnol., 26*(1), 92–118.

Medaura, M. C., & Ercoli, E. C., (2008). Bioconversion of petroleum hydrocarbons in soil using apple filter cake. *Braz. J. Microbiol., 39*(3), 427–432.

Méndez, V., Fuentes, S., Morgante, V., et al., (2017). Novel hydrocarbon-degrading and heavy metal-resistant *Acinetobacter*, *Kocuria* and *Pseudomonas* strains isolated from a crude oil-polluted soil in central Chile. *J. Soil Sci. & Plant Nutr., 17*(4), 1074–1087.

Méndez, V., Hernández, L., Salvà-Serra, F., et al., (2018). Complete genome sequence of the hydrocarbon-degrading strain *Achromobacter* sp. B7, isolated during petroleum hydrocarbon bioremediation in Valparaiso Region, *Chile. Microbiol. Resour. Announce., 7*(19), e01326–18.

Morris, J., & Jin, S., (2007). Feasibility of using microbial fuel cell technology for bioremediation of hydrocarbons in groundwater. *J. Environ. Sci. Health, Part A: Toxic/ Hazard. Subst. Environ. Eng., 43*(1), 18–23.

Mrozik, A., Piotrowska-Seget, Z., & Labuzek, S., (2003). Bacterial degradation and bioremediation of polycyclic aromatic hydrocarbons. *Pol. J. Environ. Stud., 12*(1), 15–25.

Musat, F., (2015). The anaerobic degradation of gaseous, nonmethane alkanes—from in situ processes to microorganisms. *Comput. Struct. Biotechnol. J., 13*, 222–228.

Navia, R., & Seeger, M., (2006). *Biorremediación de Suelos Contaminados con Compuestos Orgánicos Persistentes (COPs)* (Vol. 1, p. 225). Editorial Universidad de La Frontera: Temuco, Chile.

Nazmul, H., Nazir, K. H., Ichinose, H., et al., (2010). Molecular characterization and isolation of cytochrome P450 genes from the filamentous fungus *Aspergillus oryzae. Arch. Microbiol., 192*(5), 395–408.

Nealson, K. H., & Saffarini, D. A., (1994). Iron and manganese in anaerobic respiration. *Annu. Rev. Microbiol., 48*, 311–343.

Nechepurenko, I., (2020). *Russia Declares Emergency After Arctic Oil Spill*. The New York Times [Online]. https://www.nytimes.com/2020/06/04/world/europe/russia-oil-spill-arctic.html (accessed on 25 October 2021).

Nie, Y., Liang, J. L., Fang, H., et al., (2014). Characterization of a CYP153 alkane hydroxylase gene in a gram-positive *Dietzia* sp. DQ12-45-1b and its "team role" with *alkW1* in alkane degradation. *Appl. Microbiol. Biotechnol., 98*(1), 163–173.

Nikolopoulou, M., & Kalogerakis, N., (2010). Biostimulation strategies for enhanced bioremediation of marine oil spills including chronic pollution. In: Timmis, K. N., McGenity, T., Van, D. M. J. R., & De Lorenzo, V., (eds.), *Handbook of Hydrocarbon and Lipid Microbiology* (Vol. 4, pp. 2521–2530). Springer, Berlin, DE.

Notomista, E., Lahm, A., Di Donato, A., et al., (2003). Evolution of bacterial and archaeal multicomponent monooxygenases. *J. Mol. Evol., 56*(4), 435–445.

Ogburn, E. T., Jones, D. R., Masters, A. R., et al., (2010). Efavirenz primary and secondary metabolism *in vitro* and *in vivo*: Identification of novel metabolic pathways and cytochrome P450 2A6 as the principal catalyst of efavirenz 7-hydroxylation. *Drug Metab. Dispos., 38*(7), 1218–1229.

Orellana, R., Macaya, C., Bravo, G., et al., (2018). Living at the frontiers of life: Extremophiles in Chile and their potential for bioremediation. *Front. Microbiol., 9*, 2309.

Ortiz De, M. P. R., (2010). Hydrocarbon hydroxylation by cytochrome P450 enzymes. *Chem. Rev., 110*(2), 932–948.

Ossai, I. C., Ahmed, A., Hassan, A., et al., (2020). Remediation of soil and water contaminated with petroleum hydrocarbon: A review. *Environ. Technol. Inno., 17*(100526), 1–42.

Parales, R. E., Parales, J. V., Pelletier, D. A., et al., (2008). Diversity of microbial toluene degradation pathways. *Adv. Appl. Microbiol., 64*, 1–73.

Pathak, C., & Mandalia, H. C., (2012). Petroleum industries: Environmental pollution effects, management and treatment methods. *Int. J. Sep. Environ. Sci., 1*(1), 55–62.

Peng, R. H., Xiong, A. S., Xue, Y., et al., (2008). Microbial biodegradation of polyaromatic hydrocarbons. *FEMS Microbiol. Rev., 32*(6), 927–955.

Peng, T., Kan, J., Hu, J., et al., (2020). Genes and novel sRNAs involved in PAHs degradation in marine bacteria *Rhodococcus* sp. P14 revealed by the genome and transcriptome analysis. *Biotechnology, 10*(3), 1–10.

Pérez-Armendáriz, B., Loera, O., Fernández-Linares, L., et al., (2004). Biostimulation of microorganisms of sugarcane bagasse pith for hydrocarbon removal from a tropical weathered soil. *Lett. Appl. Microbiol., 38*, 373–377.

Piccolo, L. L., De Pasquale, C., Fodale, R., et al., (2011). Involvement of an alkane hydroxylase system of *Gordonia* sp. strain SoCg in degradation of solid *n*-alkanes. *Appl. Environ. Microbiol., 77*(4), 1204–1213.

Pieper, D., & Seeger, M., (2008). Bacterial metabolism of polychlorinated biphenyls. *J. Mol. Microbiol. Biotechnol., 15*(2, 3), 121–138.

Pilloni, G., Von, N. F., Engel, M., et al., (2011). Electron acceptor-dependent identification of key anaerobic toluene degraders at a tar-oil-contaminated aquifer by Pyro-SIP. *FEMS Microbiol. Ecol., 78*(1), 165–175.

Potter, M. C., (1911). Electrical effects accompanying the decomposition of organic compounds. *Proc. R. Soc. London, Ser. B, 84*(571), 260–276.

Powlowski, J., & Shingler, V., (1994). Genetics and biochemistry of phenol degradation by *Pseudomonas* sp. CF600. *Biodegradation, 5*(3, 4), 219–236.

Prince, R. C., (2010). Bioremediation of marine oil spills. In: Timmis, K. N., McGenity, T., Van, D. M. J. R., & De Lorenzo, V., (eds.), *Handbook of Hydrocarbon and Lipid Microbiology* (Vol. 4, pp. 2617–2630). Springer, Berlin.

Rabaey, K., Angenent, L., Schroder, U., et al., (2010). Bioelectrochemical systems (BES) for subsurface remediation. In: *Bioelectrochemical Systems: From Extracellular Electron Transfer to Biotechnological Application* (pp. 305–322). IWA Publishing, London, UK.

Rabus, R., Boll, M., Heider, J., et al., (2016). Anaerobic microbial degradation of hydrocarbons: From enzymatic reactions to the environment. *J. Mol. Microbiol. Biotechnol., 26*(1), 5–28.

Reisfeld, A., Rosenberg, E., & Gutnick, D., (1972). Microbial degradation of crude oil: Factors affecting the dispersion in sea water by mixed and pure cultures. *Appl. Microbiol. Biotechnol., 24*(3), 363–368.

Rock, C. O., Jackowski, S., & Cronan, J. E., (1996). Lipid metabolism in prokaryotes. In: Vance, D. E., & Vance, J. E., (eds.), *Biochemistry of Lipids, Lipoproteins and Membranes* (pp. 35–74). Elsevier Science, B.V., Amsterdam, Netherlands.

Rojas, L. A., Yáñez, C., González, M., et al., (2011). Characterization of the metabolically modified heavy metal-resistant *Cupriavidus metallidurans* strain MSR33 generated for mercury bioremediation. *PloS One, 6*(3), e17555.

Rojo, F., (2009). Degradation of alkanes by bacteria. *Environ. Microbiol., 11*(10), 2477–2490.

Ron, E. Z., (2000). Microbial life on petroleum. In: Seckbach, J., (ed.), *Journey to Diverse Microbial Worlds* (pp. 303–315). Cellular origin and life in extreme habitats, Springer Netherlands, Dordrecht.

Saavedra, M., Acevedo, F., González, M., et al., (2010). Mineralization of PCBs by the genetically modified strain *Cupriavidus necator* JMS34 and its application for bioremediation of PCB in soil. *Appl. Microbiol. Biotechnol., 87*(4), 1543–1554.

Saefong, M. P., (2020). *The Oil Market is Running Out of Storage Space and Production Cuts Loom*. MarketWatch [Online], U.S. Energy Information Administration. https://www.marketwatch.com/story/the-oil-market-is-running-out-of-storage-options-2020-04-21 (accessed on 25 October 2021).

Sakai, Y., Maeng, J. H., Tani, Y., et al., (1994). Use of long-chain *n*-alkanes (C_{13}–C_{44}) by an isolate, *Acinetobacter* sp. M-1. *Biosci., Biotechnol., Biochem., 58*(11), 2128–2130.

Santos, P. M., & Sá-Correia, I., (2007). Characterization of the unique organization and co-regulation of a gene cluster required for phenol and benzene catabolism in *Pseudomonas* sp. M1. *J. Biotechnol., 131*(4), 371–378.

Sazykin, I., Makarenko, M., Khmelevtsova, L., et al., (2019). Cyclohexane, naphthalene, and diesel fuel increase oxidative stress, CYP153, *sodA*, and *recA* gene expression in *Rhodococcus erythropolis*. *Microbiology Open, 8*, 855.

Scoma, A., Hernandez-Sanabria, E., Lacoere, et al., (2015). Primers: Bacterial genes encoding enzymes for aerobic hydrocarbon degradation. In: *Hydrocarbon and Lipid Microbiology Protocols* (pp. 23–37). Springer, Berlin-Heidelberg.

Seeger, M., Cámara, B., & Hofer, B., (2001). Dehalogenation, denitration, dehydroxylation and angular attack of substituted biphenyls and related compounds by a biphenyl dioxygenase. *J. Bacteriol., 183*, 3548–3555.

Seeger, M., González, M., Cámara, B., et al., (2003). Biotransformation of isoflavonoids by recombinant microbial enzymes. *Appl. Environ. Microbiol., 69*, 5045–5050.

Seeger, M., Rojas, L., González, M., & Yáñez, C., (2014). *Recombinant Bacterium Capable of Removing Mercury (II) Species, Cadmium (II) and Copper (II) in Presence of Other Heavy Metals from Polluted Sites, Product for the Bioremediation, Process of Obtaining the Product and Method of Bioremediation*. U.S. Patent 8,846,376B2.

Seeger, M., Saavedra, M., & Acevedo, F., (2011). *Recombinant PCB-Degrading Bacterium, Product for Bioremediation and Method of Application.* U.S. Patent 7,989,194B2.

Seeger, M., Timmis, K. N., & Hofer, B., (1995). Conversion of chlorobiphenyls into phenylhexadienoates and benzoates by the enzymes of the upper pathway for PCB degradation encoded by the *bph* locus of *Pseudomonas* sp. strain LB400. *Appl. Environ. Microbiol., 61,* 2654–2658.

Seeger, M., Timmis, K. N., & Hofer, B., (1997). Bacterial pathways for degradation of polychlorinated biphenyls. *Marine Chem., 58,* 327–333.

Seeger, M., Zielinski, M., Timmis, K. N., et al., (1999). Regiospecificity of dioxygenation of di- to pentachlorobiphenyls and their degradation to chlorobenzoates by the *bph*-encoded catabolic pathway of *Burkholderia* sp. strain LB400. *Appl. Environ. Microbiol., 65,* 3614–3621.

Seo, J., Keum, Y., & Li, Q. X., (2009). Bacterial degradation of aromatic compounds. *Int. J. Environ. Res. Public Health, 6*(1), 278–309.

Shuai, Y., Zhou, H., Mu, Q., et al., (2019). Characterization of a biosurfactant-producing *Leclercia* sp. B45 with new transcriptional patterns of *alkB* gene. *Ann. Microbiol., 69*(2), 139–150.

Sierra-Garcia, I. N., & De Oliveira, V. M., (2013). Microbial hydrocarbon degradation: Efforts to understand biodegradation in petroleum reservoirs. In: Chamy, R., & Rosenkranz, F., (eds.), *Biodegradation-engineering and Technology* (pp. 47–72). IntechOpen, London, UK.

Singh-Khadka, N., (2020). *Why the Mauritius Oil Spill is So Serious.* BBC News [Online], https://www.bbc.com/news/world-africa-53754751 (accessed on 25 October 2021).

Smits, T. H., Balada, S. B., Witholt, B., et al., (2002). Functional analysis of alkane hydroxylases from Gram-negative and Gram-positive bacteria. *J. Bacteriol. Res., 184*(6), 1733–1742.

Smits, T. H., Röthlisberger, M., Witholt, B., et al., (1999). Molecular screening for alkane hydroxylase genes in gram-negative and gram-positive strains. *Environ. Microbiol., 1*(4), 307–317.

So, C. M., Phelps, C. D., & Young, L. Y., (2003). Anaerobic transformation of alkanes to fatty acids by a sulfate-reducing bacterium, strain Hxd3. *Appl. Environ. Microbiol., 69,* 3892–3900.

Speight, J. G., (2014). In: Heinemann, H., & Speight, J. G., (eds.), *The Chemistry and Technology of Petroleum* (5th edn.). CRC Press, Boca Raton.

Speight, J. G., (2017). Removal of organic compounds from the environment. In: *Environmental Organic Chemistry for Engineers* (pp. 387–432). Butterworth-Heinemann, Oxford.

Stagars, M. H., Ruff, S. E., Amann, R., et al., (2016). High diversity of anaerobic alkane-degrading microbial communities in marine seep sediments based on (1-methylalkyl) succinate synthase genes. *Front. Microbiol., 6,* 1511.

Stams, A. J. M., De Bok, F. A. M., Plugger, C. M., et al., (2005). Exocellular electron transfer in anaerobic microbial communities. *Environ. Microbiol., 8,* 371–382.

Suenaga, H., Sato, M., Goto, M., et al., (2006). Steady-state kinetic characterization of evolved biphenyl dioxygenase, which acquired novel degradation ability for benzene and toluene. *Biosci., Biotechnol., Biochem., 70*(4), 1021–1025.

Sun, D., Wang, A., Cheng, S., et al., (2014). *Geobacter anodireducens* sp. nov., an exoelectrogenic microbe in bioelectrochemical systems. *Int. J. Syst. Evol. Microbiol., 64*(10), 3485–3491.

Syutsubo, K., Kishira, H., & Harayama, S., (2001). Development of specific oligonucleotide probes for the identification and in situ detection of hydrocarbon-degrading *Alcanivorax* strains. *Environ. Microbiol., 3*(6) 371–379.

Tan, B., Dong, X., Sensen, C. W., et al., (2013). Metagenomic analysis of an anaerobic alkane-degrading microbial culture: Potential hydrocarbon-activating pathways and inferred roles of community members. *Genome, 56*(10), 599–611.

Tan, H. M., & Fong, K. P. Y., (1993). Molecular analysis of the plasmid-borne bed gene cluster from *Pseudomonas putida* ML2 and cloning of the *cis*-benzene dihydrodiol dehydrogenase gene. *Can. J. Microbiol., 39*(4), 357–362.

Tani, A., Ishige, T., Sakai, Y., et al., (2001). Gene structures and regulation of the alkane hydroxylase complex in *Acinetobacter* sp. strain M-1. *J. Bacteriol., 183*(5), 1819–1823.

Throne-Holst, M., Wentzel, A., Ellingsen, T. E., et al., (2007). Identification of novel genes involved in long-chain *n*-alkane degradation by *Acinetobacter* sp. strain DSM 17874. *Appl. Environ. Microbiol., 73*(10), 3327–3332.

Tourova, T. P., Sokolova, D. S., Semenova, E. M., et al., (2016). Detection of *n*-alkane biodegradation genes *alkB* and *ladA* in thermophilic hydrocarbon-oxidizing bacteria of the genera *Aeribacillus* and *Geobacillus*. *Microbiology, 85*(6), 693–707.

Unimke, A. A., Mmuoegbulam, O. A., & Anika, O. C., (2018). Microbial degradation of petroleum hydrocarbons: Realities, challenges and prospects. *Biotechnol. J. Int., 22*(2), 1–10.

Urlacher, V. B., & Girhard, M., (2012). Cytochrome P450 monooxygenases: An update on perspectives for synthetic application. *Trends Biotechnol., 30*(1), 26–36.

Utkin, I. B., Yakimov, M. M., Matveeva, L. N., et al., (1991). Degradation of styrene and ethylbenzene by *Pseudomonas* species Y2. *FEMS Microbiol. Lett., 77*(2, 3), 237–241.

Van, B. J. B., & Funhoff, E. G., (2007). Alkane hydroxylases involved in microbial alkane degradation. *Appl. Microbiol. Biotechnol., 74*(1), 13–21.

Van, B. J. B., Panke, S., Lucchini, S., et al., (2001). Analysis of *Pseudomonas putida* alkane-degradation gene clusters and flanking insertion sequences: Evolution and regulation of the *alk* genes. *Microbiology, 147*(6), 1621–1630.

Van, H. W., (1965). A bacterial methane fuel cell. *J. Electrochem. Soc., 112*(3), 258.

Van, L. M., Prokop, G., Rabl-Berger, S., & Louwagie, G., (2014). *Progress in the Management of Contaminated Sites in Europe*. Publications Office of the European Union: Italy.

Vaneechoutte, M., Young, D. M., Ornston, L. N., et al., (2006). Naturally transformable *Acinetobacter* sp. strain ADP1 belongs to the newly described species *Acinetobacter baylyi*. *Appl. Environ. Microbiol., 72*(1), 932–936.

Varjani, S. J., (2017). Microbial degradation of petroleum hydrocarbons. *Bioresour. Technol., 223*, 277–286.

Varjani, S. J., Rana, D. P., Jain, A. K., et al., (2015). Synergistic ex-situ biodegradation of crude oil by halotolerant bacterial consortium of indigenous strains isolated from on shore sites of Gujarat, India. *Int. Biodeterior. Biodegrad., 103*, 116–124.

Varjani, S., Upasani, V. N., & Pandey, A., (2020). Bioremediation of oily sludge polluted soil employing a novel strain of *Pseudomonas aeruginosa* and a phytotoxicity of petroleum hydrocarbons for seed germination. *Sci. Total Environ., 737*, 139766–139774.

Venosa, A. D., Lee, K., Suidan, M. T., et al., (2002). Bioremediation and biorestoration of a crude oil contaminated freshwater wetland on St. Lawrence River. *Bioremed. J., 6*(3), 261–281.

Wang, Q., Zhang, S., Li, Y., et al., (2011a). Potential approaches to improving biodegradation of hydrocarbons for bioremediation of crude oil pollution. *J. Environ. Prot., 2*(01), 47–55.

Wang, W., & Shao, Z., (2012). Diversity of flavin-binding monooxygenase genes (*almA*) in marine bacteria capable of degradation long-chain alkanes. *FEMS Microbiol. Ecol., 80*(3), 523–533.
Wang, W., & Shao, Z., (2014). The long-chain alkane metabolism network of *Alcanivorax dieselolei. Nat. Commun., 5*(1), 1–11.
Wang, X. B., Chi, C. Q., Nie, Y., et al., (2011b). Degradation of petroleum hydrocarbons (C6–C40) and crude oil by a novel *Dietzia* strain. *Bioresour. Technol., 102*(17), 7755–7761.
Wang, Y., Li, J., & Liu, A., (2017). Oxygen activation by mononuclear nonheme iron dioxygenases involved in the degradation of aromatics. *J. Biol. Inorg. Chem., 22*(2), 395–405.
Wentzel, A., Ellingsen, T. E., Kotlar, H. K., et al., (2007). Bacterial metabolism of long-chain *n*-alkanes. *Appl. Microbiol. Biotechnol., 76*(6), 1209–1221.
Whyte, L. G., Smits, T. H. M., Labbe, D., et al., (2002). Gene cloning and characterization of multiple alkane hydroxylase systems in *Rhodococcus* strains Q15 and NRRL B-16531. *Appl. Environ. Microbiol., 68*(12), 5933–5942.
Widdel, F., & Grundmann, O., (2010). Biochemistry of the anaerobic degradation of non-methane alkanes. In: Timmis, K. N., McGenity, T., Van, D. M. J. R., & De Lorenzo, V., (eds.), *Handbook of Hydrocarbon and Lipid Microbiology* (pp. 909–924). Springer, Berlin.
Widdel, F., & Rabus, R., (2006). Anaerobic biodegradation of hydrocarbons including methane. *The Prokaryotes, 2*, 1028–1049.
Wu, C. H., Yet-Pole, I., Chiu, Y. H., et al., (2014). Enhancement of power generation by toluene biodegradation in a microbial fuel cell in the presence of pyocyanin. *J. Taiwan. Inst. Chem. Eng., 45*(5), 2319–2324.
Wu, Y., Lai, Q., Zhou, Z., et al., (2008). *Alcanivorax hongdengensis* sp. nov., an alkane-degrading bacterium isolated from surface seawater of the straits of Malacca and Singapore, producing a lipopeptide as its biosurfactant. *Int. J. Syst. Evol. Microbiol., 59*(6), 1474–1479.
Zampolli, J., Collina, E., Lasagni, M., et al., (2014). Biodegradation of variable-chain-length *n*-alkanes in *Rhodococcus opacus* R7 and the involvement of an alkane hydroxylase system in the metabolism. *AMB Express, 4*(1), 1–9.
Zedelius, J., Rabus, R., Grundmann, O., et al., (2011). Alkane degradation under anoxic conditions by a nitrate-reducing bacterium with possible involvement of the electron acceptor in substrate activation. *Environ. Microbiol. Rep., 3*(1), 125–135.
Zeinali, M., Vossoughi, M., & Ardestani, S. K., (2008). Degradation of phenanthrene and anthracene by *Nocardia otitidiscaviarum* strain TSH1, a moderately thermophilic bacterium. *J. Appl. Microbiol., 105*(2), 398–406.
Zengler, K., Richnow, H. H., Rosselló-Mora, R., et al., (1999). Methane formation from long-chain alkanes by anaerobic microorganisms. *Nature, 401*(6750), 266–269.
Zhang, J., Wallar, B. J., Popescu, C. V., et al., (2006). Methane monooxygenase hydroxylase and B component interactions. *Biochemistry, 45*(9), 2913–2926.
Zhang, T., Gannon, S. M., Nevin, K. P., et al., (2010). Stimulating the anaerobic degradation of aromatic hydrocarbons in contaminated sediments by providing an electrode as the electron acceptor. *Environ. Microbiol., 12*(4), 1011–1020.
Zhou, J., Bian, X. Y., Zhou, L., et al., (2016). Synthesis and characterization of anaerobic degradation biomarkers of n-alkanes via hydroxylation/carboxylation pathways. *Eur. J. Mass Spectrom., 22*, 31–37.

CHAPTER 4

Microbial Degradation of Spilled Oil in Aqueous Environments: Beyond C15 Hydrocarbons

FERNANDA LOPES MOTTA, NAYEREH SABORIMANESH, and HEATHER D. DETTMAN

Natural Resources Canada, CanmetENERGY Devon, 1 Oil Patch Drive, Devon, AB T9G 1A8, Canada, E-mail: heather.dettman@nrcan-rncan.ca (Heather D. Dettman)

4.1 INTRODUCTION

Petroleum is the major source of energy fueling the world economy and spills of petroleum during exploration, transport, and use can cause impacts to soil and water environments. Both governments and industry are working to reduce the risk of oil spills with the introduction of more restrictive regulations, the use of stringent operating codes, and the development of improved handling methods. However, as the production and consumption of petroleum and its products continue to increase, the number of incidents of oil spills is also likely to increase (Fingas, 2013).

Oil spill responders are familiar with the potential impacts of petroleum immediately after a spill. Volatile compounds present in crude oil are of great concern in terms of air quality for responders and the public, acute toxicity effects upon biota, and the potential hazards of explosion and fire. Over the longer term, much is unknown about the fate of the oil. Complete pathways for the natural biodegradation of petroleum organic compounds have not been elucidated. However, these processes can completely mineralize compounds up to 15 carbons (C15) in size to carbon dioxide relatively quickly (days to weeks), suggesting that these smaller compounds will not contribute to longer-term impacts. For compounds larger than C15, especially if they include poly-ring structures, less is known of their final fate. As larger organic compounds become oxidized by microbial and photooxidation processes, the chemically added molecular oxygen increases their water solubility, making aqueous dissolution a potential route for their "disappearance" from a spill site. However, it is currently not known how the potential environmental impacts of these compounds change with oxidation, and at what point, and in which environments, these larger molecules become "environmentally benign." Essentially, the question is whether natural degradation processes can turn petroleum into humic substances (Jednak et al., 2017; Ressler et al., 1999), the organic material found in soils, rivers, lakes, and sea sediments (Weber et al., 2018).

This chapter presents an overview of the current literature on the biodegradation of petroleum compounds when spilled in aqueous environments with an emphasis on relating the composition of oils in terms of their relative contents of both less polar (saturates and aromatics) and more polar (resins and asphaltenes) organic compounds, to their inherent biodegradability. Aspects such as the mechanisms of biodegradation, the types of microorganisms capable of degrading petroleum compounds, the role of biosurfactants during biodegradation, and current evidence for conversion of petroleum into humic substances are important considerations for predicting the degree to which natural attenuation can remediate a spill site. Key environmental conditions that either enable or hamper biodegradation are examined and the potential impacts of biodegradation of petroleum compounds ($\geq$ C15) on water quality are discussed.

4.2 PETROLEUM FORMATION AND COMPOSITION

It is widely believed that petroleum originates from biomass from the remains of living organisms including vegetation that were buried in the earth (Figure 4.1) (University of Waterloo, 2020). At the earth's surface, once an organism dies, lipids, carbohydrates, proteins, and lignin are converted to humic substances (humification), which are the end products of decaying organic matter. Lignin and its transformation products, as well as polyphenols, melanin, cutin, proteins, and other derived polymers, are important building blocks in this process (Stevenson, 1994). However, the exact mechanism of humic substances formation is still unknown (Van Trump et al., 2006). These humic substances, if trapped in a kerogen matrix at elevated temperatures and pressures in the subsurface may be converted into a broad range of organic compounds, which form petroleum. Lipid and the lipid-like fractions of organisms are among the chemical constituents playing a major role in the formation of petroleum (Tissot and Welte, 1984).

Hydrocarbons, organic molecules containing no heteroatoms, constitute an important fraction of any crude oil (Petrov, 1987), ranging from smaller more volatile compounds to very large compounds (Speight, 2007). The composition of crude oil varies from location to location (Tissot and Welte, 1984) and influences its properties (Ali et al., 2012; Stauffer et al., 2008). While carbon (80-87%) and hydrogen (10-15%) are the main elements in petroleum, the presence of organic molecules containing sulfur (up to 10 wt%), nitrogen (up to 1 wt%), and oxygen (up to 5 wt%) are important minor constituents present in heterocyclic compounds, and as functional groups. Sulfur can also be present as elemental sulfur. As well, crude oils contain widely varying concentrations of trace metals such as V, Ni, Fe, Al, Na, Ca, Cu, and/or U (Boesch and Rabalais, 1987).

SARA (Saturates-Aromatics-Resins-Asphaltenes) analysis methodology, which is standardized by the American Society for Testing and Materials, is a characterization method for crude oils that quantitatively separates the oil into subfractions based on polarity. The procedure used for isolating asphaltenes is ASTM D-6560-00 (ASTM, 2005). The saturates, aromatics, and resins fractions from the remaining oil (called maltenes) are isolated by using a modified version of the procedure ASTM D-2007-03 (ASTM, 2016).

The saturates fraction consists of nonpolar compounds including linear, branched, and cyclic saturated hydrocarbons. Aromatics, which contain one or more aromatic rings, and low contents of heteroatoms, are more polar than saturates. The remaining two fractions, resins and asphaltenes are more polar than the aromatics, containing aromatic rings and significant contents of heteroatoms. The operational distinction between resins and asphaltenes is that asphaltenes are insoluble in an excess of heptane (or pentane) while resins are soluble (Fan and Buckley, 2002). Asphaltenes are the heaviest and most polar fraction of crude oils, and they tend to have the highest contents of sulfur, nitrogen, oxygen, and trace metals, organized in a large variety of chemical functionalities (Sheu, 2002). Table 4.1 gives the average weight percentage contents of saturates, aromatics, and the sum of asphaltenes and resins, based on the analysis of 517 crude oils (Tissot and Welte, 1984). Examples of chemical structures for each fraction are also shown in Table 4.1. The smallest and most volatile aromatic compounds include benzene, toluene, ethylbenzene, and xylenes and are referred to as BTEX. Larger aromatic compounds, containing two or more rings are called polycyclic aromatic hydrocarbons (PAHs) and can have contents of up to 30 wt% in crude oils (Ali et al., 2012). In Figure 4.1, it is shown that humic substances were, and are, the indirect precursors to petroleum formation on Earth. Studies have proposed that once petroleum is released at the Earth's surface, degradation processes working on the organic compounds in petroleum begin to change them back into humic substances (Jednak et al., 2017; Ressler et al., 1999).

TABLE 4.1 A Summary of the Classification of Organic Compounds in Crude Oils Based on Polarity Including Their Average Contents (wt.%) and Representative Chemical Structures

Fraction	Average (wt.%) (Tissot and Welte, 1984)	Examples (Arboleda et al., 2015; Hoff and Dettman, 2012; Lee et al., 2017; Meslé et al., 2013)
Saturates	57.2	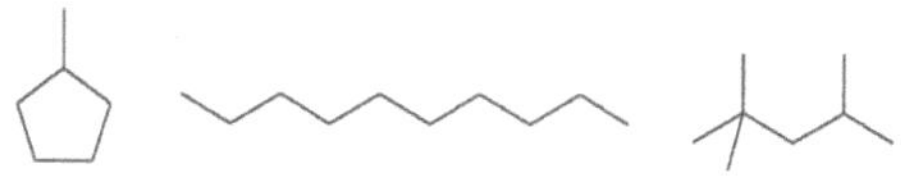
Aromatics	28.6	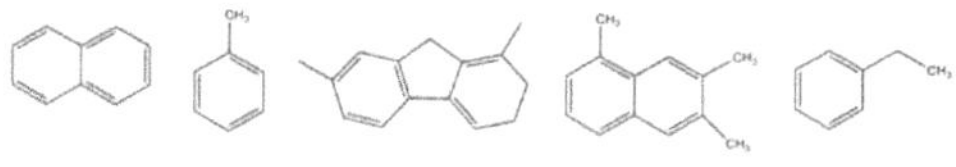
Resins	14.2*	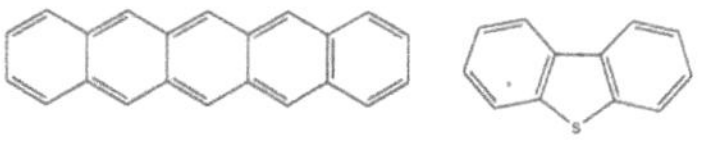
Asphaltenes		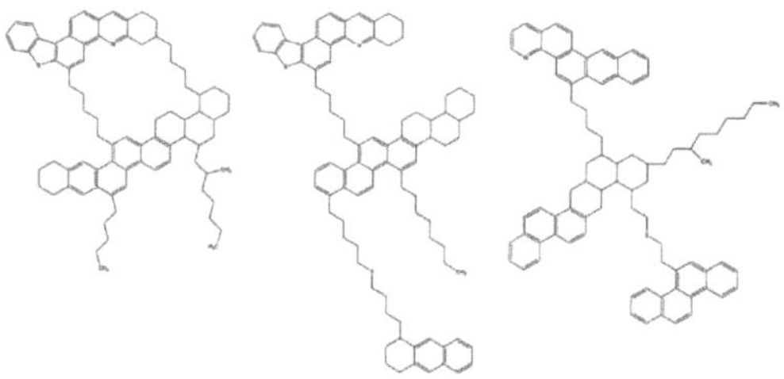

**Resins + Asphaltenes.*

Humic substances are highly varying combinations of organic molecules (carbohydrates, amino acids, and fatty acids) as well as compounds that are highly resistant to further microbial degradation (Adey, 2007). These substances have been classified into humin, humic acid and fulvic acid, based on their differing solubilities in acidic versus alkaline solutions (Trevisan et al., 2010). Humic and fulvic acids are predominantly hydrophilic, but they also contain substantial concentrations of aromatic rings, fatty acid esters, aliphatic, and other hydrophobic compounds (Schnitzer, 1978). Representative elemental compositions of samples extracted from soils formed under widely differing geographic and pedologic conditions are shown in Table 4.2. It is noted that in both humic acids and fulvic acids, C and O are the major elements (Schnitzer, 1978). Chemical structure models proposed for humic and fulvic acids are presented in Figure 4.2 (a) and Figure 4.2 (b), respectively.

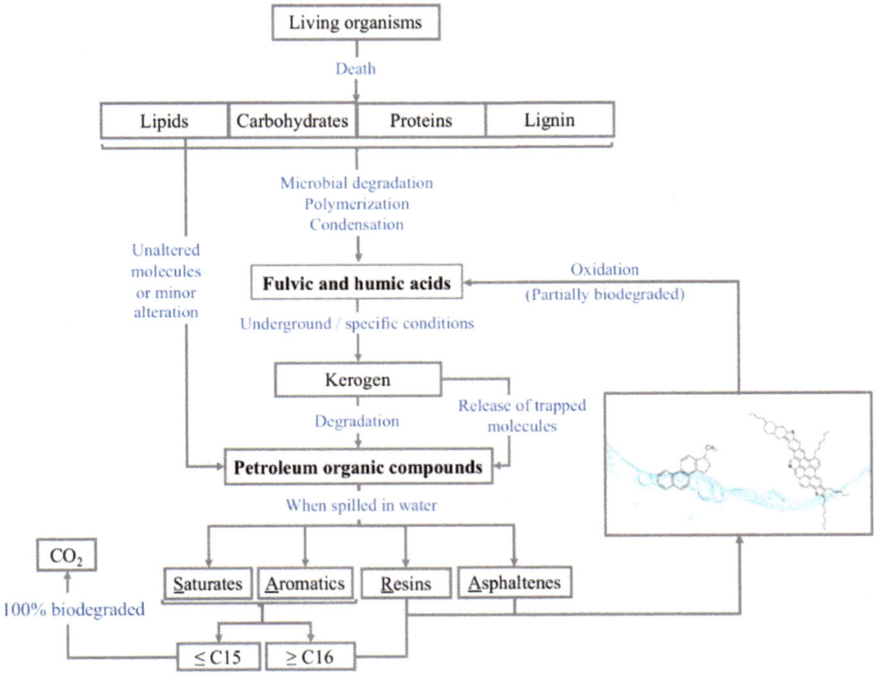

FIGURE 4.1 Schematic representation of the geological sources of petroleum organic compounds and their degradation when accessible by active microbial communities (formation scheme developed based on the work of Tissot and Welte (1984)).

TABLE 4.2 Representative Elemental Analysis Results of Humic and Fulvic Acids Samples Extracted from Soils from Differing Climates.

Elemental Analysis (wt.%)	Humic Acids	Fulvic Acids
C	56.2	45.7
H	4.7	5.4
N	3.2	2.1
S	0.8	1.9
O	35.5	44.8

Source: Reprinted with permission from Schnitzer, 1978. © Elsevier

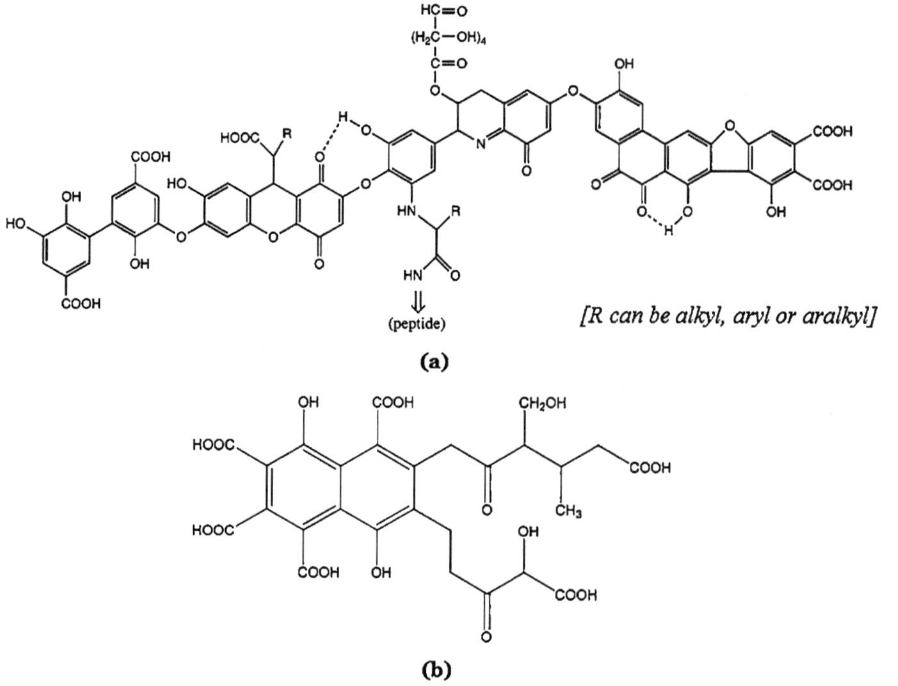

FIGURE 4.2 Model structure of humic acid (a) according to Stevenson (1982); and of fulvic acid (b) according to Buffle (1977).

Source: Reprinted with permission from a) Tissot and Welte, 1984. © 1995 American Chemical Society; b) Buffle, 1977. © 1977 American Chemical Society.

4.3 THE ROLE OF BIODEGRADATION ON THE REMEDIATION OF OIL SPILLS

Once oil is spilled into a natural water body, its composition starts to change under the influence of concurrent processes collectively termed as oil weathering (Mishra and Kumar, 2015). These processes include evaporation, photooxidation, biodegradation, dissolution, dispersion, and emulsification. Of these processes, evaporation of the most volatile oil compounds is quantitatively one of the major processes for mass loss (Prince et al., 2003) and is responsible for the quickest loss of BTEX compounds (Kotzakoulakis and George, 2018). Loss of compounds with low boiling points can be responsible for the loss of one- to two-thirds of an oil spill's mass during the first 48 hours after a spill. The oil mass remaining has higher specific gravity and viscosity, and is richer in metals (mainly nickel and vanadium), waxes, and asphaltenes than the original oil (Jafarinejad, 2017).

Although occurring at much lower rates than evaporation, natural biodegradation is also an important process for removal of the oil whereby microorganisms convert its organic molecules into other substances, such as organic acids and carbon dioxide (Hoff, 1993; Prince, 1993). Once introduced into the environment, petroleum may be biotically degraded through the oxidation of organic compounds by both aerobic and anaerobic microbes.

The natural exposure of microbial communities to petroleum in aquatic environments at the earth's surface occurs through natural seeps or spills (National Research Council, 2003). The contamination of water environments by natural seeps are phenomena that occur when crude oil flows from geological strata beneath the ground or sea floor to nearby and/or overlying water. Their ecological impacts seem to be limited because the slow but steady rate of release allows surrounding ecosystems to adapt (Burgherr, 2007). Some organisms incorporate petroleum carbon and other compounds from the releases directly into biomass with limited biotransformations (*e.g.* alkanes oxidized to carboxylic acids used for cell membranes) (Burgherr, 2007). In contrast, spills that occur during production, transportation, and utilization of oil can result in relatively short, high-volume releases in places that previously have never experienced significant exposure to these types of organic compounds. Consequently, environmental impacts can be significant, particularly immediately after the spill (Burgherr, 2007).

Studies of the fate of organic compounds such as hydrocarbons in the environment form the basis for bioremediation treatments (Hoff, 1993; Prince, 1993), in which the rates of degradation are accelerated by both encouraging natural biodegradation processes, and supplementing the existing system with microorganisms better able to degrade the contamination (Brown et al., 2017). Advantages of bioremediation processes for oil clean up are their feasibility, cost-effectiveness, and environmentally friendly properties (Gomez and Sartaj, 2014; Wu et al., 2016), such as minimal site disruption and permanent waste removal (Azubuike et al., 2020). Bioremediation approaches compatible with natural biogeochemical cycles comprise of biosurfactant amendments, biostimulation, and bioaugmentation. Biosurfactants, also called "dispersants," favour oil solubilization and oil droplet formation in water (dispersion), making organic compounds available to non-biosurfactant-producing microorganisms (Mapelli et al., 2017). After a spill, the high carbon content of the contaminant causes an imbalance in the carbon:nitrogen ratio of the environment and so biodegradation of the organic compounds can be retarded or prevented by nitrogen deficiency. Biostimulation is one of the most effective approaches to enhance the degrading activity of indigenous bacterial communities because it re-establishes the nutrient imbalance caused by the spill (Foght, 2010). Bioaugmentation, which consists of the introduction of cultured microorganisms into a contaminated environment (Kouzuma and Watanabe, 2011), might also enhance degradation (Tyagi et al., 2011). However, the effectiveness of bioaugmentation is variable, as the survival of the foreign (xenobiotic) microbes added may be variable. In addition, the oil-degrading abilities of xenobiotic microbes are highly dependent on their adaptation to the environmental conditions of the particular site. As an alternative, autochthonous bioaugmentation has been proposed to overcome these difficulties. For this bioaugmentation method, microorganisms indigenous to the contaminated site have been selected for their ability to degrade oils, enriched in the laboratory, and then re-introduced into the environment (Hosokawa et al., 2009).

A high number of potential oil-degrading microorganisms have been reported. Although bacteria are considered as primary degraders and the most active agents for petroleum pollutant degradation (Abbasian et al., 2015; Meckenstock et al., 2016), fungi and algae can also degrade petroleum pollutants (Hendrickx et al., 2006; Wilkes et al., 2016). Organic compounds can be selectively metabolized by individual strains of microorganisms or by diverse consortia of microbial strains (Varjani et al., 2015; Varjani and Upasani, 2016). Mixed microbial cultures have proven to be more effective than individual cultures for metabolizing complex hydrocarbon mixtures (Deppe et al., 2005; Varjani et al., 2015). The advantages of mixed cultures as opposed to pure cultures, for bioremediation are attributed to synergistic interactions between members of the consortia. For example, some species can remove metabolites that are toxic to other species, or they can continue the degradation of compounds that have been partially degraded by other species (Alexander, 1999).

Many studies have investigated the microbial degradation of oil (Ali et al., 2012; Bressler and Gray, 2003; Gomez and Sartaj, 2014; Hosokawa et al., 2009; Ressler et al., 1999; Wu et al., 2016). However, published work has focused mainly on the biodegradation of smaller-sized alkane and aromatic hydrocarbons (up to C40). These compounds are known to be more readily biodegraded compared to the larger organic compounds found in heavy fractions of oil that include>C40 saturates and aromatics, as well as resins and asphaltenes (Ali et al., 2012). As molecular complexity increases, susceptibility to biodegradation decreases. In general, organic compound susceptibility is in the order *n*-alkanes > branched alkanes > low molecular weight aromatics > cyclic alkanes > polyaromatics > resins and asphaltenes (Head et al., 2006). Moreover, even during studies of the biodegradation of smaller hydrocarbons, total removal is rarely reported and tends not to exceed 20-50% of the oil introduced (Venosa et al., 1996). However, with time some spill sites do recover to a point where oil is no longer detected. Consequently, more information is needed to understand the processes by which the most recalcitrant organic compounds are either eventually converted to more water-soluble compounds, or conjugated through covalent linkages to soil complexes, even if they are not completely decomposed.

4.4 CELLULAR UPTAKE OF ORGANIC COMPOUNDS

The cellular uptake of organic compounds is the first step towards the biodegradation of these chemicals. Before cellular uptake, some organic compounds can be converted into oxidized intermediates by extracellular enzymes produced by some types of microorganisms. This step facilitates the degradation of higher molecular weight organic compounds (Figure 4.3). The enzymes involved can degrade polymeric chains into shorter intermediates with modified mechanical properties that make them more accessible for microbial assimilation (Palmisano and Pettigrew, 1992). Extracellular enzymes produced by fungi are particularly suitable biodegradation catalysts due to their specificity and ability to decompose high molecular weight substrates (Fedorak et al., 1993).

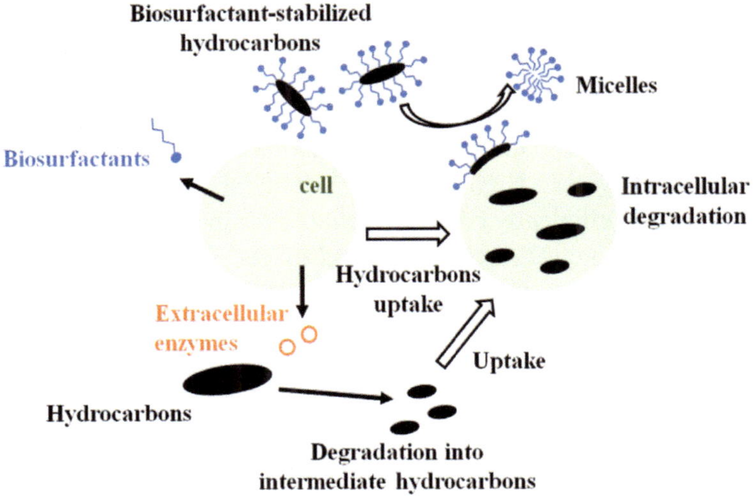

FIGURE 4.3 Probable mechanisms by which petroleum organic compounds enter a microbial cell for degradation.

Various white-rot fungi in consortia with bacteria can facilitate PAH metabolism in cells by improving the bioavailability of target compounds. Due to the lack of suitable enzymes, generally, fungi cannot degrade high molecular weight PAHs completely. However, their extracellular enzymes can transform these molecules into more polar, water-soluble metabolites, which can be further degraded by bacteria and other microbes (Ghosal et al., 2016). Van Hamme et al. (2003) studied the microbial metabolism of dibenzyl sulphide, an organosulfur compound of interest to the petroleum industry, by white-rot fungi (Van Hamme et al., 2003). Results suggested that the first oxidation step is catalyzed by extracellular enzymes and the subsequent metabolism is cytochrome P-450 mediated. Asphaltenes have been isolated from oil samples where sulfur atoms were oxidized, suggesting possible exposure to water and extrinsic enzymes in their process history (Yang et al., 2015).

Besides the production of extracellular enzymes, microorganisms have also developed strategies to modify their cell membrane fluidity to allow the transport of organic compounds across their membranes in polluted environments. Some microorganisms produce amphiphilic molecules called biosurfactants that are either released extracellularly or become part of their cell membrane surface to enhance the bioavailability of organic pollutants (Figure 4.3) (Skovhus and Whitby, 2019). Biosurfactants are produced by microorganisms (*e.g.*, bacteria, yeasts, and fungi) (Mulligan, 2005) that are classified as cationic, anionic, and neutral based on their hydrophilic head group (Mulligan, 2005), and their molecular weight (Ron and Rosenberg, 2002). The low molecular weight biosurfactants are usually glycolipids (carbohydrates attached to a long-chain aliphatic acid or lipopeptide) (Ron and Rosenberg, 2002). High molecular weight biosurfactants are composed of polysaccharides, proteins, lipopolysaccharides, lipoproteins or complex mixtures of biopolymers. High molecular weight biosurfactants (also known as bioemulsifiers) efficiently coat hydrophobic compound droplets reducing their tendency to aggregate into larger droplets. Low molecular weight biosurfactants effectively reduce liquid/gas or liquid/liquid surface and interfacial tension (Ron and Rosenberg, 2002). This is due to the interactions of biosurfactants monomers with both water and hydrophobic molecules. At biosurfactant concentrations up to their critical micelle concentration, the surface/interfacial tension of the water/hydrophobic phase decreases. At concentrations above the critical micelle concentration, biosurfactant monomers form aggregates called micelles and encapsulate hydrophobic compounds (Figure 4.3) (Muthusamy et al., 2008).

Biosurfactant-producing microorganisms have been isolated from different petroleum-contaminated environments (e.g., soil, marine waters, and sediments) (Rahman et al., 2003). Microbial biosurfactants include rhamnolipids produced by *Pseudomonas aeruginosa* (Abdel-Mawgoud et al., 2009); sophorolipids produced by *Candida bombicola* (Elshafie et al., 2015); surfactins produced by *Bacillus subtilis* (Whang et al., 2008); lipopolysaccharides produced by *Acinetobacter calcoaceticus* (Fondi et al., 2016); trehalolipids produced by species of *Mycobacterium* and *Rhodococcus erythropolis* (Kügler et al., 2015; White et al., 2013); and phospholipids produced by *Thiobacillus thiooxidans* (Mohanty et al., 2013), among others. Examples of chemical structures for rhamnolipid, surfactin, and sophorolipids are presented in Figure 4.4.

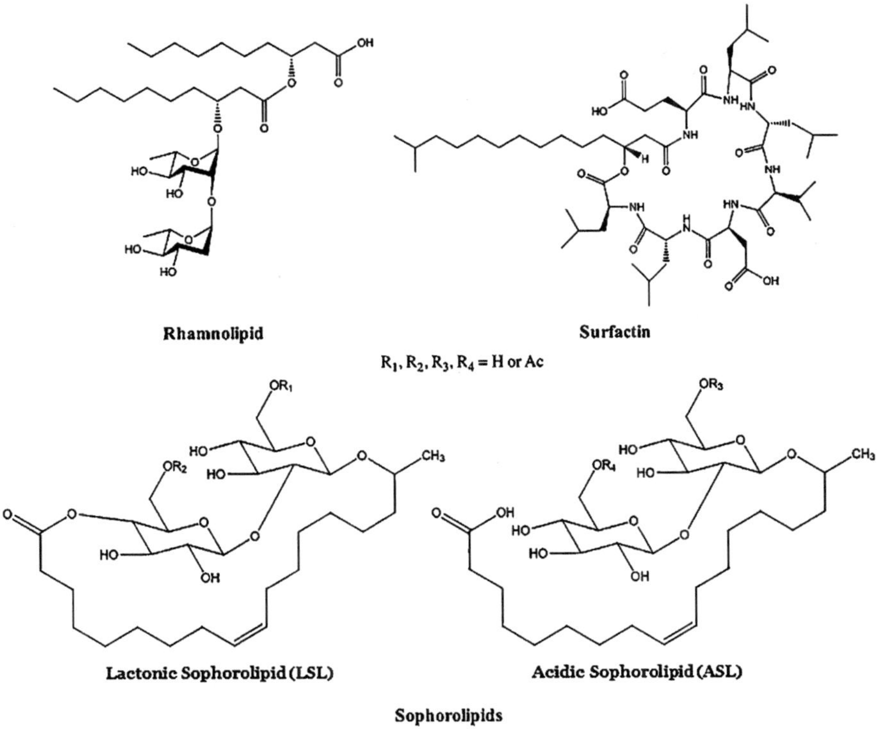

FIGURE 4.4 Chemical structures of some common biosurfactants (Abdel-Mawgoud et al., 2010; Nguyen and Sabatini, 2011; Wu et al., 2017).

Biosurfactant production involves diverse metabolic pathways and is influenced by the carbon sources available to the microorganisms (Van

Bogaert et al., 2007). Synthesis of biosurfactants is often associated with the assimilation of hydrophobic substrates (Ito et al., 1980; Noordman and Janssen, 2002). However, this is not the case in the synthesis of all biosurfactants, as some biosurfactants (*e.g.*, sophorolipids) are synthesized in the absence of hydrophobic substrates, but under nitrogen-limiting conditions (Van Bogaert et al., 2007).

Several physiological roles for biosurfactants have been proposed. Biosurfactants play multiple roles during the incorporation of hydrophobic substrates into microbial cells, for example by influencing the binding affinity of cells to *n*-alkanes (Ito et al., 1980; Kappeli et al., 1978), emulsifying water-insoluble substrates (Francy et al., 1991; Harvey et al., 1990; Zhang and Miller, 1992), promoting direct contact of cells with oil droplets (Bouchez-Naitali et al., 1999; Saborimanesh and Mulligan, 2015), interacting with emulsified droplets (Bouchez-Naitali et al., 1999; Saborimanesh and Mulligan, 2015), contributing to cell adherence to hydrophobic substrates (Bouchez-Naitali et al., 1999; Kappeli et al., 1978), and providing inhibiting effects toward other microbes in the environment (Ito et al., 1980).

There are examples of successful applications of biosurfactants for oil spill treatment in water and soil, including biosurfactant isolated from the *Gordonia* sp. strain JE-1058 (Saeki et al., 2009), surfactin synthesized by *Bacillus subtilis* ATCC 21332 and rhamnolipid synthesized by *Pseudomonas aeruginosa* J4 (Whang et al., 2008). Antoniou et al. (2015) observed that *Paracoccus marcusii* isolated from sediment can produce glycolipids and degrade PAHs (Antoniou et al., 2015). Durval et al. (2019) used *Bacillus cereus* as an *ex situ* biosurfactant producer and observed that motor oil degradation was enhanced up to 96% over 27 days when the biosurfactant was applied to contaminated seawater (Durval et al., 2019). Abalos et al. (2004) investigated the potential of rhamnolipids produced by *Pseudomonas aeruginosa* to enhance bioavailability and biodegradation of crude oil in a liquid medium (Abalos et al., 2004). The addition of these biosurfactants accelerated the biodegradation of total petroleum hydrocarbons from 32% to 61% after 10 days of incubation, particularly for isoprenoids from the aliphatic fraction and alkylated PAHs from the aromatic fraction (Abalos et al., 2004). Darvishi et al. (2011) investigated biosurfactant production by a microbial consortium including *Enterobacter cloacae* and *Pseudomonas* sp. isolated from heavy crude oil-contaminated soil under extreme environmental conditions (Darvishi et al., 2011). They observed that the isolated consortium was able to grow and produce biosurfactant and cause increased oil spreading and emulsification properties at temperatures up to 70°C, pressures up to 6000 psia, salinities up to 15% (w/v), and pH range 4-10 (Darvishi et al., 2011). In addition to these studies, bioprospecting of novel strains of microbes and their genetic engineering for enhanced efficacy have been met with variable degrees of success (Patel et al., 2018).

Molecular genetics and biochemistry studies of biosurfactant-producing microorganisms have shed light on the required enzymes, operons, and metabolic pathways of biosurfactant production (Das et al., 2008). Some microorganisms (e.g., *Candida bombicola* and *Candida apicola*) produce enzymes for the terminal oxidation of hydrophobic substrates with a chain length of C16 to C19 (e.g., alkanes, fatty acids, and derivatives) (Lottermoser et al., 1996; Van Bogaert et al., 2007) generating terminal or subterminal hydroxy fatty acids (Van Bogaert et al., 2007) for direct incorporation into their cell wall (Lottermoser et al., 1996; Van Bogaert et al., 2007). In these cases, longer-chain hydrophobic compounds, or intermediates of the terminal oxidation pathway, are shortened by β-oxidation before incorporation (Van Bogaert et al., 2007).

4.5 MECHANISMS OF BIODEGRADATION INSIDE CELLS

Once organic compounds cross the membrane into the cytoplasm, they are oxidized by enzymes and enzyme complexes by a process that can be relatively quick (Eguchi et al., 1992). Simplified pathways for the microbial degradation of the various types of petroleum organic compounds are presented in Table 4.3. Overall, the pathways of peripheral degradation transform organic contaminants into chemical intermediates of the central intermediary metabolism such as, the tricarboxylic acid (TCA) cycle, also known as the citric acid or Krebs cycle. Generally, organic compounds are oxidized to acetyl-CoA for entry into the TCA cycle, and from there, sugars are synthesized via gluconeogenesis (Al-Hawash et al., 2018; Tornheim, 2018). The tricarboxylic acid cycle is an important energy-yielding metabolic pathway in aerobic and anaerobic organisms as it provides the greater part of the reduced coenzymes that will be oxidized by the electron transport chain to ultimately drive adenosine triphosphate production via oxidative phosphorylation. In addition to its role in energy-yielding metabolism, and the oxidation of 2-carbon units, the cycle is also the major pathway for the interconversion of 4- and 5-carbon compounds in the cell, many of which arise from, or are intermediates in, the synthesis of amino acids (Bender, 2003).

TABLE 4.3 Proposed Biodegradation Pathways of Organic Compounds Reported in the Scientific Literature

Substrate	Proposed Biodegradation Pathway	References
n-Alkanes *anaerobic*	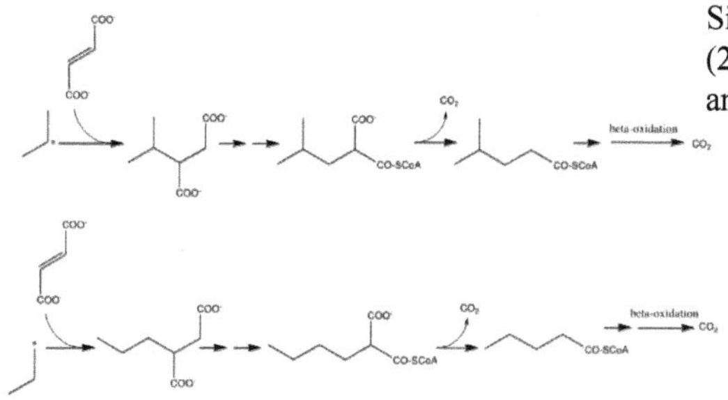	Singh et al. (2017); Skovhus and Whitby, 2019
n-Alkanes *anaerobic*	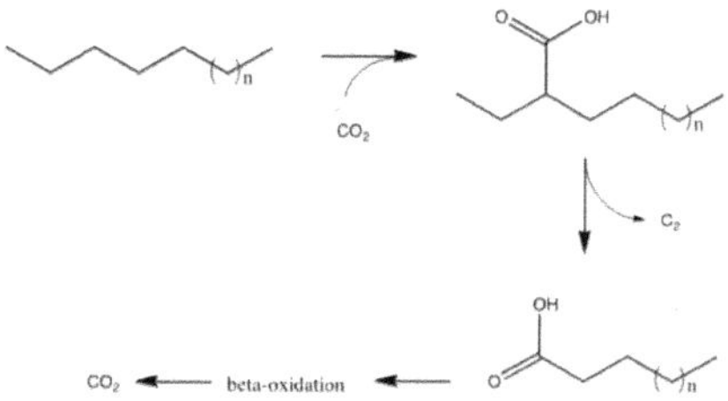	So et al. (2003)
n-Alkanes *aerobic*	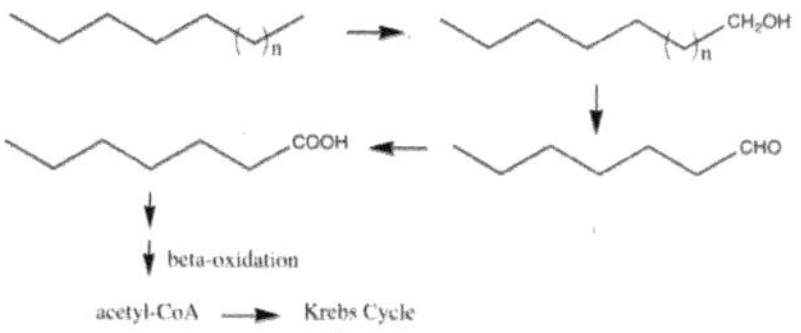	Hong et al. (2017)
Cycloalkanes: Cyclohexane	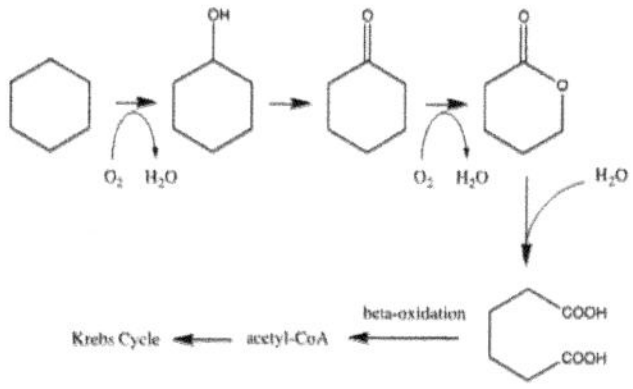	Miransari (2011)
Branched alkanes: 2-Methyloctane	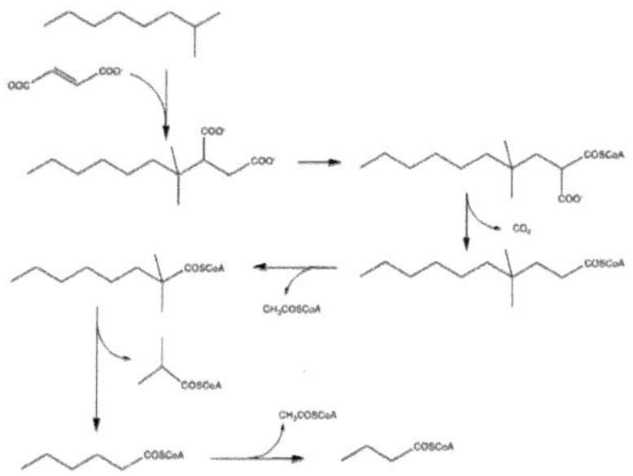	Chen et al. (2019)

TABLE 4.3 *(Continued)*

Substrate	Proposed Biodegradation Pathway	References
Branched alkanes: Pristane	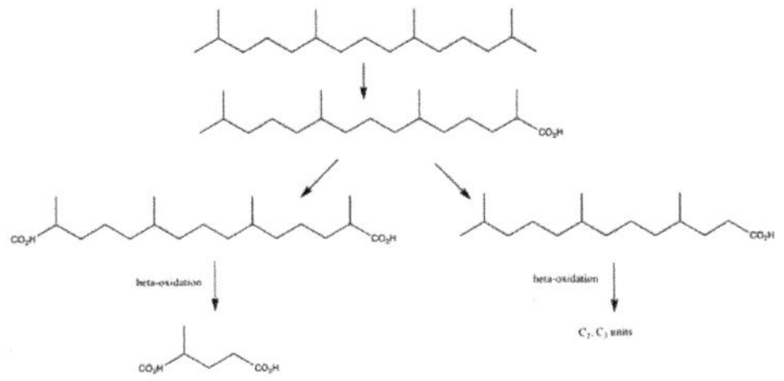	McKenna and Kallio (1971); Pirnik et al. (1974)
Mono-aromatics: Benzene	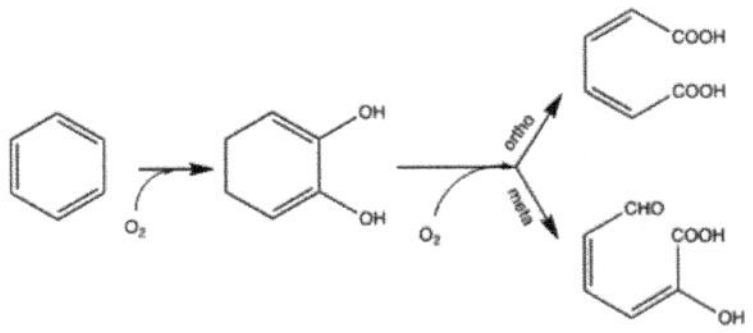	Wilson and Bouwer (1997)
Alkylated aromatics: Ethylbenzene	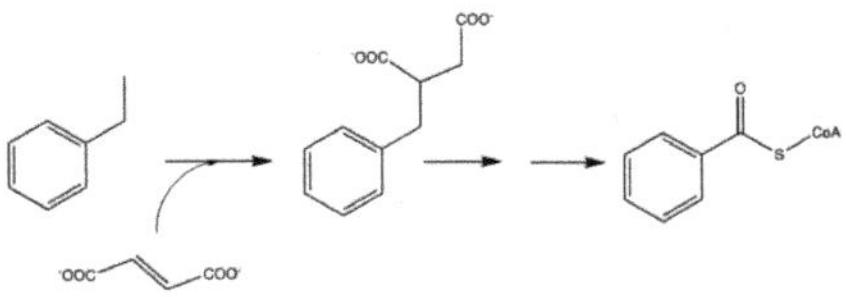	Ghattas et al. (2017)
Polycyclic aromatics: Phenanthrene	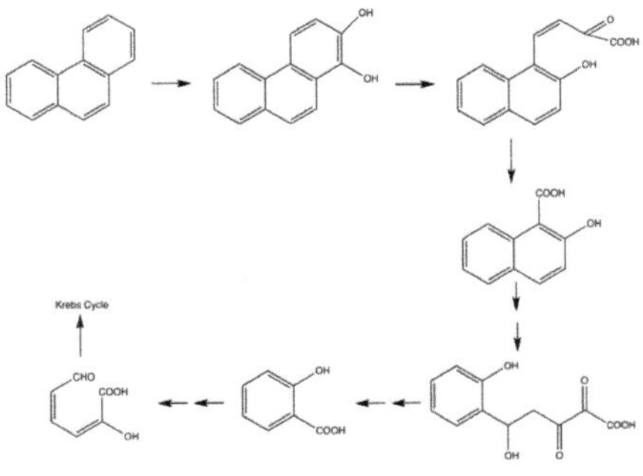	Ghosal et al. (2010)
Polycyclic aromatics: Pyrene	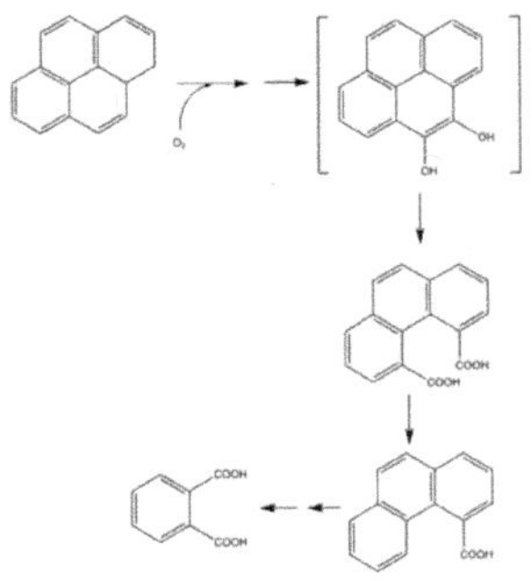	Keith (2003); Vila et al. (2001)

TABLE 4.3 *(Continued)*

Substrate	Proposed Biodegradation Pathway	References
Organosulfur compounds: Dibenzyl sulfide	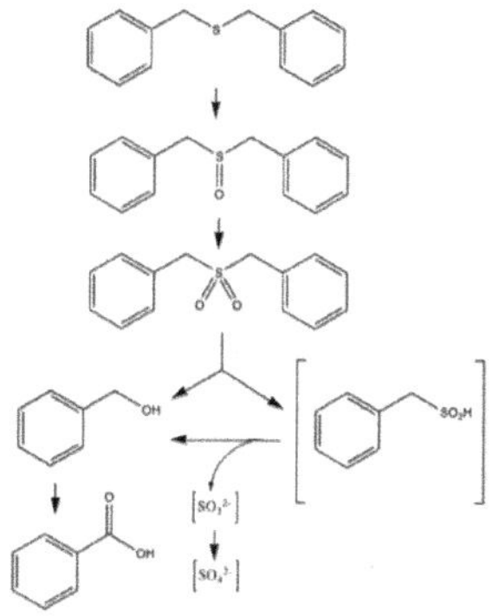	Van Hamme et al. (2003); Van Hamme et al. (2004)
Organosulfur compounds: Dibenzothiophene	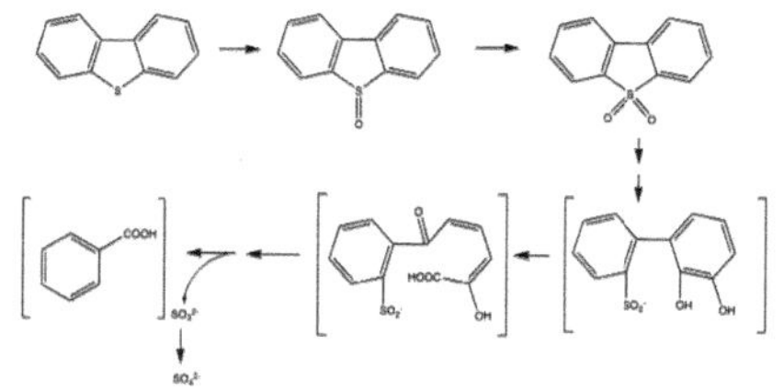	Zoveidavianpoor (2018)

Note: Two arrows represent more than one step.

Shorter chain length alkanes (up to C17) are more recalcitrant than longer chain length alkanes for anaerobic sulfate- and nitrate-reducing bacteria. This is because these hydrocarbons can accumulate in anaerobic environments due to the lack of evaporation and so exert toxic effects on microorganisms by compromising the structural and functional properties of their membrane lipid bilayer. Consequently, the degradation of these compounds becomes impaired. Furthermore, under anaerobic conditions, degradation of branched alkanes has been shown to be more efficient than for alkanes by sulfate-reducing bacteria (Hasinger et al., 2012).

Specific steps for the metabolism of alkanes, cycloalkanes, branched alkanes, single-ring aromatics, polyaromatics, and specific reactions with sulfur in organic compounds are summarized in Table 4.3. For each compound type, information for both anaerobic and aerobic degradation is given if available. The two main mechanisms of anaerobic degradation of *n*-alkanes involve biochemical reactions that are completely different from those used during aerobic metabolism (Grossi et al., 2008). The first anaerobic mechanism involves activation at the subterminal carbon of the alkane by the addition of fumarate, which is analogous to the formation of benzyl succinate during the anaerobic degradation of toluene. However subsequent reactions are completely different and involve dehydrogenation and hydration (Timmis, 2010). The second mechanism for activation of alkanes for anaerobic degradation is carboxylation (So et al., 2003). The aerobic oxidation of alkanes is initiated by the action of monooxygenases where an oxygen atom is added to a terminal or sub-terminal carbon, ultimately leading to the formation of a corresponding acid that can be further degraded via beta-oxidation. Alkanes are catalytically converted by microbial enzymes into fatty acids, that are metabolized to acetyl-CoA for entry into the Krebs cycle with the potential for converting all of the carbon and hydrogen into CO_2 and H_2O (Singh, 2006). During this process, enzymes such as alkane monooxygenase, fatty alcohol dehydrogenase, and fatty aldehyde dehydrogenase play important roles.

The biodegradation of 1-ring cycloalkanes under aerobic conditions starts with ring-opening leading to the generation of alcohol that is subsequently converted to a ketone by the action of a dehydrogenase, and then to an acid by the action of an esterase. Substituted 1-ring cycloalkanes may be attacked at the side chain prior to ring-opening (Xue et al., 2015). Kirkwood et al. (2008) describe the isolation of a strain of *Rhodococcus* sp. capable of biotransforming decalin, a dicycloalkane. These authors indicated that the pathway for decalin biotransformation would include a cyclic alcohol, a cyclic ketone and ring-opened metabolites having either an alcohol plus a carboxylic acid group, or two carboxylic acids. The anaerobic degradation of cycloalkanes was studied by Townsend et al. (2004) who concluded that unsubstituted and monosubstituted alicyclic compounds are readily biodegraded and heavily substituted alicyclic compounds are far more recalcitrant (Townsend et al., 2004). Rios-Hernandez et al. (2002) proposed a pathway for the biodegradation of ethylcyclopentane under sulphate-reducing conditions, which includes the regeneration of fumarate for the activation of ethylcyclopentane to ethylcyclopentylsuccinic acid (Rios-Hernandez et al., 2003).

Only a few strains of microorganisms have been found to utilize multibranched saturated hydrocarbons (Berekaa and Steinbüchel, 2000). In particular, *Mycobacterium* species can degrade 2,6,10,15,19,23-hexamethyltetracosane (squalane) under aerobic conditions via the β-oxidation pathway. By this pathway, squalane is initially converted to a dioic acid, then the dioic acid is oxidized to 3,7,11-trimethyldodecandioic acid. Further degradation is completed when the β-methyl group of the intermediate is converted into carbonyl oxygen (Xue et al., 2015). Studies of the biodegradation of other types of branched alkane hydrocarbons such as 2-methyloctane and the proposed pathways under anaerobic conditions have been reported and are summarized in Table 4.3.

Petroleum waxes that are complex mixtures of n-alkanes, iso-alkanes, and cycloalkanes with carbon numbers ranging approximately from 18 to 65, are regularly found on beaches worldwide but specific pathways for their degradation have not been reported. However, evidence for biodegradation of these compounds has been reported in the literature by Patel and Bhaskaran (2018), who observed the utilization of major components of paraffin wax by *Pseudomonas nitroreducens*. In particular, 100% of heneicosane, 79.74% of pentacosane and 72.50% of triacontane were metabolized in 8 days at 37°C under aerobic conditions (Patel and Bhaskaran, 2018).

Several degradation pathways have been proposed for the smaller ring aromatic hydrocarbons (*e.g.*, mono-, di, and tricyclic) (Ghosal et al., 2010; Pinyakong et al., 2000). Aromatic hydrocarbons are oxidized by the addition of two oxygen atoms, and converted to cis-dihydrodiol by bacteria, while in fungi, aromatics are oxidized and converted to trans-dihydrodiol (Xue et al., 2015). The dihydrodiol is converted to o-dihydroxybenzene, and then the ring is opened, resulting in a chain compound that is eventually metabolized to acetyl-CoA (Fuchs et al., 2011).

Higher molecular weight aromatic compounds are considered to be recalcitrant or exhibiting only very low rates of biodegradation (Dibble and Bartha, 1979; Heitkamp and Cerniglia, 1989; Perry and Cerniglia, 1973; Pineda-Flores and Mesta-Howard, 2001; Premuzic et al., 1999), even though some studies have reported their high rate of removal under optimal conditions (Leahy and Colwell, 1990). The biodegradation of PAHs with fewer than three aromatic rings has been the subject of many reviews (Fuchs et al., 2011; Hammel, 1995; Seo et al., 2009; Xue et al., 2015). However, there is currently a lack of information regarding the biodegradation pathways of higher molecular weight PAHs (Kanaly and Harayama, 2010; Kanaly and Harayama, 2000), despite research pertaining specifically to the bacterial biodegradation of PAHs composed of more than three rings is available in the literature. Kanaly et al. (2000) isolated a microbial consortium from soil, which converted benzo[*a*]pyrene to CO_2 when grown on diesel fuel (Kanaly et al., 2000). Benzo[*a*]pyrene biodegradation pathways have been proposed by Zhang et al. (2000), which are initiated by mono- or di-oxygenation at different positions, followed by dehydration of the resulting dihydrodiols, and then oxidative cleavage, usually at the meta-position (Zhang et al., 2000).

Cleavage products are then decarboxylated to form chrysene derivatives. Proposed aerobic biodegradation pathways for pyrene and phenanthrene are presented in Table 4.3. In both cases biodegradation initiates with the incorporation of oxygen atoms followed by ring cleavage, leading to central intermediates such as protocatechuates and catechols, which are further converted to tricarboxylic acid cycle intermediates (van der Meer et al., 1992).

Although high molecular weight organic compounds such as resins and asphaltenes are less susceptible to degradation, heavy fractions of global crudes include organic acids, suggesting that some level of biodegradation can occur even for the largest organic compounds in oil (Dettman et al., 2012). Studies on the biodegradation of asphaltenes have reported mass losses from 5 to 35% (Pineda-Flores and Mesta-Howard, 2001), however, the mechanisms are not well elucidated. Pathways have been proposed based upon those known for similar chemical structures (Pineda-Flores and Mesta-Howard, 2001). One potential metabolic limitation to biodegradation for high molecular weight organic compounds is their size. Even if a high molecular weight compound was able to get inside the cell, the structure may be too large to gain access and interact with enzymatic active sites (Bressler and Gray, 2003). This steric hindrance principle may explain their relative recalcitrance. Therefore, ring and chain oxidative cleavage previously described for the biodegradation of PAHs may have important roles for the reduction of molecular weight, and at least partial degradation of resins and asphaltenes.

Specific reactions with sulfur in larger organic compounds such as asphaltenes may assist with the biodegradation of these recalcitrant compounds (Table 4.3). In particular, the cleavage of aliphatic sulphide bonds in these compounds, which may be as high as 30% of the C-S bonds, would result in a significant reduction in their molecular weight and enhance their biodegradation (Pathak, 2018). Microorganisms able to cleave alkyl sulphide bonds and remove sulfur from ring structures play a major role in the transformation of sulfur-containing organic compounds in nature. Most studies on the biodegradation of sulfur-containing compounds have focused on sulfur in ring structures such as dibenzothiophene, the most abundant heterocyclic compounds in crude oil. There are two pathways, Kodama (Kodama et al., 2000) and sulfur-specific (4S) (Gallagher et al., 1993), that microbially degrade organic sulfur when dibenzothiophene is used as a model compound. The Kodama pathway uses microbial metabolism to open carbon-carbon bonds in the dibenzothiophene, releasing the sulfur. The 4S desulfurization pathway specifically cleaves the carbon-sulfur bonds in dibenzothiphene without destroying the carbon skeleton, removing the sulfur in the form of sulfate (Hu et al., 2019). Zoveidavianpoor (2018) isolated a strain of *Cunninghamella elegans* from mangrove sediments that was able to metabolize dibenzothiophene (Zoveidavianpoor, 2018). The fungus used a portion of the carbon, removed the sulfur forming sulphite and sulfate, and left benzoic acid as a product. Karimi et al. (2017) observed that sulfur can be removed from dibenzothiphene and the carbon simultaneously metabolized to meet cellular growth demands of a bacterial strain *Rhodococcus sp* IGTS8 (Mohebali and Ball, 2008; Soleimani et al., 2007). Isolation of a bacterial strain, *Rhodococcus* sp. strain JVH1 capable of removing alkyl sulfur from organic compounds was achieved by Van Hamme et al. (2004) using a novel compound, *bis*-(3-pentafluorophenylpropyl)-sulphide as a sole sulfur source.

4.6 FACTORS AFFECTING BIODEGRADATION RATES OF SPILLED OIL

4.6.1 OIL EMULSION

Oil spilled in water tends to spread and form a slick. As a result of wind and wave action, oil-in-water or water-in-oil emulsions may form (Cooney, 1984). Dispersion of organic compounds in the water column in the form of oil-in-water emulsions increases the surface area of the oil and thus its availability for microbial attack. However, water-in-oil emulsions called "mousse" form on the water surface as large masses or plates, establish low surface-to-volume ratios and inhibit biodegradation (Davis and Gibbs, 1975). Furthermore, rapid emulsion formation after a spill could diminish the rates of evaporation and dissolution, causing retention of more toxic low molecular weight aromatic compounds within the residual oil (National Research Council U.S., 2014).

Emulsion formation starts with wave and current energy forcing the entry of small water droplets, about 10-25 μm in size, into the oil. If the oil has low viscosity, the small droplets will not leave the oil quickly. On the other hand, if the oil is too viscous, water droplets will not be able to enter the oil to a significant extent. The relationship between crude oil viscosity and water-in-oil emulsion stability was studied by Hemmingsen et al. (2005). A general trend is that as the viscosity increases, there is an increase in emulsion stability and a decrease in the coalescence of oil droplets observed from light to heavy oils (10 to 1000 cP, respectively) (BP, 2011). Once water mixes into the oil, polar compounds including asphaltenes and resins in the oil interact with the water droplets to stabilize them. Depending on the quantity and type of asphaltenes and resins, an emulsion may be formed. The conditions required for stable emulsion formation may only be reached after a period of evaporation. Evaporation lowers the amount of lower molecular weight compounds and increases oil viscosity (Kim, 2010).

In addition to oil droplets becoming dispersed into the water column, smaller aromatic compounds are water-soluble. Monocyclic aromatic compounds (e.g., benzene and alkyl-substituted benzenes) with $\log K_{ow}$ values between 2.1 and 3.7, and lower molecular weight 2-3 ring PAHs with $\log K_{ow}$ values between 3.7 and 4.8 undergo dissolution (National Research Council, 2005; Payne et al., 2003). Environmental processes such as evaporation, dissolution, biodegradation, and photooxidation can be strongly affected by interactions between dissolved and dispersed oil components and sediment particles (Gong et al., 2014). Sediments have long been recognized as important factors in the transport of oil from one environmental compartment (i.e., phase) to another following oil spill events in aquatic environments. Studies have shown that interactions between oil and sediment play important roles in dispersion and degradation of spilled oil (Muschenheim and Lee, 2002). In nearshore waters, naturally dispersed oil droplets may aggregate readily with suspended particulate material such as clay-sized mineral particles or organic matter to form oil-suspended particulate material aggregates (Gong et al., 2014).

When used, chemical dispersants are sprayed on surface oil slicks to break up oil into small droplets, thus increasing its surface area by lowering the oil/water interfacial tension, and accelerating the dispersal of oil into the water column. Under optimal conditions, the application of chemical dispersants promotes the biodegradation of oil in the water column and may eliminate, or significantly reduce, the impacts of oil on sensitive shorelines and habitats (Lessard and DeMarco, 2000). Indeed, dispersants themselves can enhance the initial rate of petroleum biodegradation by being the first substrate utilized by organic compound-degrading bacteria (Varadaraj et al., 1995). However, two issues are associated with the use of dispersants including limits of their effectiveness, and the transient increase in local toxicity resulting from increased oil dispersion into the water column (Fingas, 2013).

4.6.2 MARINE SNOW FORMATION

The role of marine snow on the fate of spilled oil in the deep sea has attracted significant attention. Marine snow aggregates form and degrade in the water column. Marine snow forms naturally in the ocean and consists of living and non-living particles, involving mucus-producing phytoplankton and bacteria, detritus, feces from zooplankton and fish, and inorganic particles (Henry et al., 2018). Degradation is to a large extent due to the activity of attached microbes (Smith et al., 1992), which typically occur in aggregates in abundances that are orders of magnitude higher than in the ambient water (Alldredge and Silver, 1988; Kiørboe, 2000; Simon et al., 2002). Dissolved organic matter liberated from snow particles due to the activity of attached bacteria may become important substrates even for free-living (non-symbiotic) bacteria (Kiørboe, 2000). Thus, marine snow aggregates are unique microcosms in the water column within which material and energy flows are regulated by complex biological and physical processes (Simon et al., 2002). Grossart et al. (2007) observed that cultures of bacteria isolated from marine snow showed a 20-fold increase in enzyme activity within 2 h of particle attachment (Grossart et al., 2007). Elevated extracellular enzymatic activity on aggregates may result from bacterial quorum sensing that has been demonstrated to occur in bacterial strains isolated from marine snow (Jatt et al., 2015).

Interactions in the ocean water between marine snow suspended particles, and dispersed oil droplets can influence oil biodegradation. Within the oiled marine snow complexes, the accessibility of the oil for biodegrading bacteria may change to become either less or more favorable (Rahsepar et al., 2017). Recently, studies of the Deepwater Horizon oil spill reported that unprecedented quantities of marine snow were found floating on the sea surface prior to sinking to the seafloor (Suja et al., 2019) suggesting that marine snow is also formed in association with oil spills (Brakstad et al., 2018).

A wide variety of microorganisms, particularly cyanobacteria (Decho et al., 2005; Han et al., 2014; Kawaguchi and Decho, 2002), bacteria (Grossart et al., 2007), and eukaryotic phytoplankton (Raposo et al., 2013) produce and secrete exopolysaccharide substances in the marine environment (Suja et al., 2019). Oil-related marine snow sedimentation and flocculant accumulation are thought to be initiated by the release of microbial extracellular polymeric substances in response to the presence of oil. The sticky extracellular polymeric substances collect the oil along with sediment particles (e.g., clays) as well as (living and dead) plankton, bacterial, and fungal materials into marine oil snow particles that, given enough inorganic matter, sink through the water column to the ocean floor (Quigg et al., 2016). Therefore, biodegradation processes in the deep-water plume are suggested to be also an important factor for the transport of oil to the seabed (Brakstad et al., 2018) and seabed sedimentation of oil (Stout and Payne, 2016; Valentine et al., 2014; Valentine et al., 2010).

Depending on local oceanic conditions, the addition of chemical dispersants can result in the formation of high volumes of marine snow (Langenhoff et al., 2020). This is presumably due to the favorable nutritive conditions in marine snow compared to seawater and may be partly explained by microbial colonization through chemotactic localization of high concentrations of organic matter. Furthermore, microbial communities are regulated by growth and predator-prey interactions, and such interactions may interplay with colonization and detachment to regulate the dynamics of marine snow-associated microbial communities (Kiørboe et al., 2003). During the Deepwater Horizon incident, up to 7 million liters of chemical dispersant was applied by both spraying on the sea surface oil slicks and directly injecting at the leaking wellhead near the seafloor (National Commission on the BP Deepwater Horizon Oil Spill and Offshore Drilling, 2011). These surface and subsurface applications of dispersant were reported to result in facilitated microbial biodegradation of the oil (Brakstad et al., 2015) and to either trigger or enhance oil-related marine snow formation (Suja et al., 2017). During the Deepwater Horizon oil spill, an estimated 1200 square miles of deep ocean sediment (Adams, 2015) was covered with a thick layer of oily material including marine snow and particulates (Brooks et al., 2015).

Once on the ocean floor, oiled marine snow accumulates because of low biodegradation rates due to low oxygen availability and temperatures (Langenhoff et al., 2020). As well, under conditions with limited oxygen availability within layers of ocean sediments marine snow could limit more general oil biodegradation by consuming the dissolved oxygen relatively quickly to create anaerobic conditions (Daly et al., 2016).

4.6.3 OXYGEN AS A LIMITING NUTRIENT

Aerobic conditions are generally considered necessary for extensive degradation of petroleum in the environment since major degradative pathways for organic compounds involve oxygenases (Atlas, 1981; Cerniglia, 1992; National Research Council, 2003). Studies have shown that oxygen depletion sharply reduced biodegradation activities in marine sediments and soils (Atlas, 1981; Bossert and Bartha, 1984; Hambrick et al., 1980; Le Moigne et al., 2017). Oxygen concentrations are not normally limiting in the upper levels of the water column in marine and freshwater environments, or the surface layer of most beach environments. It can become limiting in subsurface sediments, anoxic zones of water columns, and in most fine-grained marine shorelines, freshwater wetlands, mudflats, and salt marshes. Factors affecting the availability of oxygen also include the actions of waves and water flow, the physical state of the oil, and the availability of other required nutrients such as nitrogen and phosphorus (Zhu et al., 2001).

When oxygen concentrations decrease in the water column anoxic conditions can exist (Røy et al., 2012). Thus in the absence of oxygen, some microorganisms can use other electron acceptors that are available such as Fe^{3+}, SO_4^{2-} and NO_3^- to degrade organic compounds (Timmis, 2010; Foght, 2008; Grossi et al., 2008). However, anaerobic degradation is usually considered to be several orders of magnitude slower than aerobic degradation (Heider and Schühle, 2013).

In addition, oxygen can play an important role on biodegradation rates indirectly when there is photooxidation of petroleum. For tropical waters such as in the Gulf of Mexico receiving high year-round solar flux, photochemical processes significantly enhance petroleum biodegradation (Daling et al., 1990; Thingstad and Pengerud, 1983). The mechanisms for photooxidation of petroleum organic compounds include both direct photolysis and also the reaction of reactive oxygen species produced by solar radiation acting on a variety of photosensitizers in natural waters (Mill et al., 1980; Zepp and Baughman, 1978). Photooxidation may not directly result in significant mass losses (Ward et al., 2018), but the photooxidation products may alter the subsequent weathering processes (Garrett et al., 1998; Islam et al., 2015; Thingstad and Pengerud, 1983). Photooxidation products are generally more water-soluble and bioavailable (Aeppli et al., 2018; Ray et al., 2014), which enhances biodegradation (Dutta and Harayama, 2000; Maki et al., 2001). For instance, photooxidation can oxidize alkanes of 15 to 35 carbon atoms, as readily as polyaromatic and polyheteroaromatic structures such as phenanthrene, fluorene, and dibenzothiophene (Dutta and Harayama, 2000). A recent study showed that shorter alkanes are less susceptible to photooxidation than aromatic compounds (Hounjet et al., 2017). Photooxidation results in the production of a variety of oxygenated organic compounds and sulfur species, including aliphatic and aromatic ketones, aldehydes, carboxylic acids, fatty acids, esters, epoxides, sulphoxides, sulphones, phenols, anhydrides, quinones, and aliphatic and aromatic alcohols (Burwood and Speers, 1974; Hansen, 1975; Larson et al., 1979; Payne and Phillips, 1985). Many of these photooxidized compounds are much more rapidly degraded by microbes than the parent compounds (Lee, 1980; Lee et al., 1982; Rontani et al., 1987).

4.6.4 BIODEGRADATION RATES IN THE ARCTIC OCEAN AND THE GULF OF MEXICO: EFFECT OF SURFACE TEMPERATURE

The Arctic region differs from temperate regions in ways that may influence the fate of oil pollution. Low surface temperatures, low light intensity (Atlas, 1981; Margesin and Schinner, 2001), and ice coverage (Atlas, 1981) decrease the surface remediation rates of oil pollution in the Arctic marine environment compared to those found in temperate and tropical climates (Kristensen et al., 2015). The colder temperatures influence the physicochemical properties of oil that then affects oil biodegradation. As temperature decreases, oil fluidity decreases resulting in lower organic compound solubility in water and rates of dissolution. These processes decrease the bioavailability of oil compounds to degradative microorganisms and prolong the presence of toxic organic compounds in colder environments (Atlas, 1975; Atlas and Bartha, 1972; Faksness et al., 2008). As well, photooxidation contributions to petroleum degradation will be impacted by low light intensity. Although the long sunshine exposure times during the summer months in the Arctic might similarly influence oil photooxidation and biodegradation (Ivanov et al., 2005), the year-round intensity of solar radiation is much lower in the Arctic than in temperate or tropical regions (University of Guelph, 2020).

Below the water/ice surface in Arctic waters, biodegradation has been found not to be strongly inhibited by cold water temperatures (National Research Council U.S., 2014). While the Gulf of Mexico has very different surface characteristics compared to the Arctic Ocean, their deep waters and sediments have similar temperature conditions (National Research Council U.S., 2014; Landes, 1973) and so temperature may not be the factor that determines the difference in bacterial populations between these regions. Furthermore, both have natural oil seeps that allow long-term selection for oil-degrading bacteria adapted to regional conditions (National Research Council U.S., 2014; Venosa and Holder, 2007). Measured biodegradation rates for crude oil are remarkably similar at positive and sub-zero temperatures with half-lives of 17 days at 5°C (Brakstad et al., 2015), 18 days at -1°C and 14 days at -1.7°C (Garneau et al., 2016).

Ice coverage during much of the year also creates an extra physical phase in the Arctic not present in tropical regions where oil contamination can be trapped. However, microbes have been able to adapt to even these conditions.

Hydrocarbon degraders have been detected in both the Arctic and Antarctic sea ice contaminated with fresh crude oil (Delille et al., 2004; Deppe et al., 2005; Gerdes et al., 2005). Oil degrading microbes may also inhabit the brine channels of the ice interior. Most of these are Gammaproteobacteria, including known oil degraders of the genera *Colwellia*, *Marinomonas* and *Glaciecola* (Boccadoro et al., 2018; Brakstad et al., 2017).

4.6.5 FRESH AND MARINE WATERS: EFFECT OF SALINITY

The effect of salinity on microbial degradation activity depends on the range of salt concentration and the salinity of the habitat of the microorganisms under investigation (Martins and Peixoto, 2012). Biodegradation rates of oil in freshwater, where the concentration of salts (salinity) is low (National Oceanic and Atmospheric Administration, 2021) and marine water, where average salinity is 3.5 % w/v NaCl (National Oceanic and Atmospheric Administration, 2021), may be comparable and depend on tolerance and adaptation of resident oil-degrading microorganisms. This suggestion is supported by a study where rates of hydrocarbon utilization did not start to decrease until salinity just below that of marine water, at 3.3% w/v NaCl. Degradation rates continued to decrease as salinity increased over the entire range of testing, up to 28.4% w/v NaCl (Ward and Brock, 1978). However, the decreasing biodegradation rates with increasing salt concentrations may not necessarily be caused by decreased microbial activity. Microbial communities present in these environments would likely be able to adapt to higher salinity levels. However, the physical factor that the solubility of hydrophobic molecules in water decreases as salt concentration increases would reduce the bioavailability of the hydrocarbons in the hypersaline environments, and so would reduce their degradation rates. At the same time, the availability of oxygen also decreases as salt concentration increases due to its decreasing solubility in the water (Timmis, 2010; Pfeifer et al., 1997). Martins and Peixoto (2012) suggested that under conditions of low oxygen and high salt concentrations, halophilic photoautotrophs can be important participants in the degradation of organic compounds (Martins and Peixoto, 2012; Fathepure, 2014).

More detailed knowledge of the mechanisms by which microorganisms degrade hydrocarbons in hypersaline conditions (salinity of 35% w/v NaCl) has also accumulated in the past two decades. Studies described in a review article suggest that there is a diverse microbial population existing in these environments, some of which can efficiently degrade organic compounds

(Fathepure, 2014). Investigation of biodegradation pathways has revealed that degradation of oxygenated and non-oxygenated hydrocarbons by halophilic and halotolerant microorganisms occurs by pathways similar to those found in non-halophiles (Fathepure, 2014). In addition, some organisms capable of degrading a mixture of hydrocarbons in widely fluctuating salinities have been found to produce surfactants to assist with solubility (Fathepure, 2014).

4.7 WHAT HAPPENS TO THE PETROLEUM ORGANIC COMPOUNDS THAT CANNOT BE COMPLETELY BIODEGRADED?

Biodegradation and other natural weathering processes will degrade petroleum organic compounds in the environment, but the larger molecules (>C40) can take months to years in areas of high oil concentrations (Atlas and Hazen, 2011). For petroleum products such as crude oils, removal of 100% of the oil by biodegradation is not expected (Atlas and Hazen, 2011). For example, the residues of spilled oil from the tanker vessel Exxon Valdez in 1989 are still found within the beaches of Prince William Sound (Boufadel et al., 2010). Surface and subsurface sediment samples collected from locations affected by the Hebei Spirit (South Korea) oil spill, revealed the presence of residual crudes two years after the spill (Hong et al., 2012). Based upon previous oil spills it is clear that residual oil in shoreline environments can remain for an extended period of time (Atlas and Hazen, 2011).

In Figure 4.1, the creation of petroleum is shown to be formed with the removal of oxygen from humic substances. As degradation of petroleum after a spill occurs with the addition of oxygen, the question can be asked whether there are conditions under which the recalcitrant, partially degraded petroleum compounds can become humic substances again? Humic substances are not believed to be harmful to human health (Government of Nova Scotia, 2021), and published literature elucidates the beneficial effect of these substances in decreasing toxicity caused by other chemicals (Schnitzer and Khan, 1972). Among the various organic substances occurring naturally, humic substances are the most widespread and can be found in various environments such as soils, rivers, lakes, and sea sediment (Weber et al., 2018). In addition, humic substances have important roles in soil fertility and are considered to have primal relevance for the stabilization of soil aggregates, as well as can remove toxic metals, anthropogenic organic chemicals, and other pollutants from water (Peña-Méndez et al., 2005). Humic substances, themselves, are extremely resistant to biodegradation, with a half-life measured in tens or hundreds of years. These substances are considered to be recalcitrant in the absence of dissolved oxygen (Stevenson, 1994) and are slowly degraded by microbial populations in aerobic environments (Wetzel, 1992; Wetzel et al., 1995).

A recent study conducted by Jednak et al. (2017) suggests that humification found during bioremediation of heavy residual fuel oil is likely partially due to the resistance of the larger petroleum compounds to biodegradation, where they remained to be transformed into materials analogous to humic substances (Jednak et al., 2017). Their results indicated that the production of humic substances during bioremediation of petroleum compounds resulted from a consortium of zymogenous microorganisms (Jednak et al., 2017). Ressler et al. (1999) demonstrated that the degradation of a mixture of seven PAHs compounds by a bacterial culture enriched from contaminated soil resulted in the formation of a dark-coloured residual fraction highly resistant to bacterial degradation (Ressler et al., 1999). Analysis of this fraction revealed a molecular-size distribution similar to that of natural humic acids. Walter et al. (1991) also observed the formation of recalcitrant humic acid-like substances during the bacterial degradation of anthracene oil, a complex mixture of PAHs and other organic substances (Walter et al., 1991). Salt et al. (1998) stated that root-colonizing microorganisms that form symbiotic relationships with plants may also accelerate remediation processes by increasing the humification of organic pollutants (Salt et al., 1998).

Aguelmous et al. (2019) reviewed studies of composting petroleum sludge where different sources of co-substrates (*e.g.*, soil, wood, green wastes, and manure) were used to enhance biodegradation of total petroleum hydrocarbons including different alkanes and aromatics up to C40 in size (Aguelmous et al., 2019). Although these authors concluded that no work has addressed the formation of humic substances during composting of petroleum sludge, they pointed out that El Fels et al. (2016) and Hafidi et al. (2008) studied co-composting of hydrocarbons with green substrate and observed that hydrocarbons are incorporated or adsorbed into the structure of humic substances (El Fels et al., 2016; Hafidi et al., 2008). El Fels et al. (2016) investigated the behavior of aliphatic hydrocarbons during co-composting of sewage sludge activated with palm tree waste and Hafidi et al. (2008) studied the contents and fate of sixteen PAH compounds (El Fels et al., 2016; Hafidi et al., 2008).

Based on the presented findings, humic substances can be produced during the microbial degradation of petroleum. More work is needed to determine the extent of the conversion and the conditions under which conversion is optimized. As well, the relationship between the degree of conversion and the toxicity of the resulting residue would need to be understood. Improved knowledge for the humification of petroleum residues could lead to the development of more efficient remediation strategies for sites contaminated with heavy petroleum products.

4.8 CONCLUSIONS AND FUTURE RESEARCH DIRECTIONS

Microorganisms are major degraders of petroleum organic compounds and play a key role in the removal of petroleum compounds from the environment. Literature studies suggest that biosurfactants produced by bacteria may enhance the assimilation of both petroleum compounds and other nutrients available in the environment thereby enhancing the rate of biodegradation. In addition, a better understanding of physical, chemical, and biological factors that affect oil-related marine snow sedimentation and flocculent accumulation, and the resulting impacts on biological communities is an important next step for oil spill science. More work is needed to develop knowledge for how different environmental conditions affect the performance of chemical dispersants when applied to oil spills at sea, to determine their impacts on the formation of oil-related marine snow, and subsequent impacts on the biodegradation of spilled oil.

During the biodegradation of petroleum organic compounds, the toxicity of the partially degraded products to aquatic organisms has not been identified. Consequently, the evolution of the relative toxicity of petroleum spill sites with time, and the cumulative effects of small spills on the environment are not completely understood. However, the evidence that humification processes as natural processes to promote further degradation of petroleum organic compounds beyond that possible by biodegradation alone may be indicative of how the environment eventually remediates petroleum spills. An improved understanding of the conditions needed to promote humification may offer new insights for improved strategies to optimize remediation rates for treatment of residual oil remaining at new and historical spill sites.

ACKNOWLEDGMENTS

The authors acknowledge financial support from the Government of Canada Oceans Protection Plan.

KEYWORDS

- **hydrocarbon biodegradation**
- **microbial degradation pathways**
- **oil humification**
- **oil spill behaviour**
- **oil spill in water**
- **petroleum biodegradation**

REFERENCES

Abalos, A., Viñas, M., Sabaté, J., Manresa, M. A., & Solanas, A. M. (2004). Enhanced biodegradation of Casablanca crude oil by a microbial consortium in presence of a rhamnolipid produced by *Pseudomonas Aeruginosa* AT10. *Biodegradation, 15*, 249–260.

Abbasian, F., Lockington, R., Mallavarapu, M., & Naidu, R. (2015). A comprehensive review of aliphatic hydrocarbon biodegradation by bacteria, *Appl. Biochem. Biotechnol., 176*, 670–699.

Abdel-Mawgoud, A. M., Aboulwafa, M. M., & Hassouna, N. A., (2009). Characterization of rhamnolipid produced by *Pseudomonas aeruginosa* isolate Bs20. *Appl. Biochem. Biotechnol., 157*, 329–345.

Abdel-Mawgoud, A. M., Lépine, F., & Déziel, E., (2010). Rhamnolipids: Diversity of structures, microbial origins and roles. *Appl. Environ. Microbiol., 86*, 1323–1336.

Adams, A. (2015). *Summary of Information Concerning the Ecological and Economic Impacts of the BP Deep Water Horizon Oil Spill Disaster*. https://www.nrdc.org/sites/default/files/gulfspill-impacts-summary-IP.pdf (accessed on 25 October 2021).

Adey, W. H., (2007). *Dynamic Aquaria: Building and Restoring Living Ecosystems*. Academic Press.

Aeppli, C., Swarthout, R. F., O'Neil, G. W., Katz, S. D., Nabi, D., Ward, C. P., et al. (2018). How persistent and bioavailable are oxygenated Deepwater Horizon oil transformation products? *Environ. Sci. Technol., 52*, 7250–7258.

Aguelmous, A., El Fels, L., Souabi, S., Zamama, M., Hafidi, M. (2019). The fate of total petroleum hydrocarbons during oily sludge composting: A critical review. *Rev. Environ. Sci. Biotechnol., 18*, 473–493.

Al-Hawash, A. B., Dragh, M. A., Li, S., Alhujaily, A., Abbood, H. A., Zhang, X., & Ma, F. (2018). Principles of microbial degradation of petroleum hydrocarbons in the environment. *Egypt. J. Aquat. Res., 44*, 71–76.

Alexander, M., (1999). *Biodegradation and Bioremediation*. Academic Press.

Ali, H. R., El-Gendy, N. S., Moustafa, Y. M., Roushdy, M. I., & Hashem, A. I. (2012). Degradation of asphaltenic fraction by locally isolated halotolerant bacterial strains. *ISRN Soil Sci., 2012*, 1–11.

Alldredge, A. L., & Silver, M. W., (1988). Characteristics, dynamics and significance of marine snow. *Prog. Oceanogr., 20*, 41–82.

Antoniou, E., Fodelianakis, S., Korkakaki, E., & Kalogerakis, N. (2015). Biosurfactant production from marine hydrocarbon-degrading consortia and pure bacterial strains using crude oil as carbon source. *Front. Microbiol., 6*, 274.

Arboleda, P. H., Dettman, H. D., & Lucy, C. A., (2015). Hydrocarbon group type separation of gas oil resins by high performance liquid chromatography on hyper-cross-linked polystyrene stationary phase. *Energy Fuels, 29*, 6686–6694.

ASTM, (2005). *Standard Test Method for Determination of Asphaltenes (heptane insolubles) in Crude Petroleum and Petroleum Products.* ASTM D6560-17, ASTM International, West Conshohocken, PA.

ASTM, (2016). *Standard Test Method for Characteristic Groups in Rubber Extender and Processing Oils and Other Petroleum-Derived Oils by the Clay-Gel Absorption Chromatographic Method.* ASTM D2007-19, ASTM International, West Conshohocken, PA.

Atlas, R. M., (1975). Effects of temperature and crude oil composition on petroleum biodegradation. *Appl. Microbiol., 30*, 396–403.

Atlas, R. M., (1981). Microbial degradation of petroleum hydrocarbons: An environmental perspective. *Microbiol. Rev., 45*, 180–209.

Atlas, R. M., & Bartha, R., (1972). Biodegradation of petroleum in seawater at low temperatures. *Can. J. of Microbiol., 18*, 1851–1855.

Atlas, R. M., & Hazen, T. C., (2011). Oil biodegradation and bioremediation: A tale of the two worst spills in U. S. history. *Environ. Sci. Technol., 45*, 6709–6715.

Azubuike, C. C., Chikere, C. B., & Okpokwasili, G. C., (2020). Bioremediation: An eco-friendly sustainable technology for environmental management. In: *Bioremediation of Industrial Waste for Environmental Safety: Volume I: Industrial Waste and its Management* (pp. 19–39). Springer, Singapore.

Bender, D. A., (2003). Tricarboxylic acid cycle. In: *Encyclopedia of Food Sciences and Nutrition* (pp. 5851–5856). Academic Press: Oxford.

Berekaa, M. M., & Steinbüchel, A., (2000). Microbial degradation of the multiply branched alkane 2,6,10,15,19,23-hexamethyltetracosane (squalane) by *Mycobacterium fortuitum* and *Mycobacterium ratisbonense*. *Appl. Environ. Microbiol., 66*, 4462–4467.

Boccadoro, C., Krolicka, A., Receveur, J., Aeppli, C., & Le Floch, S. (2018). Microbial community response and migration of petroleum compounds during a sea-ice oil spill experiment in Svalbard. *Mar. Environ. Res., 142*, 214–233.

Boesch, D. F., & Rabalais, N. N., (1987). *Long-Term Environmental Effects of Offshore Oil and Gas Development*. Elsevier Applied Science.

Bossert, I., & Bartha, R., (1984). The fate of petroleum in soil ecosystems. *Environ. Sci.*, 435–473.

Bouchez-Naitali, M., Rakatozafy, H., Marchal, R., Leveau, J. Y., Vandecasteele, J. P. (1999). Diversity of bacterial strains degrading hexadecane in relation to the mode of substrate uptake. *J. Appl. Microbiol., 86*, 421–428.

Boufadel, M. C., Sharifi, Y., Van Aken, B., Wrenn, B. A., & Lee, K. (2010). Nutrient and oxygen concentrations within the sediments of an Alaskan beach polluted with the Exxon Valdez oil spill. *Environ. Sci. Technol., 44*, 7418–7424.

BP, (2011). *Heavy vs Light Oils*. https://www.scribd.com/document/281867519/HRES-3-10-11-Lunch-Learn-BP-Heavy-Oil1 (accessed on 16 November 2021).

Brakstad, O. G., Almås, I. K., & Krause, D. F., (2017). Biotransformation of natural gas and oil compounds associated with marine oil discharges. *Chemosphere, 182*, 555–558.

Brakstad, O. G., Lewis, A., & Beegle-Krause, C. J., (2018). A critical review of marine snow in the context of oil spills and oil spill dispersant treatment with focus on the deep-water horizon oil spill. *Mar. Pollut. Bull., 135*, 346–356.

Brakstad, O. G., Nordtug, T., & Throne-Holst, M., (2015). Biodegradation of dispersed Macondo oil in seawater at low temperature and different oil droplet sizes. *Mar. Pollut. Bull., 93*, 144–152.

Bressler, D. C., & Gray, M. R., (2003). Transport and reaction processes in bioremediation of organic contaminants. 1. Review of bacterial degradation and transport. *Int. J. Chem. React. Eng., 1*, 1–16.

Brooks, G. R., Larson, R. A., Schwing, P. T., Romero, I., Moore, C., Reichart, G. J. (2015). Sedimentation pulse in the NE Gulf of Mexico following the 2010 DWH blowout. *PLoS One, 10*, 1–24.

Brown, L. D., & Ulrich, A. C. (2017). Bioremediation of oil spills on land. In: *Handbook of Oil Spill Science and Technology* (pp. 699–729). Gulf Professional Publishing: Cambridge.

Buffle, J. A. E., (1977). Humic substances in water and their interaction with mineral ions. *Techn. Sci. Munic., 72*, 3–10.

Burgherr, P., (2007). In-depth analysis of accidental oil spills from tankers in the context of global spill trends from all sources. *J. Hazard. Mater., 140*, 245–256.

Burwood, R., & Speers, G. C., (1974). Photo-oxidation as a factor in the environmental dispersal of crude oil. *Estuar. Coast. Mar. Sci., 2*, 117–135.

Cerniglia, C. E., (1992). Biodegradation of polycyclic aromatic hydrocarbons. *Biodegradation, 3*, 351–368.

Chen, J., Liu, Y. F., Zhou, L., Mbadinga, S. M., Yang, T., Zhou, J. (2019). Methanogenic degradation of branched alkanes in enrichment cultures of production water from a high-temperature petroleum reservoir. *Appl. Microbiol. Biotechnol., 103*, 2391–2401.

Cooney, J. J., (1984). The fate of petroleum pollutants in freshwater ecosystems. In: Petroleum Microbiology (pp. 399–433). Macmillan, New York.

Daling, P. S., Brandvik, P. J., Mackay, D., Johansen, O. (1990). Characterization of crude oils for environmental purposes. *Oil Chem. Pollut., 7*, 199–224.

Daly, K. L., Passow, U., Chanton, J., & Hollander, D. (2016). Assessing the impacts of oil-associated marine snow formation and sedimentation during and after the Deepwater Horizon oil spill. *Anthropocene, 13*, 18–33.

Darvishi, P., Ayatollahi, S., Mowla, D., Niazi, A. (2011). Biosurfactant production under extreme environmental conditions by an efficient microbial consortium, ERCPPI-2. *Colloids Surf. B., 84*, 292–300.

Das, P., Mukherjee, S., & Sen, R., (2008). Genetic regulations of the biosynthesis of microbial surfactants: An overview. *Biotechnol. Genet. Eng. Rev., 25*, 165–185.

Davis, S. J., & Gibbs, C. F., (1975). The effect of weathering on a crude oil residue exposed at sea. *Water Res., 9*, 275–285.

Decho, A. W., Visscher, P. T., & Reid, R. P., (2005). Production and cycling of natural microbial exopolymers (EPS) within a marine stromatolite. *Palaeogeogr. Palaeoclimatol. Palaeoecol., 219*, 71–86.

Delille, D., Coulon, F., & Pelletier, E., (2004). Effects of temperature warming during a bioremediation study of natural and nutrient-amended hydrocarbon-contaminated sub-Antarctic soils. *Cold Reg. Sci. Technol., 40*, 61–70.

Deppe, U., Richnow, H. H., Michaelis, W. Antranikian, G. (2005). Degradation of crude oil by an arctic microbial consortium. *Extremophiles, 9*, 461–470.

Dettman, H. D., Li, N., Wickramasinghe, D., Xu, Z., Chen, X. N., Elliott, G. R., & Luo, J. (2012). *The Influence of Naphthenic Acid and Sulfur Compound Structure on Global Crude Corrosively Under Vacuum Distillation Conditions*. Presentation on the Corrosion, NACE International, Salt Lake City, Utah.

Dibble, J. T., & Bartha, R., (1979). Effect of environmental parameters on the biodegradation of oil sludge. *Appl. Environ. Microbiol., 37*, 729–739.

Durval, I. J. B., Resende, A. H. M., Figueiredo, M. A., Luna, J. M., Rufino, R. D., Sarubbo, L. A. (2019). Studies on biosurfactants produced using *Bacillus cereus* isolated from seawater with biotechnological potential for marine oil-spill bioremediation. *J. Surfactants Deterg., 22*, 349–363.

Dutta, T. K., & Harayama, S., (2000). Fate of crude oil by the combination of photooxidation and biodegradation. *Environ. Sci. Technol., 34*, 1500–1505.

Eguchi, K., Yonezawa, M., Mitsui, Y., & Hiramatsu, Y. (1992). Developmental changes of glutamate dehydrogenase activity in rat liver mitochondria and its enhancement by branched-chain amino acids. *Neonatology, 62*, 83–88.

El Fels, L., Lemee, L., Ambles, A., Hafidi, M. (2016). Identification and biotransformation of aliphatic hydrocarbons during co-composting of sewage sludge-date palm waste using pyrolysis-GC/MS technique. *Environ. Sci. Pollut. Res., 23*, 16857–16864.

Elshafie, A. E., Joshi, S. J., Al-Wahaibi, Y. M., Al-Bemani, A. S., Al-Bahry, S. N., Al-Maqbali, D., Banat, I. M. (2015). Sophorolipids production by *Candida bombicola* ATCC 22214 and its potential application in microbial enhanced oil recovery. *Front. Microbiol., 6*, 1324.

Faksness, L. G., Brandvik, P. J., & Sydnes, L. K., (2008). Composition of the water accommodated fractions as a function of exposure times and temperatures. *Mar. Pollut. Bull., 56*, 1746–1754.

Fan, T., & Buckley, J. S., (2002). Rapid and accurate SARA analysis of medium gravity crude oils. *Energy Fuels, 16*, 1571–1575.

Fathepure, B. Z., (2014). Recent studies in microbial degradation of petroleum hydrocarbons in hypersaline environments. *Front. Microbiol., 5*, 1–16.

Fedorak, P. M., Semple, K. M. Vazquez-Duhalt, R., Westlake, D. W. S. (1993). Chloroperoxidase-mediated modifications of petroporphyrins and asphaltenes. *Enzyme Microb. Tech., 15*, 429–437.

Fingas, M. F., (2013). *The Basics of oil Spill Cleanup.* CRC.

Foght, J., (2008). Anaerobic biodegradation of aromatic hydrocarbons: Pathways and prospects. *J. Mol. Microbiol. Biotechnol., 15*, 93–120.

Foght, J., (2010). Nitrogen fixation and hydrocarbon-oxidizing bacteria. In: *Handbook of Hydrocarbon and Lipid Microbiology* (pp. 1661–1668). Springer Berlin Heidelberg: Berlin, Heidelberg.

Fondi, M., Maida, I., Perrin, E., Orlandini, V., La Torre, L., Bosi, E., Negroni, A., et al. (2016). Genomic and phenotypic characterization of the species *Acinetobacter venetianus*. *Sci. Rep., 6*, 21985.

Francy, D. S., Thomas, J. M., Raymond, R. L., Ward, C. H. (1991). Emulsification of hydrocarbons by subsurface bacteria. *J. Ind. Microbiol., 8*, 237–245.

Fuchs, G., Boll, M., & Heider, J., (2011). Microbial degradation of aromatic compounds - from one strategy to four. *Nat. Rev. Microbiol., 9*, 803–816.

Gallagher, J. R., Olson, E. S., & Stanley, D. C., (1993). Microbial desulfurization of dibenzothiophene: A sulfur-specific pathway. *FEMS Microbiol. Lett., 107*, 31–35.

Garneau, M. É., Michel, C., Meisterhans, G., Fortin, N., King, T. L., Greer, C. W., & Lee, K. (2016). Hydrocarbon biodegradation by Arctic sea-ice and sub-ice microbial communities during microcosm experiments, Northwest Passage (Nunavut, Canada). *FEMS Microbiol. Ecol., 92*, 1–18.

Garrett, R. M., Pickering, I. J., Haith, C. E., & Prince, R. C. (1998). Photooxidation of crude oils. *Environ. Sci. Technol., 32*, 3719–3723.

Gerdes, B., Brinkmeyer, R., Dieckmann, G., & Helmke, E. (2005). Influence of crude oil on changes of bacterial communities in Arctic sea-ice. *FEMS Microbiol. Ecol., 53*, 129–139.

Ghattas, A. K., Fischer, F., Wick, A., & Ternes, T. A. (2017). Anaerobic biodegradation of (emerging) organic contaminants in the aquatic environment. *Water Res., 116*, 268–295.

Ghosal, D., Chakraborty, J., Khara, P., Dutta, T. K. (2010). Degradation of phenanthrene via meta-cleavage of 2-hydroxy-1-naphthoic acid by *Ochrobactrum* sp. strain PWTJD. *FEMS Microbiol. Lett., 313*, 103–110.

Ghosal, D., Ghosh, S., Dutta, T. K., & Ahn, Y. (2016). Current state of knowledge in microbial degradation of polycyclic aromatic hydrocarbons (PAHs): A review. *Front. Microbiol., 7*, 1369–1369.

Gomez, F., & Sartaj, M., (2014). Optimization of field scale biopiles for bioremediation of petroleum hydrocarbon contaminated soil at low temperature conditions by response surface methodology (RSM). *Int. Biodeter. Biodegr., 89*, 103–109.

Gong, Y., Zhao, X., Cai, Z., O'Reilly, S. E., Hao, X., Zhao, D. (2014). A review of oil, dispersed oil and sediment interactions in the aquatic environment: Influence on the fate, transport and remediation of oil spills. *Mar. Pollut. Bull., 79*, 16–33.

Government of Nova Scotia. Humic Substances. 2021. https://novascotia.ca/nse/water/docs/Drop_on_Water_English.pdf (accessed on 25 October 2021).

Grossart, H. P., Tang, K. W., Kiorboe, T., Ploug, H. Comparison (2007). Comparison of cell-specific activity between free-living and attached bacteria using isolates and natural assemblages. *FEMS Microbiol. Lett., 266*, 194–200.

Grossi, V., Cravo-Laureau, C., Guyoneaud, R., Ranchou-Peyruse, A., Hirschler-Réa, A. (2008). Metabolism of n-alkanes and n-alkenes by anaerobic bacteria: A summary. *Org. Geochem., 39*, 1197–1203.

Hafidi, M., Amir, S., Jouraiphy, A., Winterton, P., El Gharous, M., Merlina, G., Revel, J. C. (2008). Fate of polycyclic aromatic hydrocarbons during composting of activated sewage sludge with green waste. *Bioresour. Technol., 99*, 8819–8823.

Hambrick, G. A., Delaune, R. D., & Patrick, W. H., (1980). Effect of estuarine sediment pH and oxidation-reduction potential on microbial hydrocarbon degradation. *Appl. Environ. Microbiol., 40*, 365–369.

Hammel, K. E., (1995). Mechanisms for polycyclic aromatic hydrocarbon degradation by ligninolytic fungi. *Environ. Health Perspect., 103*, 41–43.

Han, P.-P., Sun, Y., Jia, S.-r., Zhong, C., Tan, Z.-l. (2014). Effects of light wavelengths on extracellular and capsular polysaccharide production by nostoc flagelliforme. *Carbohydr. Polym., 105*, 145–151.

Hansen, H. P., (1975). Photochemical degradation of petroleum hydrocarbon surface films on seawater. *Mar. Chem., 3*, 183–195.

Harvey, S., Elashvili, I., Valdes, J. J., Kamely, D., Chakrabarty, A. M. (1990). Enhanced removal of Exxon Valdez spilled oil from Alaskan gravel by a microbial surfactant. *Nat. Biotech., 8*, 228–230.

Hasinger, M., Scherr, K. E., Lundaa, T., Bräuer, L., Zach, C., Loibner, A. P. (2012). Changes in iso- and n-alkane distribution during biodegradation of crude oil under nitrate and sulfate reducing conditions. *J. Biotechnol., 157*, 490–498.

Head, I. M., Jones, D. M., & Röling, W. F. M., (2006). Marine microorganisms make a meal of oil. *Nat. Rev. Microbiol., 4*, 173–182.

Heider, J., & Schühle, K., (2013). Anaerobic biodegradation of hydrocarbons including methane. In: *The Prokaryotes: Prokaryotic Physiology and Biochemistry* (pp. 605–634). Springer: Berlin.

Heitkamp, M. A., & Cerniglia, C. E., (1989). Polycyclic aromatic hydrocarbon degradation by a *Mycobacterium* sp. in microcosms containing sediment and water from a pristine ecosystem. *Appl. Environ. Microbiol., 55*, 1968–1973.

Hemmingsen, P. V., Silset, A., Hannisdal, A., & Sjöblom, J. (2005). Emulsions of heavy crude oils. I: Influence of viscosity, temperature, and dilution. *J. Dispers. Sci. Technol., 26*, 615–627.

Hendrickx, B., Junca, H., Vosahlova, J., Lindner, A., Rüegg, I., Bucheli-Witschel, M., et al. (2006). Alternative primer sets for PCR detection of genotypes involved in bacterial aerobic BTEX degradation: Distribution of the genes in BTEX degrading isolates and in subsurface soils of a BTEX contaminated industrial site. *J. Microbiol. Methods, 64*, 250–265.

Henry, I., Netzer, R., Davies, E., Skancke, J., McFarlin, K., & Brakstad, O. (2018). Marine snow in the context of oil spill response. *Presentation on the 41st AMOP*. Victoria, British Columbia.

Hoff, A., & Dettman, H. D., (2012). Global asphaltene content and structure related to process behavior. *Presentation on the World Heavy Oil Congress*. Aberdeen, Scotland.

Hoff, R. Z., (1993). Bioremediation: An overview of its development and use for oil spill cleanup. *Mar. Pollut. Bull., 26*, 476–481.

Hong, S., Khim, J. S., Ryu, J., Park, J., Song, S. J., Kwon, B. O., et al. (2012). Two years after the Hebei spirit oil spill: Residual crude-derived hydrocarbons and potential AhR-mediated activities in coastal sediments. *Environ. Sci. Technol., 46*, 1406–1414.

Hong, Y. H., Ye, C. C., Zhou, Q. Z., Wu, X. Y., Yuan, J. P., Peng, J., et al. (2017). Genome sequencing reveals the potential of *Achromobacter* sp. HZ01 for bioremediation. *Front. Microbiol., 8*, 1507.

Hosokawa, R., Nagai, M., Morikawa, M., Okuyama, H. (2009). Autochthonous bioaugmentation and its possible application to oil spills. *World J. Microb. Biot., 25*, 1519–1528.

Hounjet, L., Gieleciak, R., & Dettman, H., (2017). Photooxidation susceptibility of diluted bitumen on water. *Presentation on the 40th AMOP*. Calgary, Canada.

Hu, T., Yang, Y., Zhang, M., Gao, Y., Cheng, Q., & Ji, H. (2019). Biodesulfurization of coal using *Rhodococcus erythropolis* SX-12 and *Acidithiobacillus ferrooxidans* GF: A two-step approach. *Energy Sci. Eng., 7*, 162–169.

Islam, A., Kim, D., Yim, U. H., Shim, W. J., & Kim, S. (2015). Structure-dependent degradation of polar compounds in weathered oils observed by atmospheric pressure photo-ionization hydrogen/deuterium exchange ultrahigh resolution mass spectrometry. *J. Hazard. Mater., 296*, 93–100.

Ito, S., Kinta, M., & Inoue, S., (1980). Growth of yeasts on n-alkanes: Inhibition by a lactonic sophorolipid produced by *Torulopsis bombicola*. *Agri. Biol. Chem., 44*, 2221–2223.

Ivanov, B., Bezgreshnov, A., Kubyshkin, N., & Kursheva, A. (2005). Spreading of oil products in sea ice and their influence on the radiation properties of the snow-ice cover. *Presentation on the International Conference on Port and Ocean Engineering under Arctic Conditions*. New York, US.

Jafarinejad, S., (2017). *Petroleum Waste Treatment and Pollution Control*. Elsevier.

Jatt, A. N., Tang, K., Liu, J., Zhang, Z., Zhang, X. H., (2015). Quorum sensing in marine snow and its possible influence on production of extracellular hydrolytic enzymes in marine snow bacterium *Pantoea ananatis* B9. *FEMS Microbiol. Ecol., 91*, 1–13.

Jednak, T., Avdalović, J., Miletić, S., Slavković-Beškoski, L., Stanković, D., Milić, J., et al. (2017). Transformation and synthesis of humic substances during bioremediation of petroleum hydrocarbons. *Int. Biodeter. Biodegr., 122*, 47–52.

Kanaly, R. A., Bartha, R., Watanabe, K., & Harayama, S. (2000). Rapid mineralization of benzo[a]pyrene by a microbial consortium growing on diesel fuel. *Appl. Environ. Microbiol., 66*, 4205–4211.

Kanaly, R. A., & Harayama, S., (2000). Biodegradation of high-molecular-weight polycyclic aromatic hydrocarbons by bacteria. *J. Bacteriol., 182*, 2059–2067.

Kanaly, R. A., & Harayama, S., (2010). Advances in the field of high-molecular-weight polycyclic aromatic hydrocarbon biodegradation by bacteria. *Microb. Biotechnol., 3*, 136–164.

Kappeli, O., Muller, M., & Fiechter, A., (1978). Chemical and structural alterations at the cell surface of *Candida tropicalis*, induced by hydrocarbon substrate. *J. Bacteriol., 133*, 952–958.

Karimi, E., Jeffryes, C., Yazdian, F., Akhavan Sepahi, A., Hatamian, A., Rasekh, B., et al. (2017). DBT desulfurization by decorating *Rhodococcus erythropolis* IGTS8 using magnetic Fe_3O_4 nanoparticles in a bioreactor. *Eng. Life Sci., 17*, 528–535.

Kawaguchi, T., & Decho, A. W., (2002). A laboratory investigation of cyanobacterial extracellular polymeric secretions (EPS) in influencing $CaCO_3$ polymorphism. *J. Cryst. Growth, 240*, 230–235.

Keith, N. (2003). *Pyrene Pathway Map*. http://eawag-bbd.ethz.ch/pyr/pyr_map.html (accessed on 25 October 2021).

Kim, Y. C., (2010). *Handbook of Coastal and Ocean Engineering*. World Scientific.

Kiørboe, T., (2000). Formation and fate of marine snow: Small-scale processes with large-scale implications. *Sci. Mar., 65*, 57–71.

Kiørboe, T., Tang, K., Grossart, H.-P., & Ploug, H. (2003). Dynamics of microbial communities on marine snow aggregates: Colonization, growth, detachment, and grazing mortality of attached bacteria. *Appl. Environ. Microbiol., 69*, 3036–3047.

Kirkwood, K. M., Chernik, P., Foght, J. M., & Gray, M. R. (2008). Aerobic biotransformation of decalin (decahydronaphthalene) by *Rhodococcus* spp. *Biodegradation, 19*, 785–794.

Kodama, A., Matozaki, T., Fukuhara, A., Kikyo, M., Ichihashi, M., & Takai, Y. (2000). Involvement of an SHP-2-Rho small G protein pathway in hepatocyte growth factor/scatter factor-induced cell scattering. *Mol. Biol. Cell, 11*, 2565–2575.

Kotzakoulakis, K., & George, S. C., (2018). Predicting the weathering of fuel and oil spills: A diffusion-limited evaporation model. *Chemosphere, 190*, 442–453.

Kouzuma, A., & Watanabe, K., (2011). Molecular Approaches for the analysis of natural attenuation and bioremediation. In: *Comprehensive Biotechnology* (pp. 25–36). Academic Press, Burlington.

Kristensen, M., Johnsen, A. R., & Christensen, J. H., (2015). Marine biodegradation of crude oil in temperate and arctic water samples. *J. Hazard. Mater., 300*, 75–83.

Kügler, J. H., Le Roes-Hill, M., Syldatk, C., & Hausmann, R. (2015). Surfactants tailored by the class Actinobacteria. *Front. Microbiol., 6*, 212–212.

Landes, K. K., (1973). Mother nature as an oil polluter. *Am. Assoc. Pet. Geol. Bull., 57*, 637–641.

Langenhoff, A. A., Rahsepar, S., Van Eenennaam, J. S., Radović, J. R., Oldenburg, T. B., Foekema, E., & Murk, A. J. (2020). Effect of marine snow on microbial oil degradation. In: *Deep Oil Spills: Facts, Fate, and Effects* (pp. 301–311). Springer International Publishing: Cham.

Larson, R. A., Bott, T. L., Hunt, L. L., Rogenmuser, K. (1979). Photooxidation products of a fuel oil and their antimicrobial activity. *Environ. Sci. Technol., 13*, 965–969.

Le Moigne, F. A., Cisternas-Novoa, C., Piontek, J., Maßmig, M., & Engel, A. (2017). On the effect of low oxygen concentrations on bacterial degradation of sinking particles. *Sci. Rep., 7*, 16722.

Leahy, J. G., & Colwell, R. R., (1990). Microbial degradation of hydrocarbons in the environment. *Microbiol. Rev., 54*, 305–315.

Lee, R. F., (1980). Processes affecting the fate of oil in the sea. In: *Elsevier Oceanography Series* (pp. 337–351). Elsevier.

Lee, R. F., Hinga, K., & Almquist, G., (1982). Fate of radiolabeled polycyclic aromatic hydrocarbons and pentachlorophenol in enclosed marine ecosystems. In: *Marine Mesocosms: Biological and Chemical Research in Experimental Ecosystems* (pp. 123–135). Springer, New York.

Lee, S., Hong, S., Liu, X., Kim, C., Jung, D., Yim, U. H., et al (2017). Endocrine disrupting potential of PAHs and their alkylated analogues associated with oil spills. *Environ. Sci-Proc. IMP., 19*, 1117–1125.

Lessard, R. R., & DeMarco, G., (2000). The significance of oil spill dispersants. *Spill Sci. Technol. Bull., 6*, 59–68.

Lottermoser, K., Schunck, W. H., & Asperger, O., (1996). Cytochromes P450 of the sophorose lipid-producing yeast *Candida apicola*: Heterogeneity and polymerase chain reaction-mediated cloning of two genes. *Yeast, 12*, 565–575.

Maki, H., Sasaki, T., & Harayama, S., (2001). Photo-oxidation of biodegraded crude oil and toxicity of the photo-oxidized products. *Chemosphere, 44*, 1145–1151.

Mapelli, F., Scoma, A., Michoud, G., Aulenta, F., Boon, N., Borin, S., et al. (2017). Biotechnologies for marine oil spill cleanup: Indissoluble ties with microorganisms. *Trends Biotechnol., 35*, 860–870.

Margesin, R., & Schinner, F., (2001). Biodegradation and bioremediation of hydrocarbons in extreme environments. *Appl. Microbiol. Biotechnol., 56*, 650–663.

Martins, L. F., & Peixoto, R. S., (2012). Biodegradation of petroleum hydrocarbons in hypersaline environments. *Braz. J. Microbiol., 43*, 865–872.

McKenna, E. J., & Kallio, R. E., (1971). Microbial metabolism of the isoprenoid alkane pristane. *Proc. Natl. Acad. Sci. USA, 68*, 1552–1554.

Meckenstock, R. U., Boll, M., Mouttaki, H., Koelschbach, J. S., Cunha Tarouco, P. Weyrauch, P., et al. (2016). Anaerobic degradation of benzene and polycyclic aromatic hydrocarbons. *J. Mol. Microb. Biotech., 26*, 92–118.

Meslé, M., Dromart, G., & Oger, P., (2013). Microbial methanogenesis in subsurface oil and coal. *Res. Microbiol., 164*, 959–972.

Mill, T., Hendry, D. G., & Richardson, H., (1980). Free-radical oxidants in natural waters. *Science, 207*, 886, 887.

Miransari, M., (2011). *Soil Microbes and Environmental Health*. Nova Science Publisher's.
Mishra, A. K., & Kumar, G. S., (2015). Weathering of oil spill: Modeling and analysis. *Aquat. Procedia., 4*, 435–442.
Mohanty, S., Jasmine, J., & Mukherji, S., (2013). Practical considerations and challenges involved in surfactant enhanced bioremediation of oil. *Biomed. Res. Int., 2013*, 1–16.
Mohebali, G., & Ball, A. S., (2008). Biocatalytic desulfurization (BDS) of petrodiesel fuels. *Microbiology, 154*, 2169–2183.
Mulligan, C. N., (2005). Environmental applications for biosurfactants. *Environ. Pollut., 133*, 183–198.
Muschenheim, D. K., & Lee, K., (2002). Removal of oil from the sea surface through particulate interactions: Review and prospectus. *Spill Sci. Technol. Bull., 8*, 9–18.
Muthusamy, K., Gopalakrishnan, S., Ravi, T., Sivachidambaram, P. (2008). Biosurfactants: Properties, commercial production and application. *Curr. Sci., 94*, 736–747.
National Commission on the BP Deepwater Horizon Oil Spill and Offshore Drilling, (2011). *The Use of Surface and Subsea Dispersants During the BP Deep Water Horizon Oil Spill* (pp. 1–28). Washington.
National Oceanic and Atmospheric Administration. (2021). *Estuaries*. https://oceanservice.noaa.gov/education/kits/estuaries/estuaries01_whatis.html (accessed on 25 October 2021).
National Research Council, (2003). *The Oil Spill Recovery Institute Past, Present, and Future Directions*. The National Academies Press.
National Research Council, (2005). *Oil Spill Dispersants*. https://www.nap.edu/read/11283/chapter/1 (accessed on 25 October 2021).
National Research Council, (2014). *Responding to Oil Spills in the U. S. Arctic Marine Environment*. The National Academies Press.
Nguyen, T. T., & Sabatini, D. A., (2011). Characterization and emulsification properties of rhamnolipid and sophorolipid biosurfactants and their applications. *Int. J. Mol. Sci., 12*, 1232–1244.
Noordman, W. H., & Janssen, D. B., (2002). Rhamnolipid stimulates uptake of hydrophobic compounds by *Pseudomonas aeruginosa*. *Appl. Environ. Microbiol., 68*, 4502–4508.
Palmisano, A. C., & Pettigrew, C. A., (1992). Biodegradability of plastics. *Bioscience, 42*, 680–685.
Patel, D. D., & Bhaskaran, L., (2018). Study on paraffin wax degrading ability of *Pseudomonas nitroreducens* isolated from oil wells of Gujarat, India. *Petrol. Sci. Technol., 36*, 583–590.
Patel, S., Homaei, A., Patil, S., & Daverey, A. (2018). Microbial biosurfactants for oil spill remediation: Pitfalls and potentials. *Appl. Microbiol. Biotechnol., 103*, 27–37.
Pathak, M. S., (2018). *Asphaltenes Conversion by S-Alkylation and C-Alkylation Reactions*. Master's Thesis, University of Alberta, Edmonton, Canada.
Payne, J. R., & Phillips, C. R., (1985). Photochemistry of petroleum in water. *Environ. Sci. Technol., 19*, 569–579.
Payne, J. R., Clayton Jr, J. R., & Kirstein, B. E., (2003). Oil/suspended particulate material interactions and sedimentation. *Spill Sci. Technol. Bull., 8*, 201–221.
Peña-Méndez, E. M., Havel, J., & Patočka, J., (2005). Humic substance - compounds of still unknown structure: Applications in agriculture, industry, environment, and biomedicine. *J. Appl. Biomed., 3*, 13–24.
Perry, J. J., & Cerniglia, C. E., (1973). Crude oil degradation by filamentous fungi. *J. Gen. Appl. Microbiol., 19*, 151–153.

Petrov, A. A., (1987). *Petroleum Hydrocarbons*. Springer-Verlag.
Pfeifer, F., Krüger, K., Röder, R., Mayr, A., Ziesche, S., & Offner, S. (1997). Gas vesicle formation in halophilic Archaea. *Arch. Microbiol., 167*, 259–268.
Pineda-Flores, G., & Mesta-Howard, A. M., (2001). Petroleum asphaltenes: Generated problematic and possible biodegradation mechanisms. *Rev. Latinoam. Microbiol., 43*, 143–150.
Pinyakong, O., Habe, H., Supaka, N., Pinpanichkarn, P., Juntongjin, K., Yoshida, T., et al. (2000). Identification of novel metabolites in the degradation of phenanthrene by *Sphingomonas* sp. strain P2. *FEMS Microbiol. Lett., 191*, 115–121.
Pirnik, M. P., Atlas, R. M., & Bartha, R., (1974). Hydrocarbon metabolism by *Brevibacterium erythrogenes*: Normal and branched alkanes. *J. Bacteriol. Res., 119*, 868–878.
Premuzic, E. T., Bohenek, M. L. M., & Zhou; W. M. (1999). Bioconversion reactions in asphaltenes and heavy crude oils. *Energy Fuels, 13*, 297–304.
Prince, R. C., (1993). Petroleum spill bioremediation in marine environments. *Crit. Rev. Microbiol., 19*, 217–240.
Prince, R. C., Garrett, R. M., Bare, R. E., Grossman, M. J., Townsend, T., Suflita, J. M., et al. (2003). The roles of photooxidation and biodegradation in long-term weathering of crude and heavy fuel oils. *Spill Sci. Technol. Bull., 8*, 145–156.
Quigg, A., Passow, U., Chin, W. C., Xu, C., Doyle, S., Bretherton, L., et al. (2016). The role of microbial exopolymers in determining the fate of oil and chemical dispersants in the ocean. *Limnol. Oceanogr. Lett., 1*, 3–26.
Rahman, K. S. M., Rahman, T. J., Lakshmanaperumalsamy, P., Marchant, R., & Banat, I. M. (2003). The potential of bacterial isolates for emulsification with a range of hydrocarbons. *Acta Biotechnol., 23*, 335–345.
Rahsepar, S., Langenhoff, A. A., Smit, M. P., Van Eenennaam, J. S., Murk, A. J., & Rijnaarts, H. H. (2017). Oil biodegradation: Interactions of artificial marine snow, clay particles, oil and corexit. *Mar. Pollut. Bull., 125*, 186–191.
Raposo, M. F. J., Morais, R. M. S. C., & Morais, A. M. M. B., (2013). Bioactivity and applications of sulfated polysaccharides from marine microalgae. *Mar. Drugs, 11*, 233–252.
Ray, P. Z., Chen, H., Podgorski, D. C., McKenna, A. M., & Tarr, M. A. (2014). Sunlight creates oxygenated species in water-soluble fractions of Deepwater Horizon oil. *J. Hazard. Mater., 280*, 636–643.
Ressler, B. P., Kneifel, H., & Winter, J., (1999). Bioavailability of polycyclic aromatic hydrocarbons and formation of humic acid-like residues during bacterial PAH degradation. *Appl. Microbiol. Biotechnol., 53*, 85–91.
Rios-Hernandez, L. A., Gieg, L. M., & Suflita, J. M., (2003). Biodegradation of an alicyclic hydrocarbon by a sulfate-reducing enrichment from a gas condensate-contaminated aquifer. *Appl. Environ. Microbiol., 69*, 434–443.
Ron, E. Z., & Rosenberg, E., (2002). Biosurfactants and oil bioremediation. *Curr. Opin. Biotechnol., 13*, 249–252.
Rontani, J. F., Bonin, P., & Giusti, G., (1987). Mechanistic study of interactions between photo-oxidation and biodegradation of n-nonylbenzene in seawater. *Mar. Chem., 22*, 1–12.
Røy, H., Kallmeyer, J., Adhikari, R. R., Pockalny, R., Jørgensen, B. B., & D'Hondt, S. (2012). Aerobic microbial respiration in 86-million-year-old deep-sea red clay. *Science, 336*, 922–925.

Saborimanesh. N., & Mulligan, C. N. (2015). Effect of sophorolipid biosurfactant on oil biodegradation by the natural oil-degrading bacteria on the weathered biodiesel, diesel and light crude oil. *J. Bioremediat. Biodegrad.* *6*, 1-8.

Saeki, H., Sasaki, M., Komatsu, K., Miura, A., & Matsuda, H. (2009). Oil spill remediation by using the remediation agent JE1058BS that contains a biosurfactant produced by *Gordonia* sp. strain JE-1058. *Bioresour. Technol.*, *100*, 572–577.

Salt, D. E., Smith, R. D., & Raskin, I., (1998). Phytoremediation. *Annu. Rev. Plant Physiol.*, *49*, 643–668.

Schnitzer, M., (1978). Humic substances: Chemistry and reactions. In*: Soil Organic Matter* (pp. 1–64). Elsevier: Amsterdam.

Schnitzer, M., & Khan, S. U., (1972). *Humic Substances in the Environment.* Marcel Dekker.

Seo, J. S., Keum, Y. S., & Li, Q. X., (2009). Bacterial degradation of aromatic compounds. *Int. J. Environ. Res. Public Health, 6*, 278–309.

Sheu, E. Y., (2002). Petroleum asphaltene properties, characterization, and issues. *Energy Fuels, 16*, 74–82.

Simon, M., Grossart, H. P., Schweitzer, B., & Ploug, H., (2002). Microbial ecology of organic aggregates in aquatic ecosystems. *Aquat. Microb. Ecol., 28*, 175–211.

Singh, H., (2006). *Mycoremediation: Fungal Bioremediation.* Wiley-Interscience.

Singh, R., Guzman, M. S., & Bose, A., (2017). Anaerobic oxidation of ethane, propane, and butane by marine microbes: A mini review. *Front. Microbiol., 8*, 2056–2063.

Skovhus, T. L., & Whitby, C., (2019). *Oilfield Microbiology.* CRC Press.

Smith, D. C., Simon, M., Alldredge, A. L., & Azam, F. (1992). Intense hydrolytic enzyme activity on marine aggregates and implications for rapid particle dissolution. *Nature, 359*, 139–142.

So, C. M., Phelps, C. D., & Young, L. Y., (2003). Anaerobic transformation of alkanes to fatty acids by a sulfate-reducing bacterium, Strain Hxd3. *Appl. Environ. Microbiol., 69*, 3892–3900.

Soleimani, M., Bassi, A., & Margaritis, A., (2007). Biodesulfurization of refractory organic sulfur compounds in fossil fuels. *Biotechnol. Adv., 25*, 570–596.

Speight, J. G., (2007). *The chemistry and technology of petroleum.* CRC Press/Taylor & Francis.

Stauffer, E., Dolan, J. A., & Newman, R., (2008). Flammable and combustible liquids. In: *Fire Debris Analysis* (pp. 199–233). Academic Press: Burlington.

Stevenson, F. J., (1982). *Humus Chemistry: Genesis, Composition, Reactions.* Wiley.

Stevenson, F. J., (1994). *Humus Chemistry: Genesis, Composition, Reactions.* Wiley.

Stout, S. A., & Payne, J. R., (2016). Macondo oil in deep-sea sediments: Part 1 - sub-sea weathering of oil deposited on the seafloor. *Mar. Pollut. Bull., 111*, 365–380.

Suja, L. D., Chen, X., Summers, S., Paterson, D. M., & Gutierrez, T. (2019). Chemical dispersant enhances microbial exopolymer (EPS) production and formation of marine oil/dispersant snow in surface waters of the subarctic northeast Atlantic. *Front. Microbiol., 10*, 553.

Suja, L. D., Summers, S., & Gutierrez, T., (2017). Role of EPS, dispersant and nutrients on the microbial response and MOS formation in the subarctic northeast Atlantic. *Front. Microbiol., 8*, 676.

Thingstad, T., & Pengerud, B., (1983). The formation of "chocolate mousse" from Statfjord crude oil and seawater. *Mar. Pollut. Bull., 14*, 214–216.

Timmis, K. N., (2010). *Handbook of Hydrocarbon and Lipid Microbiology*. Springer.
Tissot, B. P., & Welte, D. H., (1984). *Petroleum Formation and Occurrence*. Springer-Verlag.
Tornheim, K., (2018). Glucose metabolism and hormonal regulation. In: *Encyclopedia of Endocrine Diseases* (pp. 87–94). Academic Press, Oxford.
Townsend, G. T., Prince, R. C., & Suflita, J. M., (2004). Anaerobic biodegradation of alicyclic constituents of gasoline and natural gas condensate by bacteria from an anoxic aquifer. *FEMS Microbiol. Ecol., 49*, 129–135.
Trevisan, S., Francioso, O., Quaggiotti, S., & Nardi, S. (2010). Humic substances biological activity at the plant-soil interface. *Plant Signal. Behav., 5*, 635–643.
Tyagi, M., Fonseca, M. M. R., & Carvalho, C. C. C. R., (2011). Bioaugmentation and biostimulation strategies to improve the effectiveness of bioremediation processes. *Biodegradation, 22*, 231–241.
University of Guelph, (2020). Day Length. https://animalbiosciences.uoguelph.ca/aquacentre/files/aquatalk/Aquatalk%208.pdf (accessed on 25 October 2021).
University of Waterloo, (2020). *The Formation of Oil and Other Elements Required to Produce a Petroleum Accumulation*. https://uwaterloo.ca/wat-on-earth/news/formation-oil-and-other-elements-required-produce-petroleum (accessed on 25 October 2021).
Valentine, D. L., Fisher, G. B., Bagby, S. C., Nelson, R. K., Reddy, C. M., Sylva, S. P., & Woo, M. A. (2014). Fallout plume of submerged oil from Deepwater Horizon. *PNAS, 111*, 15906–15911.
Valentine, D. L., Kessler, J. D., Redmond, M. C., Mendes, S. D., Heintz, M. B., Farwell, C., Hu, L., et al. (2010). Propane respiration jump-starts microbial response to a deep oil spill. *Science, 330*, 208–211.
Van Bogaert, I. N., Saerens, K., De Muynck, C., Develter, D., Soetaert, W., & Vandamme, E. J. (2007). Microbial production and application of sophorolipids. *Appl. Microbiol. Biotechnol., 76*, 23–34.
Van Der Meer, J. R., De Vos, W. M., Harayama, S., & Zehnder, A. J. (1992). Molecular mechanisms of genetic adaptation to xenobiotic compounds. *Microbiol. Rev., 56*, 677–694.
Van Hamme, J. D., Fedorak, P. M., Foght, J. M., Gray, M. R., & Dettman, H. D. (2004). Use of a novel fluorinated organosulfur compound to isolate bacteria capable of carbon-sulfur bond cleavage. *Appl. Environ. Microbiol., 70*, 1487–1493.
Van Hamme, J. D., Wong, E. T., Dettman, H. D., Gray, M. R., & Pickard, M. A. (2003). Dibenzyl sulfide metabolism by white rot fungi. *Appl. Environ. Microbiol., 69*, 1320–1324.
Van Trump, J. I., Sun, Y., & Coates, J. D., (2006). Microbial interactions with humic substances 1. *Adv. Appl. Microbiol., 60*, 55–96.
Varadaraj, R., Robbins, M. L., Bock, J., Pace, S., & MacDonald, D. (1995). Dispersion and biodegradation of oil spills on water. *Presentation on the International Oil Spill Conference*. Long Beach, California.
Varjani, S. J., Rana, D. P., Jain, A. K., Bateja, S., & Upasani, V. N. (2015). Synergistic ex-situ biodegradation of crude oil by halotolerant bacterial consortium of indigenous strains isolated from on shore sites of Gujarat, India. *Int. Biodeterior. Biodegr., 103*, 116–124.
Varjani, S. J., & Upasani, V. N., (2016). Biodegradation of petroleum hydrocarbons by oleophilic strain of *Pseudomonas aeruginosa* NCIM 5514. *Bioresour. Technol., 222*, 195–201.
Venosa, A. D., & Holder, E. L., (2007). Biodegradability of dispersed crude oil at two different temperatures. *Mar. Pollut. Bull., 54*, 545–553.

Venosa, A. D., Suidan, M.T., Wrenn, B.A., Strohmeier, K.L., Haines, J.R., Eberhart, B.L., et al. (1996). Bioremediation of an experimental oil spill on the shoreline of Delaware bay. *Environ. Sci. Technol., 30*, 1764–1775.

Vila, J., López, Z., Sabaté, J., Minguillón, C., Solanas, A. M., & Grifoll, M. (2001). Identification of a novel metabolite in the degradation of pyrene by *Mycobacterium* sp. strain AP1: Actions of the isolate on two- and three-ring polycyclic aromatic hydrocarbons. *Appl. Environ. Microbiol., 67*, 5497–5505.

Walter, U., Beyer, M., Klein, J., & Rehm, H. J. (1991). Degradation of pyrene by *Rhodococcus* sp. UW1. *Appl. Microbiol. Biotechnol., 34*, 671–676.

Ward, C. P., Sharpless, C. M., Valentine, D. L., French-McCay, D. P., Aeppli, C., White, H. K., et al. (2018). Partial photochemical oxidation was a dominant fate of Deepwater Horizon surface oil. *Environ. Sci. Technol., 52*, 1797–1805.

Ward, D. M., & Brock, T. D., (1978). Hydrocarbon biodegradation in hypersaline environments. *Appl. and Environ. Microbiol., 35*, 353–359.

Weber, J., Chen, Y., Jamroz, E., & Miano, T. (2018). Preface: Humic substances in the environment. *J. Soil. Sediment., 18*, 2665–2667.

Wetzel, R. G., (1992). Gradient-dominated ecosystems: Sources and regulatory functions of dissolved organic matter in freshwater ecosystems. *Hydrobiologia, 229*, 181–198.

Wetzel, R. G., Hatcher, P. G., & Bianchi, T. S., (1995). Natural photolysis by ultraviolet irradiance of recalcitrant dissolved organic matter to simple substrates for rapid bacterial metabolism. *Limnol. Oceanogr., 40*, 1369–1380.

Whang, L. M., Liu, P. W. G., Ma, C. C., & Cheng, S. S. (2008). Application of biosurfactants, rhamnolipid, and surfactin, for enhanced biodegradation of diesel-contaminated water and soil. *J. Hazard. Mater., 151*, 155–163.

White, D. A., Hird, L. C., & Ali, S. T., (2013). Production and characterization of a trehalolipid biosurfactant produced by the novel marine bacterium *Rhodococcus* sp., strain PML026. *J. Appl. Microbiol., 115*, 744–755.

Wilkes, H., Buckel, W., Golding, B. T., & Rabus, R. (2016). Metabolism of hydrocarbons in *n*-alkane-utilizing anaerobic bacteria. *J. Mol. Microb. Biotech., 26*, 138–151.

Wilson, L. P., & Bouwer, E. J., (1997). Biodegradation of aromatic compounds under mixed oxygen/denitrifying conditions: A review. *J. Ind. Microbiol. Biotechnol., 18*, 116–130.

Wu, M., Dick, W. A., Li, W., Wang, X., Yang, Q., Wang, T., et al. (2016). Bioaugmentation and biostimulation of hydrocarbon degradation and the microbial community in a petroleum-contaminated soil. *Int. Biodeter. Biodegr., 107*, 158–164.

Wu, Y. S., Ngai, S. C., Goh, B. H., Chan, K. G., Lee, L. H., & Chuah, L. H. (2017). Anticancer activities of surfactin and potential application of nanotechnology assisted surfactin delivery. *Front. Pharmacol., 8*, 761.

Xue, J., Yu, Y., Bai, Y., Wang, L., & Wu, Y. (2015). Marine oil-degrading microorganisms and biodegradation process of petroleum hydrocarbon in marine environments: A review. *Curr. Microbiol., 71*, 220–228.

Yang, F., Tchoukov, P., Dettman, H., Teklebrhan, R. B., Liu, L., Dabros, T., et al. (2015). Asphaltene subfractions responsible for stabilizing water-in-crude oil emulsions. Part 2: Molecular representations and molecular dynamics simulations. *Energy Fuels, 29*, 4783–4794.

Zepp, R. G., & Baughman, G. L., (1978). Prediction of photochemical transformation of pollutants in the aquatic environment. In: *Aquatic Pollutants* (pp. 237–263). Pergamon Press: Oxford.

Zhang, E., Schrom, J., & Dodge, T. (2000). *Benzo(a)pyrene Pathway Map*. http://eawag-bbd.ethz.ch/bap/bap_map.html (accessed on 25 October 2021).

Zhang, Y., & Miller, R. M., (1992). Enhanced octadecane dispersion and biodegradation by a *Pseudomonas* rhamnolipid surfactant (biosurfactant). *Appl. Environ. Microbiol., 58*, 3276–3282.

Zhu, X., Venosa, A. D., Suidan, M. T., & Lee, K. (2001). *Guidelines for the Bioremediation of Marine Shorelines and Freshwater Wetlands.* Environmental Protection Agency.

Zoveidavianpoor, M., (2018). *Recent Insights in Petroleum Science and Engineering*. IntechOpen.

CHAPTER 5

Functional Diversity of Microbial Communities in Hydrocarbon-Polluted Ecosystems

R. M. M. ABED,[1] H. MAHMOUD,[2] and N. SIVAKUMAR[1]

[1]*Sultan Qaboos University, College of Science, Biology Department, P.O. Box–36, Postal Code 123, Al Khoud, Sultanate of Oman*

[2]*Kuwait University, Department of Biological Sciences, P.O. Box–5969, Safat–13060, Kuwait*

5.1 INTRODUCTION

Contamination with crude oil, refined products, or individual hydrocarbons (HCs) affects not only the diversity and distribution of microorganisms but also the activity of microorganisms in impacted ecosystems (Das and Chandran, 2011; Hassanshahian, 2014; Alvarez et al., 2017). Culture-dependent and culture-independent techniques have been employed for decades to study the composition of bacterial communities in oil-contaminated sites, to follow community changes upon exposure to pollutants, and to isolate and characterize oil-degrading microorganisms (Kostka et al., 2011; Mason et al., 2012; Al-Mailem et al., 2015; Ali et al., 2016a; Al Kindi and Abed, 2016b; Handley et al., 2017). In contrast, real-time measurement of microbial activity in oil-contaminated sites is a more recent development. The introduction of microsensors to the field of microbial ecology has facilitated *in situ* measurements of rates of metabolic process, including photosynthesis, respiration, and sulfate reduction, and measurements of the interactions and responses of mixed bacterial communities to oil pollution at the microscale (Pringault et al., 2015; Müller et al., 2016; Abed and Al-Kindi, 2017). Microsensors have been successfully applied to study the role of cyanobacterial mats from hypersaline environments and from hydrothermal vents in the degradation of HCs (Grötzschel et al., 2002; Benthien et al., 2004; Musat et al., 2004; Felden et al., 2013; Mishra et al., 2017). The activity of oil-degrading bacteria has been followed by quantifying catabolic gene expression, using, for example, real time-quantitative polymerase chain reaction (RT-qPCR), microarrays, and transcriptomics, offering valuable information about the degradation capability of the microorganisms and the various aliphatic and aromatic degradation pathways they employ (Aitken et al., 2004; Verde et al., 2013).

Cultivation-dependent methods are incapable of evaluating or accessing the uncultivable majority of environmental microorganisms, including those in oil-contaminated environments (Amann et al., 1995; Orphan et al., 2000; Stefani et al., 2015; Zhang et al., 2018). Research on functional diversity of microbial communities using cultivation-independent tools has increased in the era of "-omics" and "meta-omics" (meta meaning "beyond"). Several "meta-omics" platforms have been established to investigate whole community DNA (metagenomics), mRNA transcripts levels (metatranscriptomics), protein identity and abundance (metaproteomics), and low-molecular-weight metabolites (metabolomics) (Stenuit et al., 2008; Kumavath and Pratap, 2012; Uhlik et al., 2013; Gruner et al., 2017). All of these approaches have been recently employed to gain more insight into the functional diversity of microorganisms in oil-contaminated areas.

5.2 MICROSENSORS FOR MEASUREMENT OF MICROBIAL ACTIVITY

5.2.1 MICROSENSORS AND PROCESS RATE MEASUREMENTS

Microsensors are glass electrodes that are needle-shaped with a tip diameter ranging from 1 to 100 μm. They were introduced into the field of microbial ecology almost five decades ago (Revsbech and Jørgensen, 1983). Microsensors are mainly used to measure the concentration of specific chemicals in mixed microbial communities at different depths, generating a concentration-depth profile. Microsensors are available that allow the measurement of O_2, CO_2, CO_3^{2-}, NO_3^-, NO_2^-, N_2O, NH_4^+, Ca^{2+}, pH, H_2S, S^{2-}, CH_4, temperature, diffusivity, and irradiance. These microsensors have been used to investigate carbon, sulfur, and nitrogen cycling in sediments, biofilms, and microbial mats (Jeroschewski et al., 1996; Wieland and Kühl, 2000; Wieland et al., 2005; Abed et al., 2006, 2007, 2013, 2015; Revsbech et al., 2016). Microbial processes, including respiration, photosynthesis (de Beer et al., 2000; Köhler-Rink and Kühl, 2000), sulfate reduction, sulfide oxidation (Kühl and Jørgensen, 1992; Klatt et al., 2016; de Beer et al., 2017), nitrification, and denitrification (Gieseke et al., 2001; Schutte et al., 2018) have been measured. From the measured depth concentration profiles, net specific consumption and production rates can be estimated. Microsensors have become a useful tool in microbial ecology due to their fine spatial resolution and ability to measure several microbial processes simultaneously.

Although a wide array of microsensors are available, most oil biodegradation studies have used only O_2, pH, and H_2S sensors. O_2 microsensors of the Clark-type that have been converted to operate on the microscale have a gold-coated cathode, at which O_2 is reduced, behind a silicon membrane immersed in an electrolyte solution (Revsbech, 1989). O_2 sensors are stable, lasting for more than a year, and have a detection limit of approximately 0.1 μM. Using fast O_2 microsensors (90% response time < 0.5 s) and the light-dark-shift method (Revsbech and Jørgensen, 1986), rates of gross photosynthesis can be calculated. H_2S microsensors are also Clark-type sensors, but with a platinum anode behind a silicon membrane. H_2S diffuses into the sensor, is oxidized to elemental sulfur by ferricyanide, with the resulting ferrocyanide being re-oxidized at the anode generating a current in the pA range. pH measurements are then used to calculate total sulfide from H_2S profiles according to the equation $[S^{2-}_{tot}] = [H_2S] \times (1+10^{-pK1}/10^{-pH})$ (Kühl et al., 1998). Unlike O_2 sensors, H_2S sensors have a relatively short life, lasting only a few weeks. They are light-sensitive and have a detection limit of 1 μM. pH microsensors are similar to well-known commercial pH electrodes but with tip sizes of 25–200 μm. Measurements are based on quantifying the electrical potential difference across a pH-sensitive glass membrane (Hinke, 1969). For more details on the construction and working principles of these sensors, the reader is referred to Kühl and Revsbech (2001); Gieseke and de Beer (2007); and Abed and de Beer (2010).

5.2.2 MICROSENSORS AND BIODEGRADATION STUDIES

The application of microsensors in oil biodegradation studies has been limited to studying the effect of oil pollution on key bacterial processes; for example, photosynthesis, aerobic respiration, and sulfate reduction in cyanobacterial mats. Cyanobacterial mats are stratified agglomerations of microorganisms, with a top 1–3 mm oxic layer dominated by cyanobacteria and aerobic heterotrophic bacteria, and a bottom anoxic layer dominated by sulfate-reducing bacteria (SRB) (Wong et al., 2015; Abed et al., 2018). Their growth on oily sediments after the oil spill in Kuwait in 1991 suggests a role in oil biodegradation (Höpner et al., 1996). The exact role of the physiologically different bacterial groups in these mats with respect to oil degradation is unknown, and the effect of oil pollution on the diversity and activity of mat bacteria was never investigated. In microbial mats, cyanobacteria are able to perform oxygenic photosynthesis, generating oxygen that can be taken up by oil-degrading aerobic bacteria. The rates of both respiration and photosynthesis in these mats can be measured using O_2 microsensors (Abed and de Beer, 2010). When a thick viscous oil layer is formed on mats and sediments, oxygen diffusion and light penetration become limited, resulting in the creation of anoxic conditions. These conditions stimulate the growth of anaerobic SRB, some of which are capable of anaerobically degrading HCs (Widdel and Rabus, 2001; Kleindienst et al., 2014; Jaekel et al., 2015). The activity of SRB leads to the release of H_2S, as measured using H_2S sensors, and changes in pH, a parameter that generally decreases due to respiratory processes and increases due to photosynthesis, can be measured using pH sensors. Microsensor measurements have been coupled with molecular analysis of bacterial communities and chemical analysis of contaminants in order to monitor changes in the diversity and activity of microorganisms upon HC contamination, and in an effort to identify bacteria involved in HC degradation (Grötzschel et al., 2002; Benthien et al., 2004).

Microsensors have been used to study the effect of the addition of phenanthrene, dibenzothiophene (DBT), pristine, and *n*-octadecane on bacterial processes in a pristine hypersaline cyanobacterial mat (Grötzschel et al., 2002). In this mat, photosynthetic and respiration rates remained unchanged throughout the experiment, and pH and H_2S profiles did not show dramatic changes. Nevertheless, the four compounds were partially degraded, with residues still detectable after 18 weeks. Molecular analysis of the bacterial community in this mat did not show any significant shifts in composition. In contrast to this study, contamination of a pristine hypersaline cyanobacterial mat from France with crude oil resulted in stimulation of aerobic respiration, particularly in the non-photosynthetic zone, and inhibition of photosynthesis (Benthien et al., 2004). Photosynthesis recovered and increased after 3 months, and H_2S sensors revealed that total sulfide decreased over time in the uncontaminated and in the heavily contaminated mats incubated in the dark, whereas it increased slightly at the beginning of the experiment and remained stable over 90 days in the light-incubated mats.

The enrichment of microbial populations associated with oil was simulated in aquaria under laboratory settings by contaminating a tidal sediment with crude oil (Musat et al., 2004). Microsensor measurements showed that the addition of oil significantly promoted subsurface oxygen consumption and microbial sulfate reduction. There was no indication that oil stimulated cyanobacterial growth or oxygenic phototrophy during the first two months of mat development. The pronounced depletion of oxygen and the increase in the activity of SRB in the subsurface layer were attributed to oxygen scavenging by aerobic oil-degrading bacteria, the supply of substrates from the oil to SRB, and to the oil layer severely limiting oxygen diffusion. Indeed, SRB have been demonstrated to utilize HCs (Rueter et al., 1994; Widdel and Rabus, 2001; Liu et al., 2015; Müller et al., 2016), utilize intermediate products resulting from incomplete mineralization of HCs by aerobic bacteria (Nazina et al., 1985), and grow on cyanobacterial photosynthetic and fermentation products. A follow-up experiment using the same mats (Abed et al., unpublished) showed a stimulation of sulfate-reduction and the enrichment of sulfate-reducing bacterial populations with the potential to degrade HCs. The activity of SRB was stimulated by direct oil degradation, and was independent of organic compounds provided by phototrophic and other heterotrophic activities.

Microsensors have been used to study the activity of microbial communities in natural petroleum seepages and mud volcanos (Felden et al., 2013; Mishra et al., 2017). For instance, the microbial community response to petroleum seepage in a sediment core collected close to a natural HC seepage in the Caspian Sea was investigated in the laboratory for 190 days (Mishra et al., 2017). In this sediment, HCs were degraded under methanogenic conditions, and H_2S microsensors showed an increase in sulfide concentrations, indicating that petroleum degradation may be coupled to sulfate reduction. In another study, O_2, H_2S, and pH microsensors were used to measure *in situ* profiles in microbial mats in the Amon mud volcano (Felden et al., 2013). These measurements showed that the muds surrounding the central methane gas and fluid ejected from the volcano hosted active anaerobic methane-degrading microbial communities. In contrast, active aerobic methane-degrading bacteria developed in the highly disturbed areas at the sulfur band, as indicated by high oxygen consumption rates and low sulfide production. These studies demonstrate the strong potential of microsensors for studying microbial communities in deep-sea research, despite the associated difficulties in deployment, maintenance, and calibration.

The application of microsensors in oil-polluted samples comes with several challenges. Microsensors break easily in sediments, calibration must be done before, during, and after experiments, and care must be taken to ensure that oil does not affect sensor sensitivity. Of course, microsensors disrupt the physicochemical structure of microbial communities through the consumption of solutes, compression of the local matrix, and by disturbing flow regimes. Microsensors generate point measurements, so extrapolating process rates to larger heterogeneous samples must be carefully considered, as diffusion coefficients are sediment dependent and vary with porosity, physical constituents, and pollutant type.

5.3 FUNCTIONAL GENES

The analysis of functional bacterial groups using metabolic genes of key proteins as targets has brought a new dimension to microbial community studies. An increasing number of studies in diversity analysis using functional genes receive significant support from the genome and metagenome sequence collections that form the basis for the application of functional genes in environmental studies (Imhoff, 2016). Real-time quantitative PCR (RT-qPCR) is a very reliable and sensitive technique, that is useful for the quantification of specific nucleic acid sequences in real-time through PCR product detection. This technique is useful for analyzing the quantity of functional genes found in different bacterial genera. In this section, we briefly discuss the genes involved in HC degradation and the quantification of functional genes using RT-qPCR.

5.3.1 GENES INVOLVED IN HYDROCARBON (HC) DEGRADATION

Alkanes and polycyclic aromatic hydrocarbons (PAHs) are common petroleum HC contaminants, constituting approximately 40–70% of crude oil depending on oil density (Jauhari et al., 2014). Investigating well-characterized HC metabolism genes from bacteria and fungi provides insight into the distribution of hydrocarbonoclastic microorganisms and their biodegradation potential (Yang et al., 2015).

5.3.1.1 ALKANE DEGRADATION GENES

Under aerobic conditions, alkane hydroxylases oxidize alkanes to alcohols that are further oxidized to fatty acids and catabolized via β-oxidation (Rojo, 2009). Integral membrane alkane monooxygenases, i.e., AlkB-related alkane hydroxylases, are commonly found in alkane-degrading bacteria (Smits et al., 2002). From a biotechnological perspective, alkane hydroxylases are versatile biocatalysts that carry out a wide range of useful oxidation reactions. *Alk* genes are generally chromosomal and many strains capable of consuming alkanes as carbon and energy source have one or more *alk* genes (Shahi et al., 2017). *Pseudomonas putida* GPo1 has only one alkane hydroxylase, while *Acinetobacter* sp. strain M1 contains two *alkB*-related alkane hydroxylases, *alkMa*, and *alkMb*, which are regulated based on the alkane available in the medium (Tani et al., 2001). When a number of alkane hydroxylases coexist in a bacterial strain, they are normally located at different sites in the chromosome.

The growth of anaerobic microbes on *n*-alkanes is usually slower (Widdel and Grundmann, 2010), although many examples of anaerobic bacteria metabolizing *n*-alkanes with six or more carbons with sulfate or nitrate as electron acceptors are available (Grossi et al., 2008; Widdel and Grundmann, 2010). The dominant mechanism for activating alkanes for metabolism appears to be the addition of fumarate yielding alkylsuccinates. Alkylsuccinic acids are important intermediates of anaerobic alkane oxidation originating from C3 to C11 alkanes (Gieg and Suflita, 2002) and their presence in oil reservoir fluids is an indication of *in situ* microbial degradation (Callaghan et al., 2010; Mbadinga et al., 2012). An anaerobic glycyl radical enzyme, alkylsuccinate synthase, succinate synthase or (1-methylalkyl) succinate synthase (Ass or Mas), is responsible for the synthesis of alkylsuccinates in the alkane degrading sulfidogenic bacteria *Desulfococcus alkenivoras* AK-01 (Callaghan et al., 2008), in *Desulfoglaeba alkanedexens* ALDCT (Callaghan et al., 2010), and in nitrate-reducing strains HxN1 (Grundmann et al., 2008) and OcN1 (Zedelius et al., 2010) in the Proteobacteria phylum (Mbadinga et al., 2011). Moreover, fumarate addition has been proposed to be important during methanogenic alkane degradation based on the presence of *assA* and *masD* genes in enrichments (Aitken et al., 2013; Cheng et al., 2013).

5.3.1.2 POLYCYCLIC AROMATIC HYDROCARBON (PAH) DEGRADATION GENES

Naphthalene and phenanthrene have long been used as model compounds to investigate the ability of bacteria to degrade polycyclic aromatic HCs. Naphthalene dioxygenase genes (*nah*) are well characterized PAH-metabolic genes responsible for the biodegradation of low molecular weight PAHs with two to three rings (Zhou et al., 2006). In *Pseudomonas putida* strain G7, naphthalene catabolic genes are organized in three operons on the 83-kb NAH7 plasmid. The three operons code for the upper pathway enzymes involved in the conversion of naphthalene to salicylate, while the lower-pathway enzymes are involved in the conversion of salicylate to a tricarboxylic acid

(TCA) cycle intermediate (Yen and Gunsalus, 1985). A trans-acting positive control regulator encoded by the *nahR* gene regulates both upper and lower operons; and NahR is essential for *nah* gene expression via induction by salicylate (Schell, 1985). The naphthalene catabolic plasmids present in different *Pseudomonas* strains were very similar to the NAH7 plasmid from strain G7 (Habe and Omori, 2003). Several *Pseudomonas* strains were reported to have genes encoding the naphthalene upper-catabolic enzymes such as *ndo* genes from *P. putida* strain NCIB 9816 (Kurkela et al., 1988), *nah* genes from *P. putida* strain G7, BS202, NCIB 9816-4 and *P. stutzeri* strain AN10 (Simon et al., 1993; Eaton, 1994; Bosch et al., 1999), *dox* genes from *Pseudomonas* sp. strain C18 (Denome et al., 1993), and *pah* genes from *P. putida* strain OUS82 and *P. aeruginosa* strain PaK1 (Takizawa et al., 1994,1999).

Bacteria degrade phenanthrene, a low-molecular-weight PAH, by one of two distinct routes. Bacteria able to grow on both phenanthrene and naphthalene mineralize these substrates via the salicylate pathway, while those growing only on phenanthrene degrade phenanthrene via the protocatechuate pathway (Kiyohara et al., 1982). The *phd* genes are responsible for the degradation of phenanthrene to *o*-phthalate. *Alcaligenes faecalis* AFK2 strain utilizes phenanthrene as a carbon source using the *o*-phthalate pathway (Kiyohara et al., 1982). *A. faecalis* AFK2 prefers phenanthrene to naphthalene due to the presence of 3,4-dihydroxyphenanthrene dioxygenase (*phnC*), 2′-carboxybenzaldehyde dehydrogenase (*phnI*), and trans-2-carboxybenzal-pyruvate hydratase aldolase (*phnH*) genes (Nagao et al., 1988; Kiyohara et al., 1990; Peng et al., 2008). *Nocardioides* sp. strain KP7 grows on phenanthrene but not on naphthalene, and has *phdA*, *phdB*, *phdC*, and *phdD* genes, encoding α and β subunits of the oxygenase component, a ferredoxin and a ferredoxin reductase, respectively, of phenanthrene dioxygenase (Saito et al., 2000).

5.3.2 QUANTIFICATION OF FUNCTIONAL GENES BY RT-QPCR

RT-qPCR was used to quantify the proportion of microorganisms harboring alkane monooxygenase genes and used to follow the alterations in the microbial community in HC-contaminated Antarctic soil during a bioremediation field trial (Powell et al., 2006). Assays for the *alkB* and *rpoB* (ribosomal polymerase B subunit) genes were validated and found to be both sensitive and reproducible. The quantity of these *alkB*-like genes was positively correlated with *n*-alkane concentrations in petroleum-contaminated soils. The proportion of *alkB*-harboring microorganisms was higher in the treatments with the most *n*-alkanes present, indicating that the alkane monooxygenase pathway is essential for the degradation of *n*-alkanes in Antarctic soils. When the concentration of *n*-alkanes in the soil decreased, the proportion of alkane-degrading microbes decreased, and the proportion of total HC-degrading microbes increased, demonstrating another shift in the microbial community structure and ongoing biodegradation.

RT-qPCR was used to estimate the number of transcripts related to *alkB* and *ndoB* from *Pseudomonas* strains and the *alkB1* and *alkB2* genes from *Rhodococcus* in order to understand the functional dynamics of indigenous bacterial communities in diesel-contaminated soils at two Canadian high-Arctic sites (Yergeau et al., 2012). The expression of degradation-related genes measured by RT-qPCR exhibited significant positive correlations with other traditional measures of HC degradation and with the mineralization of hexadecane and naphthalene. This indicates that RT-qPCR is a reliable method for the evaluation of degradation of HCs. At both locations, a decrease in soil HC concentrations within a month of treatment was observed.

The diversity and relative abundance of *alkB* genes were studied in sediments of the Timor Sea, Australia, where natural leaks are the source of petroleum HCs (Wasmund et al., 2009). RT-qPCR primers were designed to target the most abundant *alkB* gene sequences identified in Timor Sea clone libraries, and studies showed significantly higher copy numbers of *alkB* gene clusters in sediments associated with an actively venting seep. Approximately 2.9×10^5 and 1.1×10^5 copies g^{-1} of the *alkB* gene were recovered from the sediment for primer sets targeting two *alkB* gene clusters, A, and C, respectively.

The metabolic potential of oil-degrading microorganisms was evaluated using RT-qPCR to quantify the abundance of *alkB* and *nah* genes in three different soils (Liu et al., 2015). The number of copies of *alkB* ranged between 9.1×10^5 and 1.9×10^7 copies g^{-1} dry soil and were positively correlated with total concentrations of petroleum HCs and alkanes (C33 ~ C40). The *nah* gene was distributed evenly between the sampling zones, ranging between 1.9×10^7 and 1.1×10^8 copies g^{-1} dm^{-1} soil, and it was negatively correlated with concentrations of total aromatic HCs. Numbers of copies of the *nah* gene were approximately 10-fold greater than the numbers of copies of *alkB*. Hence, RT-qPCR quantification of the numbers of copies of genes involved in the degradation of petroleum HCs can be applied to biomarkers to determine the bioremediation potential of microorganisms in oil-contaminated soils. Oil-degrading microorganisms containing the *alkB* gene were a good indicator of long-term exposure to alkanes in contaminated soils due to the significant, positive correlation between the number of copies of the *alkB* gene and concentrations of alkanes.

Using RT-qPCR assays to target both *alkB* and *nah* genes in the soils of oil exploration areas may help to characterize the abundance and diversity of the petroleum HC-degrading bacterial communities. For the determination of the diversities of *alkB* and *nah* genes, the primer pairs alkB-F/alkB-R and NAH-F/NAH-R were used. The number of copies of the *alkB* gene varied from 2.56×10^6 to 9.37×10^7 copies per gram dry soil. The *nah* gene copy number varied from 2.76×10^6 to 2.29×10^7 copies per gram dry soil in oil field soil samples (Yang et al., 2015). Primer selection plays a key role in the quantification of catabolic genes, and therefore, developing the best possible primer sets is essential. Good RT-qPCR assays must produce high levels of accuracy, consistency, reproducibility, dynamic range, sensitivity, and specificity. A list of RT-qPCR primers, selected based on their extensive coverage of bacterial PAH catabolic genes, can be found in Table 5.1.

RT-qPCR was adopted to detect and quantify characteristic functional genes of PAH-degrading bacteria in oil fields of the largest oil producer worldwide, Saudi Aramco (Nasser et al., 2017). Oil field water from an injection water treatment facility, soil, and sludge were analyzed for the presence of PAH-degrading bacteria. The naphthalene dioxygenase-encoding gene *nahA* was used for RT-qPCR assays conducted with nahAF (5′-GGGYCTKTCCGCYATCAAYAC-3′) and nahAR (5′-GCACATGTC-GAGGAAGTCTTC-3′) primers. The quantified copy numbers of the *nahA* gene in the soil and sludge samples (60,540 copies g^{-1}) were higher than in the water samples (3,846 copies ml^{-1}). The variance in the gene copy numbers may reflect temporal and spatial variations between the samples. In another study, PAH ring-hydroxylating dioxygenase genes in soil and sediment samples with different contamination levels were quantified using RT-qPCR (Cébron et al., 2008). The RT-qPCR assays were based on the quantification of the copy number of the gene that encodes the alpha subunit of the PAH ring-hydroxylating dioxygenases (PAH-RHDα) from different Gram-positive and Gram-negative bacteria. Gene copy numbers of approximately 4.4×10^4 to 4.7×10^7 copies per gram soil or sediment were detected in the highly contaminated samples. It was found that the PAH biodegradation ability was positively correlated with the level of PAH contamination in the soil and sediment samples. The RT-qPCR assays developed in these studies could be used to monitor PAH-degrading bacterial population density in contaminated environments or during bioremediation.

RT-qPCR was used to detect a variety of *nagAc*-like gene sequences, responsible for conversion of naphthalene to central metabolites via gentisate (Dionisi et al., 2004). An amplification efficiency of 0.97 without any PCR inhibition was obtained from contaminated sediment total DNA (1 to 100 ng) in 25 μl reaction mixtures. The assay was applied to surficial freshwater sediment samples obtained in, or in close proximity to, a coal tar-contaminated Superfund site in the United States. The assay for *nagAc*-like sequences indicated the presence of $4.1 \pm 0.7 \times 10^3$ to $2.9 \pm 0.3 \times 10^5$ copies of *nagAc*-like dioxygenase genes per μg of DNA extracted from sediment samples. A positive correlation was observed between naphthalene concentrations and *nagA*c-like gene copy numbers.

The assessment of functional genes numbers, together with quantification of total bacterial and fungal community numbers, provides valuable information about the PAH degradation potential of microbial communities during bioremediation. When ring-hydroxylating dioxygenase genes from Gram-negative and Gram-positive bacteria (PAH-RHD GNF/R and PAH-RHD GPF/R) are above 10^6 per g dry weight of soil, bioremediation is more likely to be successful (Shahsavari et al., 2016). RT-qPCR studies revealed a significant increase in hydroxylating dioxygenase gene copies in Gram-positive bacteria after 12 weeks of bioremediation. On the other hand, the number of copies of the pyrene degradation gene (*nidA*) decreased, suggesting that bacteria having the *nidA* gene may have had only a limited impact (Shahsavari et al., 2016). The biodegradation potential of Lake Erie *Mycobacterium* populations was evaluated through quantification of pyrene dioxygenase genes (*nidA*) and mycobacteria 16S rRNA genes using RT-qPCR (Debruyn et al., 2009). The *nidA* gene was detected at all sampling sites, with abundances ranging from 2.09 to 70.4×10^6 copies g^{-1} sediment, with the highest abundance at the most PAH-contaminated site. In contrast, *nahAc* was not detected, and *nagAc* was detected at only one site in the Lake Erie study, despite the dominance of Proteobacteria. The results suggest a role for PAH-degrading mycobacteria in natural attenuation of PAH contamination in Lake Erie.

5.3.3 FUNCTIONAL GENE ARRAYS

Many culture-independent methods introduce biases (Lueders and Friedrich, 2003). Further, designing conserved PCR primers for some functional genes is difficult because of sequence variability and availability in databases (Van Nostrand et al., 2012). Hence, sometimes there is a need to perform PCR with

TABLE 5.1 Primers Targeting Hydrocarbon Degradation Genes

Primer Name	Target Gene	Annealing Temperature	Sequence	Reference
TOD-F/TOD-R	Toluene dioxygenase	47	F: CAAAA(A/G)CACCTGATT(C/T)ATGG R: A(C/T)(A/G)CG(A/G)G(C/G)GACTTCTTTCAA	Baldwin et al. (2003)
AlkB	Alkane monooxygenases	50	F: AACTACMTCGARCAYTACGG R: TGAMGATGTGGTYRCTGTTCC	Powell et al. (2010)
alkB-F/alkB-R	Alkane monooxygenases	60	F:AAYACIGCICAYGARCTIGGICAYAA R: GCRTGRTGRTCIGARTGICGYTG	Yang et al. (2015)
NAH-F/NAH-R	Naphthalene dioxygenase	47	F: CAAAA(A/G)CACCTGATT(C/T)ATGG R: A(C/T)(A/G)CG(A/G)G(C/G)GACTTCTTTCAA	Baldwin et al. (2003)
nahAF/nahAR	Naphthalene dioxygenase gene	55	F: GGGYCTKTCCGCYATCAAYAC R: GCACATGTCGAGGAAGTCTTC	Nasser et al. (2017)
Nah	Naphthalene dioxygenase	57	F: ACTTGGTTCCGGAGTTGATG R: CAGGTCAGCATGCTGTTGTT	Park and Crowley (2006)
TOL-F/TOL-R	Xylene monooxygenase	55	F: TGAGGCTGAAACTTTACGTAGA R: CTCACCTGGAGTTGCGTAC	Baldwin et al. (2003)
PAH-RHD GNF/R	PAH-ring hydroxylating dioxygenase genes of G– bacteria	57	F: GAGATGCATACCACGTKGGTTGGA R: AGCTGTTGTTCGGGAAGAYWGTGCMGTT	Cébron et al. (2008)
PAH-RHD GPF/R	PAH-ring hydroxylating dioxygenase genes of G+ bacteria	54	F: CGGCGCCGACAAYTTYGTNGG R: GGGGAACACGGTGCCRTGDATRAA	Cébron et al. (2008)

TABLE 5.1 *(Continued)*

Primer Name	Target Gene	Annealing Temperature	Sequence	Reference
BPH2-F/BPH2-R	Biphenyl dioxygenase	63	F: GACGCCCGCCCCTATATGGA R: AGCCGACGTTGCCAGGAAAAT	Baldwin et al. (2003)
CATA-F/CATA-R	Catechol-1,2-dioxygenase genes	58	F: ACVCCVCGHACCATYGAAGG R: CGSGTNGCAWANGCAAAGT	El Azhari et al. (2010)
C23O-F/C23O-R	Catechol-2,3-dioxygenase genes	58	F: AAGAGGCATGGGGGCGCACCGGTTCGATCA R: CCAGCAAACACCTCGTTGCGGTTGCC	Sei et al. (1999)
RMO-F/RMO-R	Toluene monooxygenase	53	F: TCTC(A/C/G)AGCAT(C/T)CAGAC(A/C/G)GACG R: TT(G/T)TCGATGAT(C/G/T)AC(A/G)TCCCA	Baldwin et al. (2003)
PHE-F/PHE-R	Phenol monooxygenase	49	F: GTGCTGAC(C/G)AA(C/T)CTG(C/T)TGTTC R: CGCCAGAACCA(C/T)TT(A/G)TC	Baldwin et al. (2003)
phnAc	PAH initial dioxygenases	55	F: TTCGAGCTGGAATGTGAGC R: AATAACCGGCGATTCCAAAC	Laurie and Lloyd-Jones (2000)

different primer sets targeting the same gene. Microarrays provide the possibility of investigating thousands of genes at one time without the need for individual PCR amplification of each gene. Since microarrays are valuable for the study of microbial communities, several types have been designed to examine microbial communities, including functional gene arrays, designed to examine multiple functional genes at one time.

To better understand the diversity and functions of microbial communities, GeoChip, a robust functional gene array containing 24,243 oligonucleotide probes covering more than 10,000 genes in more than 150 functional groups involved in nitrogen, carbon, and sulfur cycling, metal reduction and resistance, and organic contaminant degradation, was developed (He et al., 2007). A functional gene array study was conducted with soils collected along a contamination gradient at an oil field in northeast China to study the *in situ* microbial functional diversity (Liang et al., 2009). The results revealed that the total functional gene abundance and diversity decreased along an increasing gradient of oil contamination. The abundance of a number of functional genes in the families *pgl*, *rbcL*, *nifH*, and *nor* and those encoding cellulase, laccase, chitinase, urease, and other enzymes involved in the metabolism of organic compounds significantly decreased with high oil contamination stress. However, a few genes encoding key enzymes for catechol, protocatechuate, and biphenyl degradation, and the gene families *nir*, *rbcL*, and *pgl*, significantly increased with moderate oil contamination. Oil content and soil available nitrogen were the key factors that influenced the microbial community structure. Hence, microarray studies provide potential information for on-site management and remediation measures.

Bacterial community structure and function were evaluated using microarrays targeting the 16S rRNA genes of bacteria living in Arctic soil (Yergeau et al., 2009). A rapid reorganization of the bacterial community structure and functional potential was observed. Similarly, an increase in the expression of alkane monooxygenases and PAH ring-hydroxylating dioxygenases (RHDs) were observed. This could be due to the differences in bioremediation treatments, HC weathering, native microbial communities and other environmental factors. This shows the value of the microarray technique for monitoring bacteria in polar soil and their functions during bioremediation.

Both culture and community DNA fingerprinting by functional gene arrays were used to characterize the dynamics of microbial community structure within a fluidized bed reactor used to bioremediate diesel-contaminated groundwater in a tropical environment (Rodríguez-Martínez et al., 2006).

Sequence analysis of 16S rRNA genes was used to examine microbial diversity, with most bacteria belonging to the α, β and γ subdivisions of the Proteobacteria, Bacilli, and Actinobacteria groups. In addition, the genetic constitution of the microbial community was studied at multiple time points with a GeoChip 1.0 functional gene array containing over 12,000 probes for organic contaminant degradation and major biogeochemical cycling genes. A total of 270 genes for organic contaminant degradation (including naphthalene, toluene [aerobic and anaerobic], octane, biphenyl, pyrene, xylene, phenanthrene, and benzene), and 333 genes involved in other metabolic activities, were detected. This functional gene array study demonstrated the early establishment of a diverse microbial community with concurrent aerobic and anaerobic processes contributing to the bioremediation process.

The effect of different bioremediation treatments on microbial communities was examined using laboratory-scale bioremediation systems (Liang et al., 2009). The study was conducted with sediment from contaminated oil fields that was inoculated with oil-degrading enrichment cultures and treated with ozone after 242 days of incubation. With GeoChip 2.0, many HC degradation genes (for benzene, benzoate, catechol, PAH, protocatechuate, and phthalate) were detected. Due to ozonation, the number of functional genes decreased by 50%, but after a recovery period, the gene numbers increased, and oil degradation continued.

To study the impact of the 2010 Gulf of Mexico oil spill, GeoChip 4.0 was used to compare microbial communities in oil-contaminated water with that of uncontaminated water (Lu et al., 2012). After 40 days, the presence of the HC plume (1,100 m depth) caused a major shift in the microbial community's functional structure and diversity. Many genes involved in HC degradation were significantly enriched in plume samples, including naphthalene 1,2-dioxygenase, cyclohexanone 1,2-monooxygenase, and alkene monooxygenase (AMO) (Hazen et al., 2010; Lu et al., 2012). These findings suggest that the microbial communities in the Gulf of Mexico were capable of intrinsic bioremediation, and oil stimulated the oil-degrading community members. Hence, the functional gene array studies provide information about microbial community dynamics and the role of functional genes in oil remediation. Although DNA microarray methods are rapid and facilitate monitoring of a variety of samples (Desai et al., 2010), this method has recently become less utilized due to issues related to chip sensitivity, specificity, and quantitative ability. In addition, microarray techniques are not able to identify novel sequences, overlooking the diverse reservoir of uncharacterized sequences that are common in natural environments (He et al., 2013).

5.4 OMICS APPROACHES

Omics approaches (genomics, proteomics, metabolomics, and transcriptomics) can be applied to study specific organisms under defined conditions, while meta-omics techniques can be adopted for investigating the structure and function of microbial communities in natural ecosystems. Recently, scientists have adapted integrated multi-omic approaches to their research on oil- and HC-polluted ecosystems, both at the single-cell (cytomic) and the environmental level, resulting in comprehensive datasets describing the functional role of microorganisms in oil-polluted sites. In this section, we summarize these studies.

5.4.1 METAGENOMICS

Metagenomics is typically applied to describe the taxonomic profile and metabolic potential of a microbial community (Stenuit et al., 2008). Unlike amplicon sequencing (An et al., 2013; Aquilar et al., 2014), which targets specific gene markers such as ribosomal RNA or a subset of functional genes, metagenomic approaches endeavor to capture all DNA sequences from an environment for cloning into libraries for functional screening, or for whole metagenome shotgun sequencing (WMS).

Function-based screening of clones in metagenome libraries has been used to identify DNA sequences with unknown functions detected in oil-polluted environments. The method involves introducing vectors, such as cosmid, fosmid, and bacterial artificial chromosomes (BAC) (Brennerova et al., 2009; Sierra-Garcia et al., 2014; Bouhajja et al., 2017), containing inserts from environmental DNA into suitable microbial hosts, and screening the hosts for certain activities. This method has helped to discover clusters of functional genes and link them to specific metabolic pathways. For instance, this method was used to describe a previously undescribed metabolic pathways involved in aromatic HC degradation in oil reservoir microbial communities (Sierra-Garcia et al., 2014).

Functional metagenomics is technically challenging and requires automation of the screening processes to deal with the low number of positive hits. For example, <2% of clones in a fosmid clone library showed positive activity against HCs (Brennerova et al., 2009; Sierra-Garcia et al., 2014). The presence of novel toluene monooxygenase (TMO) genes in HC-polluted sediments was revealed after screening 845,000 BAC clones and 53,760 fosmid clones (Bouhajja et al., 2017).

Conversely, continual developments in high-throughput sequencing analysis have made direct WMS more practical. For samples with low levels of DNA, WMS may require a random amplification step (Khodakova et al., 2014), or a PCR-independent multiple displacement approaches to amplify DNA using phi29 DNA polymerase (Abulencia et al., 2006). Following sequencing, reads may be mapped to a reference genome, or *de novo* assembly approaches can be used in an effort to construct contigs and scaffolds of groups of microbial genes (Aguiar-Pulido et al., 2016). This approach provides a high-level picture of the microbial ecosystem, instead of focusing on specific genes, pathways, or individual organisms. The WMS approach has been used to characterize microbial communities with diverse metabolic potential in various oil-contaminated terrestrial and marine environments (Mason et al., 2012, 2014; Joshi et al., 2014; Jung et al., 2016; Abbasian et al., 2016; Salam et al., 2017). The genetic diversity in these environments was found to be dominated by genes related to amino acid, carbohydrate, cofactor, vitamin, prosthetic group and pigment metabolism (Salam et al., 2017), motility, chemotaxis, aliphatic HC degradation (Mason et al., 2012), and genes involved in denitrification (Mason et al., 2014). The dominance of these genes suggested an adaptation of the microbial community in polluted environments to stressors including HC hydrophobicity, heavy metal toxicity oxidative stress, imbalances in carbon, nitrogen, and phosphorous ratios, and nutrient starvation. In diesel-contaminated microcosms, WMS analysis showed significant changes in a large proportion of functional gene categories, especially those related to HC degradation and tetracycline biosynthesis, and microbial diversity decreased in the environments where alkane monooxygenase and P450 genes increased between 2 to 13-fold (Jung et al., 2016).

Despite the importance of the metagenomic approach in revealing the potential functional diversity of microbes in oil-polluted environments, this approach is limited by its inability to measure functional activities under specific conditions. This limitation is overcome by investigating products of transcription instead of DNA.

5.4.2 METATRANSCRIPTOMICS

Metatranscriptomics is the study of the active functional diversity of microbial communities at a specific time by sequencing total community mRNA transcripts (Aguiar-Pulido et al., 2016). The transcripts can be obtained by extracting and sequencing total RNA (Mason et al., 2012; Lamendella et al., 2014), or by separating the more abundant rRNA from mRNA prior to sequencing. The latter method involves either enriching mRNA from the total RNA pool or depleting rRNA (Militon et al., 2016). Studies have shown metatranscriptomic datasets are sometimes less complex than metagenomic datasets, and can be generated with less sequencing depth. Furthermore, metatranscriptomic methods have a demonstrated ability to detect large numbers of novel sequences from low copy number genes that may be overlooked by DNA sequencing (Zarraonaindia et al., 2013).

Metatranscriptomics studies have demonstrated that transcripts provide a more precise understanding of what is occurring in an ecosystem compared to what can be inferred from DNA sequences. For example, the metagenome showed an abundance of both aromatic and aliphatic HC degradation genes in the plume, but the transcriptomics approach showed that only the aliphatic HC degradation genes were being expressed at the time of sampling (Mason et al., 2012). Using total RNA sequencing, the microbial community structure, diversity, and function were profiled, and it was found that the key species in HC degradation of the DWHOS that reached the beaches were Alphaproteobacteria and Gammaproteobacteria, and that xenobiotic degradation transcripts from the samples matched PAH, *n*-alkane, toluene, and benzoate degradation pathways (Lamendella et al., 2014). However, a study that screened mRNA from DWHOS bathypelagic bacterioplankton found β-oxidation and other key HC metabolism pathways in the plume, and that Gammaproteobacteria dominated the HC-degrading community (Rivers et al., 2013).

Recently, studies have been undertaken to couple mRNA sequencing with stable isotope probing (SIP), which facilitates the tracking of isotopically labeled substances in functionally informative biomarkers with the expectation of linking microbial metabolic capability to mRNA (Huang et al., 2009). SIP has also been coupled with metagenomics and proteomics methods (Uhlik et al., 2013).

Metatranscriptomics is a powerful tool because inventories of mRNA pools are informative about ongoing processes (Ismael et al., 2014). This is true in a stable environment, but can be challenging in a naturally fluctuating environment where the abundance of mRNAs from a functional gene may change quickly, and does not guarantee translation to functional proteins (Moran et al., 2013). That is, transcriptomics is unable to provide information regarding the activity, arrangement, or final destination of the gene products; metaproteomics and metabolomics can be used to overcome these obstacles.

5.4.3 METAPROTEOMICS AND INTERACTOMICS

When applied to microbial communities, proteomics can provide a picture of the protein content in cells, including information on protein abundance, stability, turnover rate, post-translational modifications, and protein-protein interactions. Both bottom-up protein analysis techniques (i.e., analyzing peptide fragments from the proteins) and top-down techniques (i.e., analyzing whole proteins) are made possible with tandem mass spectrometry (MS/MS) and matrix-assisted laser desorption ionization time-of-flight mass spectrometry (MALDI-ToF) techniques. MALDI-ToF has generally been used to identify proteins excised from two-dimensional gel electrophoresis (2DE) or separated in some other way. Liquid chromatography (LC) coupled with tandem mass spectrometry (LC/MS/MS) led to the development of high-throughput gel-free profiling procedures (van Straalen and Roelofs, 2006; Kumavath and Pratap, 2012), allowing for detection of a wide dynamic range of thousands of proteins in single run. While less popular now, 2DE has provided considerable information on aromatic HC metabolism (see review by Kim et al., 2009).

It is now well-established that the one gene-one enzyme (or one function) notion is not valid, and that a single gene can produce different proteins due to variations in splicing and translational factors, as well as through post-transcriptional modification of mRNA. Therefore, the phenotypic traits of microbes are better reflected by protein content, rather than the gene sequences (Wang et al., 2016). Metaproteomic analysis of oil-contaminated environments has provided information on the overall physiological adaptations and regulatory mechanisms operating at one time in response to oil contamination (Desai et al., 2010). Guazzaroni et al. (2013) investigated the effect of naphthalene exposure on the microbial community in PAH-polluted soil in northern Spain. Metaproteome analysis revealed that PAH exposure induced production of proteins with functions involved in biodegradation of aromatics via di- and tri-hydroxylated intermediates. Metaproteome analyzes of chronically oil-polluted marine sediments from three sites along the coastline of Italy indicated dominance of anaerobic metabolic reactions not associated with oil biodegradation (Bargiela et al., 2015), suggesting that in oxygen-limited Mediterranean Sea sediments, oil biodegradation is suppressed. Bastida et al. (2015) used metaproteomics to describe novel functional HC degradation processes in contaminated semiarid soil upon compost amendment during bioremediation.

Meta-interactomics is a sub-discipline of metaproteomics for investigating complex protein interaction networks that result from protein aggregation in multicomponent protein complexes (Kiemer and Cesareni, 2007; Singh and Nagaraj, 2006). The application of meta-interactomics for investigating the functional activity of microbial communities in oil-polluted sites is still in its infancy.

5.4.4 METABOLOMICS

Metabolomics involves the identification and quantification of low-molecular-weight intra- and extra-cellular constituents (<1,000 Da) of microbial cells or communities (Villas-Boas and Bruheim, 2007). Since metabolites are functional entities in microbial cells, investigating them provides a powerful way to discover, for example, biomarker profiles of toxicant exposure and to identify the metabolic pathways involved in such processes, linking the phenotype-genotype gap (Fiehn, 2002). Metabolomic data signify physiological responses to environmental stressors, as well as developmental and genetic changes throughout time (Weckwerth and Fiehn, 2002). An investigation of HC biodegradation via metabolic footprint analysis of extracellular molecules showed that the Gulf of Mexico deep sediment affected by the DWHOS was dominated by anaerobic Deltaproteobacteria, and the presence of benzylsuccinate indicated active anaerobic HC metabolism (Kimes et al., 2013). Anaerobic degradation of alkanes via fumarate addition was detected in Chinese oil reservoirs as alkylsuccinate dominated the metabolome pool (Bian et al., 2015). Using signature metabolites, it was suggested that microbial degradation of HCs at 80°C in oil reservoir formation water produced (poly)alkylated succinic acids during the process of HC transformation and intracellular catabolism of fatty acids (Gruner et al., 2017).

Recent advances in metabolomics lead to the development of "fluxomics," which deals with real-time flux analysis of metabolites inside cells over time (Wiechart et al., 2007), thus revealing metabolome functional interactions with the environment and the genome (Dettmer et al., 2007). However, metabolomic approaches have many limitations, including the inability to detect all cellular molecules even when a combination of mass spectrometry (MS), nuclear magnetic resonance (NMR), and vibrational spectroscopy is used. The challenges with environmental metabolomics are greater than those of single organism metabolomics because of the high level of metabolic noise, which is a direct result of the interactions between microorganisms in the environment. It has become clear that solely applying metabolomic approaches will not reveal the linkages between metabolites and genes, enzymes, or specific pathways in cellular or community samples. For this reason, integrated multi-omics approaches have gained popularity in recent years in which different omics approaches are executed on the same sample.

5.4.5 INTEGRATIVE MULTI-OMICS APPROACH

To provide a holistic view of the functional activities of microbes in their environment and to find the relationship between different data while filling the mechanistic gaps, various "–omics" methods can be applied simultaneously. The results of integrating several "-omics" approaches have been found to be dramatically different from those obtained by applying a single technique, which have inherent limitations (Zhang et al., 2010; Franszosa et al., 2015) (Table 5.2). Combining metagenome with metatranscriptome and metaproteome analyzes can help draw linkages between the functional potential and the functional activities of the system investigated. Metagenome and metatranscriptome data were combined with cytomic analysis of *Oceanospirillales*, the most active member of the DWHOS plume indigenous microbial community, to provide in-depth knowledge regarding the role of this bacterium in the degradation of aliphatic HCs in the early stages of the plume (Mason et al., 2012). Combining metagenome and metaproteome analyzes revealed the mechanism involved in naphthalene degradation in soil (Guazzaroni et al., 2013).

Combined "-omics" have also been adopted to provide stronger support for the acceptance or rejection of hypotheses when different meta-omics methods independently reach the same conclusion. As such, an integrated approach has been demonstrated to be effective, not only in descriptive modeling but also in predictive modeling studies. Bargiela et al. (2015) detected the minimal signature of active biodegradation pathways in oil-polluted sediments under limited oxygen concentrations using both metaproteome and metabolome data. This study represents an example of how integrated methods are now being applied with different meta-omics approaches complementing each other; revealing novel biological insights that usually cannot be obtained using a single technique. However, to gain the maximum benefit from integrating "-omics" methods, a proper strategy should be applied to determine the type of questions under investigation and therefore the number of samples to be analyzed, the number of assays to be performed, and the integrative analysis method to be used (Quinn et al., 2016).

5.5 CONCLUSIONS AND FUTURE RESEARCH DIRECTIONS

Recent advancements in microsensor and molecular tools have made it possible to link the diversity of bacterial communities in oil-polluted environments to their function, which has been very challenging for several decades. Using microsensors, the rates of different microbial processes can now be instantly measured in contaminated sediments and any changes in these processes can be monitored. The expression of genes can be quantified using RT-qPCR, and with the recent development of "–omics" tools, an extensive knowledge of the diversity and function of microorganisms in oil-polluted environments can now be obtained. Despite the importance of these tools, their application can be associated with several challenges. For instance, the presence of a high concentration of oil in the studied samples may change the porosity of sediments and affect the sensitivity of microsensors. Oil may hamper nucleic acid extractions, resulting in an incomplete picture of microbial diversity. Although, the discovery of new pathways and genes using "-omics" approaches has now become possible, the validation of these pathways can be tedious and time consuming. In spite of all the pitfalls, these techniques have dramatically increased our understanding of the microbial ecology of oil-impacted ecosystems. Future research should focus on optimizing these techniques, increasing their application, and combining them with the newly emerging single-cell techniques in order to study the functional diversity of microorganisms in oil-polluted sites from the single-cell level to the whole community level.

TABLE 5.2 Advantages and Disadvantages of Various Omics Approaches

Meta-Omics Approach	Advantages	Limitations
Metagenomics	• Reveal the taxonomic profile of the microbial community based on DNA analysis • When coupled with microarray analysis, function-based screening of clones or whole nucleic acid shotgun sequencing it provides information regarding the potential functional diversity of the microbial community.	• Unable to measure functional activities under specific conditions.

TABLE 5.2 *(Continued)*

Meta-Omics Approach	Advantages	Limitations
Metatranscriptomics	• Disclose functional diversity microbial communities at specific time points by providing information about the gene expression. • Data are less complex than from the metagenomics approach and requires less sequencing depth to achieve the same level of coverage. • Able to produce large numbers of novel sequences from low copy number genes, which may be overlooked by DNA sequencing.	• The abundance of mRNAs from a functional gene is not a reliable proxy for the microbial community functions in unstable environments. • The approach fails to account for post-transcriptional events. • Unable to provide information regarding the activity, arrangement, or final destination of the gene products.
Metaproteomics	• Enables a snapshot of the protein repertoire in microbial communities from environmental sources, where the production of proteins varies with environmental conditions. • Provides a picture of the protein characteristics in cells including protein abundance, stability, turnover rate, and protein-protein interactions.	• Protein extraction, separation, and identification may be limited in complex environments. • Databases are underdeveloped.
Metabolomics	• Reflect cellular processes through measurements of low molecular weight metabolites.	• Not useful for assessing microbial community structure. • No single method can measure and analyze the thousands of different chemical compounds that may be present. • Measurement from environmental samples include metabolic noise • Databases are underdeveloped.

KEYWORDS

- **functional diversity**
- **metabolomics**
- **metagenomics**
- **metaproteomics**
- **metatranscriptomics**
- **microarrays**
- **microsensors**

REFERENCES

Abbasian, F., et al., (2016). Microbial diversity and hydrocarbon-degrading gene capacity of a crude oil field soil as determined by metagenomics analysis. *Biotechnol. Prog., 32*(3), 638–648.

Abed, R. M. M., & Al-Kindi, S., (2017). Effect of disturbance by oil pollution on the diversity and activity of bacterial communities in biological soil crusts from the sultanate of Oman. *Appl. Soil Ecol., 110,* 88–96.

Abed, R. M. M., & De Beer, D., (2010). Microsensor techniques to study in situ bacterial metabolic processes in hydrocarbon-polluted sediments. In: Timmis, K. N., (ed.), *Handbook of Hydrocarbon and Lipid Microbiology* (pp. 3656–3667). Springer-Verlag, Berlin, Heidelberg.

Abed, R. M. M., De Beer, D., & Stief, P., (2015). Functional-structural analysis of nitrogen cyclebacteria in a hypersaline mat from the Omani desert. *Geomicrobiol., 32*(2), 119–129.

Abed, R. M. M., et al., (2013). High rates of denitrification and nitrous oxide emission in arid biological soil crusts from the Sultanate of Oman. *ISME J, 7*(9), 1862–1875.

Abed, R. M. M., et al., (2018). Abundance and diversity of aerobic heterotrophic microorganisms and their interaction with cyanobacteria in the oxic layer of an intertidal hypersaline cyanobacterial mat. *FEMS Microbiol. Ecol., 94*(2), 183.

Abed, R. M. M., Kohls, K., & De Beer, D., (2007). Effect of salinity changes on the bacterial diversity, photosynthesis and oxygen consumption of cyanobacterial mats from an intertidal flat of the Arabian Gulf. *Environ. Microbiol., 9*(6), 1384–1392.

Abed, R. M. M., Polerecky, L., Al-Najjar, M., & De Beer, D., (2006). Effect of temperature on photosynthesis, oxygen consumption, and sulfide production in an extremely hypersaline cyanobacterial mat. *Aquat. Microb. Ecol., 44*(1), 21–30.

Abulencia, C. B., Keller, M., et al., (2006). Environmental whole-genome amplification to access microbial populations in contaminated sediments. *Appl. Environ. Microbiol, 72*(5), 3291–3301.

Aguiar-Pulido, V., et al., (2016). Metagenomics, metatranscriptomics, and metabolomics approaches for microbiome analysis. *Evol. Bioinform., 12*(S1), 5–16.

Aitken, C. M., Head, I. M., et al., (2013). Evidence that crude oil alkane activation proceeds by different mechanisms under sulfate-reducing and methanogenic conditions. *Geochim. Cosmochim. Acta, 109*, 162–174.

Aitken, C. M., Jones, D. M., & Larter, S. R., (2004). Anaerobic hydrocarbon biodegradation in deep subsurface oil reservoirs. *Nature, 431*(7006), 291–294.

Ali, N., et al., (2016a). Dynamics of bacterial populations during bench-scale bioremediation of oily seawater and desert soil bioaugmented with coastal microbial mats. *Microb. Biotechnol., 9*(2), 157–171.

Al-Kindi, S., & Abed, R. M. M., (2016b). Effect of biostimulation using sewage sludge, soybean meal, and wheat straw on oil degradation and bacterial community composition in a contaminated desert soil. *Front. Microbiol., 7*, 240.

Al-Mailem, D. M., Kansour, M. K., & Radwan, S. S., (2015). Moderately thermophilic, hydrocarbonoclastic bacterial communities in Kuwaiti desert soil: Enhanced activity via Ca^{2+} and dipicolinic acid amendment. *Extremophiles, 19*(3), 573–583.

Álvarez, L. M., et al., (2017). Bioremediation of hydrocarbon-contaminated soils in cold regions: Development of a pre-optimized biostimulation biopile-scale field assay in Antarctica. *Sci. Total Environ., 590*, 194–203.

Amann, R. I., Ludwig, W., & Schleifer, K. H., (1995). Phylogenetic identification and in situ detection of individual microbial cells without cultivation. *Microb. Reviews, 59*(1)143–169.

An, D., et al., (2013). Microbial community and potential functional gene diversity involved in anaerobic hydrocarbon degradation and methanogenesis in an oil sands tailings pond. *Genome, 56*(10), 612–618.

Aquilar, M., et al., (2014). *Next-Generation Sequencing of Protists as a Measure of Microbial Community in Oil Sands Tailings Ponds: Amplicon Versus Metagenomic Approaches.* OSRIN Report No.

Bae, J. W., & Park, Y. H., (2006). Homogeneous versus heterogeneous probes for microbial ecological microarrays. *Trends Biotechnol., 24*(7), 318–323.

Baldwin, B. R., Nakatsu, C. H., & Nies, L., (2003). Detection and enumeration of aromatic oxygenase genes by multiplex and real-time PCR. *Appl. Environ. Microbiol., 69*(6), 3350–3358.

Bargiela, R., et al., (2015). Metaproteomics and metabolomics analyses of chronically petroleum-polluted sites reveal the importance of general anaerobic processes uncoupled with degradation. *Proteomics, 15*(20), 3508–3520.

Bastida, F., et al., (2016). The ecological and physiological responses of the microbial community from a semiarid soil to hydrocarbon contamination and its bioremediation using compost amendment. *J. Proteom., 135*, 162–169.

Benthien, M., et al., (2004). Oil-contamination effects on a hypersaline microbial mat community (Camargue, France) as studied with microsensors and geochemical analysis. *Ophelia, 58*(3), 135–150.

Bian, X. Y., et al., (2015). Insights into the anaerobic biodegradation pathway of n-alkanes in oil reservoirs by detection of signature metabolites. *Sci. Rep., 5*, 9801.

Bosch, R., Garcia-Valdáes, E., & Moore, E. R. B., (1999). Genetic characterization and evolutionary implications of a chromosomally encoded naphthalene degradation upper pathway from *Pseudomonas stutzeri* AN10. *Gene, 236*(1), 149–157.

Bouhajja, E., et al., (2017). Identification of novel toluene monooxygenase genes in a hydrocarbon-polluted sediment using sequence-and function-based screening of metagenomic libraries. *Appl. Microbiol. Biotechnol., 101*, 797–808.

Brennerova, M. V., et al., (2009). Metagenomics reveals diversity and abundance of meta-cleavage pathways in microbial communities from soil highly contaminated with jet fuel under air-sparging bioremediation. *Environ. Microbiol., 11*(9), 2216–2227.

Callaghan, A. V., et al., (2008). Anaerobic alkane-degrading strain AK-01 contains two alkylsuccinate synthase genes. *Biochem. Biophys. Res. Commun., 366*(1), 142–148.

Callaghan, A. V., et al., (2010). Diversity of benzyl- and alkylsuccinate synthase genes in hydrocarbon-impacted environments and enrichment cultures. *Environ. Sci. Technol., 44*(19), 7287–7294.

Cébron, A., et al., (2008). Real-time PCR quantification of PAH-ring hydroxylating dioxygenase (PAH-RHDa) genes from gram-positive and gram-negative bacteria in soil and sediment samples. *J. Microbiol. Methods, 73*(2), 148–159.

Cheng, L., et al., (2013). Enrichment and dynamics of novel syntrophs in a methanogenic hexadecane-degrading culture from a Chinese oilfield. *FEMS Microbiol. Ecol., 83*(3), 757–766.

Das, N., & Chandran, P., (2011). Microbial degradation of petroleum hydrocarbon contaminants: An overview. *Biotechnol. Res. Int.* Article ID 941810.

De Beer, D., et al., (2000). A microsensor study of light enhanced Ca^{2+} uptake and photosynthesis in the reef-building hermatypic coral *Favia sp. Mar. Ecol. Prog. Ser., 194, 75–85.*

De Beer, D., et al., (2017). Oxygenic and anoxygenic photosynthesis in a microbial mat from an anoxic and sulfidic spring. *Environ. Microbiol., 19(3),* 1251–1265.

Debruyn, J. M., et al., (2009). PAH biodegradative genotypes in Lake Erie sediments: Evidence for broad geographical distribution of pyrene-degrading mycobacteria. *Environ. Sci. Technol., 43*(10), 3467–3473.

Denome, S. A., et al., (1993). Metabolism of dibenzothiophene and naphthalene in *Pseudomonas* strains: Complete DNA sequence of an upper naphthalene catabolic pathway. *J. Bacteriol., 175*(21), 6890–6901.

Desai, C., Pathak, H., & Madamwar, D., (2010). Advances in molecular and "-omics" technologies to gauge microbial communities and bioremediation at xenobiotic/anthropogen contaminated sites. *Bioresour. Technol., 101*(6), 1558–1569.

Dettmer, K., Aronov, P. A., & Hammock, B. D., (2007). Mass spectrometry-based metabolomics. *Mass Spectrom. Rev., 26*(1), 51–78.

Dionisi, H. M., et al., (2004). Abundance of dioxygenase genes similar to *Ralstonia* sp. strain U2 *nagAc* is correlated with naphthalene concentrations in coal tar-contaminated freshwater sediments. *Appl. Environ. Microbiol., 70*(7), 3988–3995.

Eaton, R. W., (1994). Organization and evolution of naphthalene catabolic pathways: Sequence of the DNA encoding 2-hydroxychromene-2-carboxylate isomerase and trans-o-hydroxybenzylidenepyruvate hydratase-aldolase from the NAH7 plasmid. *J. Bacteriol., 176*(24), 7757–7762.

El Azhari, N., et al., (2010). Molecular analysis of the catechol-degrading bacterial community in a coal wasteland heavily contaminated with PAHs. *J. Hazard. Mater., 177*(1–3), 593–601.

Felden, J., et al., (2013). Limitations of microbial hydrocarbon degradation at the Amon mud volcano (Nile deep-sea fan). *Biogeosciences, 10*(5), 3269–3283.

Fiehn, O., (2002). Metabolomics—the link between genotypes and phenotypes. *In: Functional Genomics* (pp. 155–171) Springer, Dordrecht.

Franzosa, E. A., et al., (2015). Sequencing and beyond: integrating molecular 'omics' for microbial community profiling. *Nat. Rev. Microbiol., 13*(6), 360.

Gieg, L. M., & Suflita, J. M., (2002). Detection of anaerobic metabolites of saturated and aromatic hydrocarbons in petroleum-contaminated aquifers. *Environ. Sci. Technol., 36*(17), 3755–3762.

Gieseke, A., & De Beer, D., (2004). Use of microelectrodes to measure in situ microbial activities in biofilms, sediments, and microbial mats. In: Akkermans, A. D. L., & Elsas, D., (eds.), *Molecular Microbial Ecology Manual.* Kluwer, Dordrecht (NL).

Gieseke, A., et al., (2001). Community structure and activity dynamics of nitrifying bacteria in a phosphate-removing biofilm. *Appl. Environ. Microbiol., 67*(3), 1351–1362.

Grossi, V., et al., (2008). Metabolism of n-alkanes and n-alkenes by anaerobic bacteria: A summary. *Org. Geochem., 39*(8), 1197–1203.

Grötzschel, S., et al., (2002). Degradation of petroleum model compounds immobilized on clay by a hypersaline microbial mat. *Biodegradation, 13*(4), 273–283.

Grundmann, O., et al., (2008). Genes encoding the candidate enzyme for anaerobic activation of n-alkanes in the denitrifying bacterium, strain HxN1. *Environ. Microbiol., 10*(2), 376–385.

Gruner, A., et al., (2017). Tracing microbial hydrocarbon transformation processes in a high temperature petroleum reservoir using signature metabolites. *Org. Geochem., 108*, 82–93.

Guazzaroni, M. E., et al., (2013). Metaproteogenomic insights beyond bacterial response to naphthalene exposure and bio-stimulation. *ISME J., 7*(1), 122.

Habe, H., & Omori, T., (2003). Genetics of polycyclic aromatic hydrocarbon metabolism in diverse aerobic bacteria. *Biosci. Biotechnol. Biochem., 67*(2), 22–243.

Handley, K. M., et al., (2017). Metabolic and spatio-taxonomic response of uncultivated seafloor bacteria following the Deepwater Horizon oil spill. *ISME J., 11*(11), 2569–2583.

Hassanshahian, M., Zeynalipour, M. S., & Musa, F. H., (2014). Isolation and characterization of crude oil-degrading bacteria from the Persian Gulf (Khorram Shahr provenance). *Mar. Pollut. Bull., 82*(1, 2), 39–44.

Hazen, T. C., et al., (2010). Deep-sea oil plume enriches indigenous oil-degrading bacteria. *Science, 330*(6001), 204–208.

He, Y., Xiao, X., & Wang, F., (2013). Metagenome reveals potential microbial degradation of hydrocarbon coupled with sulfate reduction in an oil-immersed chimney from Guaymas Basin. *Front. Microbiol., 4*, 148.

He, Z., Gentry, T. J., et al., (2007). GeoChip: A comprehensive microarray for investigating biogeochemical, ecological and environmental processes. *ISME J., 1*(1), 67–77.

Hinke, J., (1969). Glass microelectrodes for the study of binding and compartmentalization of intracellular ions. In: Lavallee, M., Schanne, O. F., & Herbert, N. C., (eds.), *Glass Microelectrodes* (pp. 349–375). Wiley.

Höpner, T., et al., (1996). Cyanobacterial mats on oil-polluted sediments—start of a promising self-remediation process? In: Krupp, F., Abuzinada, A. H., & Nader, I. A., (eds.), *A Marine Wildlife Sanctuary for the Arabian Gulf: Environmental Research and Conservation Following the 1991 Gulf War Oil Spill* (pp. 85–95). NCWCD, Riyad, Saudi Arabia, and Senckenberg Research Institute, Frankfurt, Germany.

Huang, W. E., et al., (2009). Resolving genetic functions within microbial populations: In situ analyses using rRNA and mRNA stable isotope probing coupled with single-cell Raman-fluorescence in situ hybridization. *Appl. Environ. Microbiol., 75*(1), 234–241.

Humphery-Smith, I., (1999). Replication-induced protein synthesis and its importance to proteomics. *Electrophoresis, 20*(4, 5), 653–659.

Imhoff, J. F., (2016). New Dimensions in microbial ecology-functional genes in studies to unravel the biodiversity and role of functional microbial groups in the environment microorganisms. *Microorganisms, 4*(2), 19.

Ismail, W. M., Ye, Y., & Tang, H., (2014). Gene finding in metatranscriptomic sequences. *BMC Bioinformatics, 15*(Suppl 9), S8.

Jaekel, U., et al., (2015). Anaerobic degradation of cyclohexane by sulfate-reducing bacteria from hydrocarbon-contaminated marine sediments. *Front. Microbiol., 6, 116.*

Jauhari, N., et al., (2014). Bacteria-mediated aerobic degradation of hexacosane *in vitro* conditions. *Bioresour Technol., 170*, 62–68.

Jeroschewski, P., Steuckart, C., & Kühl, M., (1996). An amperometric microsensor for the determination of H_2S in aquatic environments. *Anal. Chem., 68*(24), 4351–4357.

Joshi, M. N., et al., (2014). Metagenomics of petroleum muck: Revealing microbial diversity and depicting microbial syntrophy. *Arch. Microbiol., 196*(8), 531–544.

Jung, J., Philippot, L., & Park, W., (2016). Metagenomic and functional analyses of the consequences of reduction of bacterial diversity on soil functions and bioremediation in diesel-contaminated microcosms. *Sci. Rep., 6*(1), 1–10.

Khodakova, A. S., et al., (2014). Random whole metagenomic sequencing for forensic discrimination of soils. *PLoS One, 9*(8), 104996.

Kiemer, L., & Cesareni, G., (2007). Comparative interactomics: Comparing apples and pears? *Trends Biotechnol., 25*(10), 448–454.

Kim, S. J., et al., (2007). Complete and integrated pyrene degradation pathway in *Mycobacterium vanbaalenii* PYR-1 based on systems biology. *J. Bacteriol., 189*(2), 464–472.

Kim, S. J., Kweon, O., & Cerniglia, C. E., (2009). Proteomic applications to elucidate bacterial aromatic hydrocarbon metabolic pathways. *Curr. Opin. Microbiol., 12*(3), 301–309.

Kimes, N. E., et al., (2013). Metagenomic analysis and metabolite profiling of deep-sea sediments from the Gulf of Mexico following the Deepwater Horizon oil spill. *Front. Microbiol., 4*(50).

Kiyohara, H., et al., (1982). Phenanthroene-degrading phenotype of *Alcaligenes faecalis* AFK2. *Appl. Environ. Microbiol., 43*(2), 454–457.

Kiyohara, H., et al., (1990). Characterization of a phenanthrene degradation plasmid from *Alcaligenes faecalis* AFK2. *J. Ferm. Bioeng., 69*(1), 54–56.

Klatt, J. M., et al., (2016). Structure and function of natural sulfide-oxidizing microbial mats under dynamic input of light and chemical energy. *ISME J., 10*(4), 921–933.

Kleindienst, S., et al., (2014). Diverse sulfate-reducing bacteria of the *Desulfosarcina/ Desulfococcus* clade are the key alkane degraders at marine seeps. *ISME J., 8*(10), 2029–2044.

Köhler-Rink, S., & Kühl, M., (2000). Microsensor studies of photosynthesis and respiration in larger symbiotic foraminifera. I. The physicochemical microenvironment of *Marginopora vertebralis, Amphistegina lobifera* and *Amphisorus hemprichii. Mar. Biol., 137*(3), 473–486.

Kostka, J. E., et al., (2011). Hydrocarbon-degrading bacteria and the bacterial community response in Gulf of Mexico beach sands impacted by the Deepwater Horizon oil spill. *Appl. Environ. Microbiol., 77*(22), 7962–7974.

Kühl, M., & Jørgensen, B. B., (1992). Microsensor measurements of sulfate reduction and sulfide oxidation in compact microbial communities of aerobic biofilms. *Appl. Environ. Microbiol., 58*(4), 1164–1174.

Kühl, M., & Revsbech, N. P., (2001). Biogeochemical microsensors for boundary layer studies. In: Boudreau, B., & Jorgensen, B. B., (eds.), *The Benthic Boundary Layer* (pp. 180–210). Oxford University Press, New York.

Kühl, M., et al., (1998). A H_2S microsensor for profiling biofilms and sediments: Application in an acidic lake sediment. *Aquat. Microb. Ecol., 15*(2), 201–209.

Kumavath, R. N., & Pratap, D., (2012). Omics in bioremediation: A consolidated overview. *Biotechnol. Ind. J., 6*, 242–253.

Kurkela, S., et al., (1988). Cloning, nucleotide sequence and characterization of genes encoding naphthalene dioxygenase of *Pseudomonas putida* strain NCIB9816. *Gene, 73*(2), 355–362.

Lamendella, R., et al., (2014). Assessment of the Deepwater Horizon oil spill impact on Gulf coast microbial communities. *Front. Microbiol., 5*(130).

Laurie, A. D., & Lloyd-Jones, G., (2000). Quantification of *phnAc* and *nahAc* in contaminated New Zealand soils by competitive PCR. *Appl. Environ. Microbiol., 66*(5), 1814–1817.

Liang, Y., et al., (2009). Microarray-based functional gene analysis of soil microbial communities during ozonation and biodegradation of crude oil. *Chemosphere, 75*(2), 193–199.

Liang, Y., Li, G., et al., (2009). Microarray-based analysis of microbial functional diversity along an oil contamination gradient in oil field. *FEMS. Microbiol. Ecol., 70*(2), 324–333.

Liu, H., Yu, T., & Liu, Y., (2015). Sulfate-reducing bacteria and their activities in oil sands process-affected water biofilm. *Sci. Total. Environ., 536*, 116–122.

Liu, Q., et al., (2015). Distribution of petroleum degrading genes and factor analysis of petroleum-contaminated soil from the Dagang Oilfield, China. *Sci. Rep., 5*(1), 1–12.

Lu, Z. M., et al., (2012). Microbial gene functions enriched in the Deepwater Horizon deep-sea oil plume. *ISME J., 6*(2), 451–460.

Lueders, T., & Friedrich, M. W., (2003). Evaluation of PCR amplification bias by terminal restriction fragment length polymorphism: Analysis of small-subunit rRNA and mcrA genes by using defined template mixtures of methanogenic pure cultures and soil DNA extracts. *Appl. Environ. Microbiol., 69*(1), 320–326.

Mason, O. U., et al., (2012). Metagenome, metatranscriptome and single-cell sequencing reveal microbial response to Deepwater Horizon oil spill. *ISME J., 6*(9), 1715–1727.

Mason, O. U., et al., (2014). Metagenomics reveals sediment microbial community response to Deepwater Horizon oil spill. *ISME J., 8*(7), 1464–1475.

Mbadinga, S. M., et al., (2011). Microbial communities involved in anaerobic degradation of alkanes. *Inter. Biodeterior Biodegradation, 65*(1), 1–13.

Mbadinga, S. M., et al., (2012). Analysis of alkane-dependent methanogenic community derived from production water of a high-temperature petroleum reservoir. *Appl. Microbiol. Biotechnol., 96*(2), 531–542.

Militon, C., et al., (2016). Metatranscriptomes of oil-contaminated marine coastal sediment affected by oil addition and/or by the bioturbating activity of the marine polychaete *Hediste diversicolor*: Who are the microbial players?. *Mar. Genomics, 29*, 55–59.

Mishra, S., et al., (2017). Hydrocarbon degradation in Caspian Sea sediment cores subjected to simulated petroleum seepage in a newly designed sediment-oil-flow-through system. *Front. Microbiol., 8*, 763.

Moran, M. A., et al., (2013). Sizing up metatranscriptomics. *ISME J., 7*(2), 237–243.

Müller, H., et al., (2016). Long-distance electron transfer by cable bacteria in aquifer sediments. *ISME J., 10*(8), 2010–2019.

Musat, F., Wieland, A., & Widdel, F., (2004). Marine sediment with surface contamination by oil in microcosms for microbiological studies. *Ophelia, 58*(3), 217–222.

Nagao, K., Takizawa, N., & Kiyohara, H., (1988). Purification and properties of cis-phenanthrene dihydrodiol dehydrogenase in *Alcaligenes faecalis* AFK2. *Agr. Biol. Chem., 52*(10), 2621–2623.

Nasser, B., Ramadan, A. R., Hamzah, R. Y., Mohamed, M. E., & Ismail, W. A., (2017). Detection and quantification of sulfate-reducing and polycyclic aromatic hydrocarbon-degrading bacteria in oilfield using functional markers and quantitative PCR. *J. Pet. Environ. Biotechnol., 8*(348), 2.

Nazina, T. N., Rozanova, E. P., & Kuznetsov, S. I., (1985). Microbial oil transformation processes accompanied by methane and hydrogen-sulfide formation. *Geomicrobiol. J., 4*(2), 103–130.

Orphan, V. J., et al., (2000). Culture-dependent and culture-independent characterization of microbial assemblages associated with high-temperature petroleum reservoirs. *Appl. Environ. Microbiol., 66*(2), 700–711.

Ozsolak, F., et al., (2009). Direct RNA sequencing. *Nature, 461*(7265), 814–818.

Park, J. W., & Crowley, D. E., (2006). Dynamic changes in nahAc gene copy numbers during degradation of naphthalene in PAH-contaminated soils. *Appl. Microbiol. Biotechnol., 72*(6), 1322–1329.

Peng, R. H., et al., (2008). Microbial biodegradation of polyaromatic hydrocarbons. *FEMS Microbiol. Rev., 32*(6), 927–955.

Powell, S. M., et al., (2006). Using real-time PCR to assess changes in the hydrocarbon-degrading microbial community in Antarctic soil during bioremediation. *Microb. Ecol., 52*(3), 523–532.

Powell, S. M., et al., (2010). The importance of soil characteristics to the structure of alkane degrading bacterial communities on sub-Antarctic Macquarie Island. *Soil Biol. Biochem., 42*(11), 2012–2021.

Pringault, O., et al., (2015). Contrasted effects of natural complex mixtures of PAHs and metals on oxygen cycle in a microbial mat. *Chemosphere, 135*, 189–201.

Quinn, R. A., et al., (2016). From sample to multi-omics conclusions in under 48 hours. *mSystems, 1*(2).

Revsbech, N. P., & Jørgensen, B. B., (1983). Photosynthesis of benthic microflora measured with high spatial resolution by the oxygen micro profile method: Capabilities and limitations of the method. *Limnol. Oceanogr., 28*(4), 749–756.

Revsbech, N. P., & Jørgensen, B. B., (1986). Microelectrodes: Their use in microbial ecology. *Adv. Microb. Ecol., 9*, 293–352.

Revsbech, N. P., (1989). An oxygen microelectrode with a guard cathode. *Limnol. Oceanogr., 34*, 474–478.

Revsbech, N. P., et al., (2016). In situ hydrogen dynamics in a hot spring microbial mat during a diel cycle. *Appl. Environ. Microbiol., 82*(14), 4209–4217.

Rivers, A. R., et al., (2013). Transcriptional response of bathypelagic marine bacterioplankton to the Deepwater Horizon oil spill. *ISME J., 7*(12), 2315–2329.

Rodríguez-Martínez, E. M., et al., (2006). Microbial diversity and bioremediation of a hydrocarbon-contaminated aquifer (VegaBaja, Puerto Rico). *Int. J. Environ. Res. Public Health., 3*(3), 292–300.

Rojo, F., (2009). Degradation of alkanes by bacteria. *Environ. Microbiol., 11*(10), 2477–2490.

Rueter, P., et al., (1994). Anaerobic oxidation of hydrocarbons in crude oil by new types of sulfate-reducing bacteria. *Nature, 372*(6505), 455–458.

Saito, A., Iwabuchi, T., & Harayama, S., (2000). A novel phenanthrene dioxygenase from *Nocardioides* sp. Strain KP7: Expression in *Escherichia coli. J. Bacteriol., 182*(8), 2134–2141.

Salam, L. B., et al., (2017). Metagenomic insights into effects of spent engine oil perturbation on the microbial community composition and function in a tropical agricultural soil. *Environ. Sci. Pollut. Res., 24*(8), 7139–7159.

Schell, M. A., (1985). Transcriptional control of the *nah* and *sal* hydrocarbon-degradation operons by the *nahR* gene product. *Gene, 36*(3), 301–309.

Schutte, C. A., Teske, A., MacGregor, B. J., Salman-Carvalho, V., Lavik, G., Hach, P., & De Beer, D., (2018). Filamentous giant beggiatoaceae from Guaymas basin are capable of both denitrification and dissimilatory nitrate reduction to ammonium (DNRA). *Appl. Environ. Microbiol., 84*(15), e02860–17.

Sei, Y., et al., (1999). Design of PCR primers and gene probes for the general detection of bacterial populations capable of degrading aromatic compounds via catechol cleavage pathways. *J. Biosci. Bioeng., 88*(5), 542–550.

Shahi, A., et al., (2017). Assessment of the horizontal transfer of functional genes as a suitable approach for evaluation of the bioremediation potential of petroleum-contaminated sites: A mini-review. *Appl. Microbiol. Biotechnol., 101*(11), 4341–4348.

Shahsavari, E., et al., (2016). A quantitative PCR approach for quantification of functional genes involved in the degradation of polycyclic aromatic hydrocarbons in contaminated soils. *MethodsX, 3*, 205–211.

Sierra-García, I. N., et al., (2014). New hydrocarbon degradation pathways in the microbial metagenome from Brazilian petroleum reservoirs. *PLoS One, 10*(1371).

Simon, M. J., et al., (1993). Sequences of genes encoding naphthalene dioxygenase in *Pseudomonas putida* strains G7 and NCIB9816-4. *Gene, 127*(1), 31–37.

Singh, O. V., & Nagaraj, N. S., (2006). Transcriptomics, proteomics and interactomics: Unique approaches to track the insights of bioremediation. *Brief Func. Genomics, 4*(4), 355–362.

Smits, T. H. M., et al., (2002). Functional analysis of alkane hydroxylases from gram-negative and gram-positive bacteria. *J. Bacteriol., 184*(6), 1733–1742.

Stefani, F. O., et al., (2015). Culture-dependent and independent methods capture different microbial community fractions in hydrocarbon-contaminated soils. *PLoS One, 10*(6).

Stenuit, B., et al., (2008). Emerging high-throughput approaches to analyze bioremediation of sites contaminated with hazardous and/or recalcitrant wastes. *Biotechnol. Adv., 26*, 561–575.

Takizawa, N., et al., (1994). Identification and characterization of genes encoding polycyclic aromatic hydrocarbon dioxygenase and polycyclic aromatic hydrocarbon dihydrodiol dehydrogenase in *Pseudomonas putida* OUS82. *J. Bacteriol., 176*(8), 2444–2449.

Takizawa, N., et al., (1999). Nucleotide sequences and characterization of genes encoding naphthalene upper pathway of *Pseudomonas aeruginosa* PaK1 and *Pseudomonas putida* OUS82. *J. Biosci. Bioeng., 87*(6), 721–731.

Tani, A., et al., (2001). Gene structures and regulation of the alkane hydroxylase complex in *Acinetobacter* sp. strain M-1. *J. Bacteriol., 183*(5), 1819–1823.

Uhlik, O., et al., (2013). Stable isotope probing in the metagenomics era: A bridge towards improved bioremediation. *Biotechnol. Adv., 31*(2), 154–165.

Van, N. J. D., Zhili, H., & Jizhong, Z., (2012). Use of functional gene arrays for elucidating in situ biodegradation. *Front. Microbiol., 3*, 339.

Van, S. N. M., & Roelofs, D., (2006). *An Introduction to Ecological Genomics*. OUP Oxford.

Verde, L. C. L., et al., (2013). Diversity of hydrocarbon-related catabolic genes in oil samples from Potiguar basin (Rn, Brazil). *J. Phylogenetics Evol. Biol., 4*(138), 2.

Vilchez-Vargas, R., et al., (2013). Analysis of the microbial gene landscape and transcriptome for aromatic pollutants and alkane degradation using a novel internally calibrated microarray system. *Environ. Microbiol., 15*(4), 1016–1039.

Villas-Bôas, S. G., & Bruheim, P., (2007). The potential of metabolomics tools in bioremediation studies. *OMICS, 11*(3), 305–313.

Wang, D. Z., et al., (2016). Environmental microbial community proteomics: Status, challenges and perspectives. *Int. J. Mol. Sci., 17*(8), 1275.

Wasmund, K., et al., (2009). Novel alkane hydroxylase gene (*alkB*) diversity in sediments associated with hydrocarbon seeps in the Timor Sea, Australia. *Appl. Environ. Microb., 75*(23), 7391–7398.

Weckwerth, W., & Fiehn, O., (2002). Can we discover novel pathways using metabolomic analysis? *Curr. Opin. Biotechnol., 13*(2), 156–160.

Widdel, F., & Grundmann, O., (2010). Biochemistry of the anaerobic degradation of non-methane alkanes. In: Timmis, K. N., (ed.), *Handbook of Hydrocarbon and Lipid Microbiology*. Berlin, Heidelberg: Springer Berlin Heidelberg.

Widdel, F., & Rabus, R., (2001). Anaerobic biodegradation of saturated and aromatic hydrocarbons. *Curr. Opin. Biotechnol., 12*(3), 259–276.

Wiechert, W., et al., (2007). Fluxomics: Mass spectrometry versus quantitative imaging. *Curr. Opin. Plant Biol., 10*(3), 323–330.

Wieland, A., & Kühl, M., (2000). Short-term temperature effects on oxygen and sulfide cycling in a hypersaline cyanobacterial mat (Solar Lake, Egypt). *Mar. Ecol. Prog. Ser., 196*, 87–102.

Wieland, A., et al., (2005). Biogeochemistry of an iron-rich hypersaline microbial mat (Camargue, France). *Microb. Ecol., 49*(1), 34–49.

Wong, H. L., et al., (2015). Niche differentiation of bacterial communities at a millimeter-scale in Shark Bay microbial mats. *Sci. Rep., 5*(15607).

Yang, Y., et al., (2015). Abundance and diversity of soil petroleum hydrocarbon-degrading microbial communities in oil exploring areas. *Appl. Microbiol. Biotechnol., 99*(4), 1935–1946.

Yen, K. M., & Gunsalus, I. C., (1985). Regulation of naphthalene catabolic genes of plasmid NAH7. *J. Bacteriol., 162*(3), 1008–1013.

Yergeau, E., Arbour, M., & Brousseau, R., (2009). Microarray and real-time PCR analyses of the response of high-arctic soil bacteria to hydrocarbon pollution and bioremediation treatments. *Appl. Environ. Microb., 75*(19), 6258–6267.

Yergeau, E., et al., (2012). Metagenomic analysis of the bioremediation of diesel-contaminated Canadian high arctic soils. *PloS One, 7*(1).

Zarraonaindia, I., Smith, D. P., & Gilbert, J. A., (2013). Beyond the genome: Community-level analysis of the microbial world. *Biol. Philos., 28*(2), 261–282.

Zedelius, J., et al., (2010). Alkane degradation under anoxic conditions by a nitrate-reducing bacterium with possible involvement of the electron acceptor in substrate activation. *Environ. Microbiol. Rep., 3*(1), 125–135.

Zhang, S., Hu, Z., & Wang, H., (2018). A retrospective review of microbiological methods applied in studies following the Deepwater Horizon oil spill. *Front. Microbiol., 9*(520).

Zhang, W., Li, F., & Nie, L., (2010). Integrating multiple 'omics' analysis for microbial biology: Application and methodologies. *Microbiology, 156*(2), 287–301.

Zhou, H. W., et al., (2006). Genetic diversity of dioxygenase genes in polycyclic aromatic hydrocarbons-degrading bacteria isolated from mangrove sediments. *FEMS Microbiol. Lett., 262*(2), 148–157.

CHAPTER 6

Bacterial Chemotaxis, Current Knowledge and Future Need for Application in Hydrocarbon Remediation Technologies

NISENBAUM MELINA,[1] GEORGINA CORTI-MONZÓN,[1] and SILVIA E. MURIALDO[2]

[1]*Institute of Food Science and Technology and Environment (INCITAA), National University of Mar del Plata (UNMDP), CONICET. J.B. Justo–4302, Mar del Plata, Bs. As., Argentina*

[2]*Institute of Food Science and Technology and Environment (INCITAA), National University of Mar del Plata (UNMDP), CIC. J.B. Justo–4302, Mar del Plata, Bs. As., Argentina*

6.1 INTRODUCTION

Responding to changes in the environment is especially important for unicellular organisms, which directly interact with it (Pandey et al., 2002a). Motile bacteria have well-established sensing mechanisms to convert chemical stimuli to mechanical energy, which allows them to approach attractants and avoid repellents (Celani et al., 2010). This essential behavior is called chemotaxis. Bacterial chemotaxis allows microorganisms to locate to more advantageous niches for growth and survival (Pandey et al., 2002a; Parales et al., 2002; Ford et al., 2007). In many cases, positive chemotaxis serves to bring them closer to favorable attractive substances (chemoattractants), while negative chemotaxis moves them away from compounds that are toxic to bacteria (chemo-repellents) (Pandey et al., 2002a). Chemotaxis generally occurs against metabolizable compounds, which generates a concentration gradient of the carbon source. In some cases, chemotaxis towards non-metabolizable compounds has been observed. The ability of bacteria to modulate movement direction is the outcome of either controlled changes in the direction of flagellar rotation (Lacal et al., 2010), or by stopping and starting the flagellar motor at intervals. It is mediated by chemoreceptors expressed on the cell surface that can sense chemical species and their concentrations through specific receptor-ligand interactions, that in turn activates intracellular signaling cascades (Bray et al. 1998; Gestwicki and Kiessling, 2002). Some studies have shown that increased number of chemoreceptors in many free-living bacteria enables them to respond to a wider range of compounds, suggesting their potential to detect and move toward a wide variety of different pollutants (Sampedro et al., 2015). Although our current knowledge of chemoreceptor specificities is limited, recent research of receptor ligands promises a better understanding of the relation between the chemoeffect and bacterial habitats (Bi et al., 2018).

Our entire planet is under great stress, and pollution, and hydrocarbons (HCs), remains one of the major threats to environmental sustainability. In 2002, the annual global input of petroleum oils to the sea was estimated to be 1.3 million metric tons (Transportation Research Board, 2003), the majority of which is derived from anthropogenic sources. Therefore, the main challenge for remediation processes is to find fast, specific, low-cost, and low-impact methods to remove contaminants of concern. Crude oil is estimated to be composed of up to 20,000 chemical components (Marshall et al., 2004), consisting mainly of alkanes with different chain lengths and branching patterns, cycloalkanes, mono-aromatic, and polycyclic aromatic hydrocarbons (PAHs). Crude oil composition varies widely, reflected by different physicochemical properties, bioavailability, and toxicity. Because of the abundance of these pollutants in the environment, and due to their adverse effects on health and ecosystems, it is important to understand how microorganisms degrade HCs, and to continue to improve tools to mitigate ecosystem damage by these potentially toxic compounds (McGenity et al., 2012). Biodegradation is the process by which organic substances are broken down into smaller compounds by enzymes produced by native living microbial organisms, and represents one of the primary mechanisms by which petroleum and other HC pollutants are eliminated from the environment (Leahy et al., 1990). Some bacteria also have the ability to locate HCs by chemotaxis, which gives them a selective advantage to survive and grow in contaminated sites and, as a consequence, for reducing pollution in these areas (Grimm et al., 1997; Samanta et al., 2000a; Pandey et al., 2002a; Parales et al., 2002). In this context, chemotaxis may play an important role in terrestrial and aquatic ecosystems (Ford et al., 2007), and chemotaxis has been demonstrated to enhance bioremediation of a number of contaminants (Grimm et al., 1997; Samanta et al., 2000a; Pandey et al., 2002a). The use of chemotactic bacteria in bioremediation may improve the efficiency, and decrease the cost, of restoration of contaminated ecosystems (Wang et al., 2018).

Certain HC-degrading bacteria exhibit positive chemotaxis towards both halogenated and non-halogenated HC substrates (Grimm et al., 1997; Law et al., 2003; Bhushan et al., 2004). Chemoattraction has been observed for biphenyl, benzoic acid and chlorobenzoic acids (Tremaroli et al., 2010; Corti-Monzón et al., 2018), toluene, and its derivatives (Parales et al., 2000; Lacal et al., 2011), naphthalene, and its derivatives (Grimm et al., 1997; Lacal et al., 2011), nitroaromatics (Iwaki et al., 2007), chloroaromatics (Gordillo et al., 2007; Okada et al., 2017), chloronitroaromatics (Pandey et al., 2012), aminoaromatics (Parales, 2004), explosives (Leungsakul et al., 2005), aliphatic HCs (Lanfranconi et al., 2003; Nisenbaum et al., 2013), kerosene (Nisenbaum et al., 2013), and herbicides (Liu et al., 2009). Given the toxic potential of most pollutants, it is conceivable that bacteria have also evolved chemorepellent responses (Krell et al., 2013) to enhance survival. For example, negative chemotaxis away from anthracene and pyrene (Ortega-Calvo et al., 2003), and 1-nonene, 1-decene and 1-undecene (Nisenbaum et al., 2016), has been reported. Specific chemicals may even be chemoattractants for one bacterial species and repellents for another, and probably depends on the underlying physiology and metabolic potential of the organism (Parales et al., 2000; Shitashiro et al., 2003; Vardar et al., 2005). The physical state of the chemical also seems to influence the type of response. For example, *Pseudomonas putida* G7 is repelled by naphthalene in the vapor phase, while it is attracted by this compound in the aqueous phase (Hanzel et al., 2010). That said, an observed chemotaxis phenotype may be the result of the action of several potential antagonist chemoreceptors that differ in their sensitivity to a given compound (Lacal, 2017). This suggests that understanding how chemotactic responses are orchestrated may be important to enhance HC biodegradation or design chemotaxis-assisted biodegradation strategies. This chapter summarizes the main methods of detecting chemotaxis, the main HC chemotactic strains already reported, and the potential of chemotaxis to increase HC bioavailability and biodegradation efficiency.

6.2 CHEMOTAXIS DETECTION METHODS

Both qualitative and quantitative methods to measure bacterial chemotaxis have been developed to monitor either the movement of populations of cells or individual cells (Table 6.1; Jin and Herald, 2009). The detection of positive chemotactic response of bacterial populations often relies on visualization of the accumulation of cells near an attractant source. Traditional qualitative assays are generally simple, economic, and easy to use, while others require considerably more sophisticated instrumentation. Some tests are more appropriate for the analysis of HC taxis than others, since many of these compounds are volatile. In this section we provide a brief review of the standard assays in current use.

Swarm agar plates consist of laboratory glass plates (Petri dishes) filled with chains of extended polymers permeated by water-filled channels (0.4–0.8% -semisolid agar) (Jin et al., 2009), which contains a metabolizable attractant. A drop of cell culture is placed on the center of the plate and, as the colony grows, the concentration of attractant decreases. A spatial concentration gradient results, and the cells migrate outward in response to that gradient, forming a sharp ring where the cells congregate. This response can be seen by trans-illuminating the plates with white light (Parkinson, 2007) or with a higher sensitivity dynamic speckle laser (Murialdo et al., 2009; Nisenbaum et al., 2013, 2014). Problems can arise with this qualitative assay when using volatile attractants that evaporate and re-dissolve in the medium, interfering with gradient formation (Timmis, 2010).

The gradient plate test is a variation of the swarm agar plate (Pham et al., 2011). Cells grow on an alternative carbon source present in the soft agar as they respond to the attractant that diffuses from an agar plug placed in a spot on soft agar. This test is qualitative and does not require metabolism of the compound being tested.

In the drop assay (Grimm et al., 1997), cells are suspended in a viscous solution in a small Petri dish and the attractant is dropped in the center of the cell suspension. A gradient is formed by diffusion from the drop, and a ring of cells forms around the attractant in a short time. This is the main difference with the swarm assay in which cell growth over hours or days is required prior to ring detection.

In the capillary assay (Adler, 1966), a 1-mm capillary sealed at one end is filled from the open end with an attractant prior to being inserted into a chamber loaded with a suspension of highly motile bacteria in chemotaxis buffer. Chemotactic cells respond by swimming up the gradient of attractant diffusing out of the tube. In the quantitative assay, cells that enter into the capillary are enumerated by plating on nutrient agar and calculated by colony count. In the qualitative capillary assay, a solidifying agent is included in the capillary, preventing the cells from entering. This technique does not require cell growth, and works well with both volatile and non-metabolizable compounds.

With the qualitative agarose in-plug test (Yu et al., 1997) the accumulation of bacteria is visualized as a ring around an agarose plug, usually in a few minutes. A drop of attractant in agar is placed between a microscope slide and a coverslip, and a suspension of motile cells surrounds the plug. This test does not require cell growth and is useful for testing responses to volatile compounds such as low molecular weight HCs, because the response is fast and the chamber is almost completely closed (Parales et al., 2000).

Video processing analysis methods, recently presented by Nisenbaum et al. (Nisenbaum et al., 2016; Okada et al., 2017), can be used by combining microscopy with traditional qualitative methods such as the agarose insertion assay, the capillary assay and the drop assay, to capture a range of qualitative parameters. This allows for analysis of low cell numbers thanks to high sensitivity image processing of only moving objects.

In recent years, chemotaxis studies with microfluidic techniques have greatly advanced. Compared to traditional detection methods, microfluidic assays may allow for manipulation and characterization of gradients with high accuracy (Ahmed et al., 2010), afford precise control over the bacterial microenvironment, increase detection sensitivity, and can be used qualitatively and quantitatively. With this technique, Stocker et al. (2008) demonstrated bacterial migration in response to different nutrients in a microfluidic device with environmentally realistic dimensions and dynamics. Classified into two categories, microfluidic devices are for either continuous flow or static diffusion assays. The continuous flow-based methods allow the user to generate different types of chemo-effector gradients using laminar flow and diffusion mixing (Englert et al., 2009; Jeong et al., 2010, 2013b; Si et al., 2012; Wu et al., 2013b). However, bacterial motility can be disturbed by shear forces induced by high flow (Marcos et al., 2012), requiring the use of external pumps and flow regulators. Diffusion-based methods provide chemical gradients without flow, require more time to deploy, but rely on less external equipment and do not interfere with bacterial motility. Recently, Jeong et al. (2013) presented a pump-less microfluidic device that solves the problem of slow diffusion of chemoeffectors and possibly the disturbance of bacterial motility driven by high flow rate. Microfluidic devices can provide a convenient tool for laboratory experiments and could play important roles in real applications in the near future (Zang et al., 2017).

Except for the microfluidic techniques, the chemotaxis protocols described so far require simple equipment to observe chemotactic responses of bacterial populations. Temporal assays, along with computer-assisted motion analysis, allow quantitative evaluation of individual cell motility and chemotaxis (Harwood et al., 1990). Advances in quantitative analysis methods have allowed for the development of dynamic imaging capabilities at the scales relevant to microbial processes ranging from single-cell growth to population motility (Son et al., 2015). The combination of optical microscopy, digital imaging and image analysis has led to mechanistic discoveries and has enabled quantification of microbial processes.

The tracking of individual swimming bacteria was first tested by Berg (2000). In this case, bacteria are tracked in two dimensions for relatively short periods, and the rate of change of direction can be measured in response to chemicals or light. Valderrama-Gómez et al. (2020) recently developed motion analysis software, called TaxisPy, for the quantitative analysis of bacterial chemotaxis. The software can be easily accessed through Docker, an open-source project, on any operating system. Using a video of freely swimming cells as input, TaxisPy estimates the culture's average tumbling frequency over time. Even though multiple studies have used two-dimensional tracking of individual bacterial cells (Thar et al., 2000; Davis et al., 2011; Meijering et al., 2012; Vater et al., 2014; Taute et al., 2015), to date, no computer software for such analyzes is freely available.

Another method for evaluation of individual cell chemotaxis is the Tethered assay, where cells have conventionally been attached to a surface with antibodies raised against flagellar filaments (Silverman et al., 1974). The fraction of time spent in clockwise or counterclockwise rotation provides a quantitative measure of the rate of change of cell direction, visualized under an optical microscope. This assay provides a method for studies of bacterial chemotaxis mechanisms and for testing of theoretical models. A computerized image processing system can be used in real-time to quantify rotations (Poole et al., 1988; Lopez-de-Victoria et al., 1993; Cai et al., 2016).

The standard chemotaxis assays do not take into account sorption of the pollutants to soils and sediments, a detail that must be taken into account in natural environments that contain mixtures of xenobiotics. Research-based on close to *in situ* conditions has shown that pollutant chemotaxis can occur in contaminated soil (Paul et al., 2006) and in a reconstituted bench-scale microcosms (Strobel et al., 2011). These investigations contribute to understanding the potential of chemotaxis to improve the transport of bacteria to contaminating sources in more complex and larger-scale systems. Bench-scale experiments are an important intermediate step towards quantifying the impact of chemotaxis on the overall effectiveness of *in situ* bioremediation. Furthermore, including the chemotaxis in mathematical bioremediation models can improve our time scale predictions in bioremediation scenarios, allowing a more accurate assessment of the natural attenuation potential as a remedial alternative as well as optimized management of contaminated sites (Strobel et al., 2011). So far, studies that examine the differences in chemotactic responses in the presence of xenobiotic mixtures are lacking, which actually occur in natural environments and by anthropogenic effect.

TABLE 6.1 Standard Chemotaxis Detection Methods

	Method	Qualitative	For Volatiles	Need Cell Growth	References
Population Analysis	Swarming assay	No	Not adequate	Yes	Jin et al. (2009)
	Gradient plate assay	No	Yes	No	Pham et al. (2011)
	Drop assay	No	Yes	No	Grimm et al. (1997)
	Capillary assay	Yes	Yes	No	Adler (1966)
	Agarose in-plug	No	Yes	No	Yu et al. (1997)
	Microfluidic techniques	Yes	Yes	No	Si et al. (2012); Jeong et al. (2013a); Wu et al. (2013a)
Single Cell Analysis	Tracking	Yes	Yes	No	Berg (2000); Meijering et al. (2012)
	Tethered assay	Yes	No	No	Silverman et al. (1974); Cai et al. (2016)

6.3 CHEMOTACTIC BACTERIA

Many free-living bacteria have been found to have a diversity of chemoreceptor types able to mediate chemotaxis to a wide range of compounds (Kato et al., 2008; Krell et al., 2011). In some cases, chemotaxis has only been observed towards biodegradable compounds, whereas structurally similar non-substrate compounds were not found to be chemoattractants (Lacal et al., 2013), suggesting a strong link between chemotaxis and catabolism. On a molecular level, Adler (1969) described two categories of chemotaxis on the basis of signal transduction strategies. One type of chemotaxis is independent of the metabolism of the chemoeffector molecule, whereas in metabolism-dependent chemotaxis, the degradation of the chemoeffector is a prerequisite. The sensors systems for metabolism-independent chemotaxis can be the binding of the chemoeffector by the transmembrane chemoreceptors or the transport of the chemoeffector by the phosphoenolpyruvate (PEP)-dependent carbohydrate phosphotransferase system (PTS). The following facts have been reported by Adler (1969) for transmembrane chemoreceptor bindings: (i) metabolizable attractants and their non-metabolizable analogs function as attractants, (ii) the chemotactic response is unaffected if the metabolic pathway of a chemical attractant is mutated, and (iii) bacteria are attracted to a certain non-metabolizable chemoattractant, even in the presence of other metabolizable compounds. Transport, but not metabolism, of the PTS substrates is required for a behavioral response in this last sensor system (Alexandre et al., 2001). Metabolism-independent chemical sensing is found in several bacterial species, including *E. coli*, *Salmonella* sp., *Bacillus subtilis, and Pseudomonas* sp. (Pandey et al., 2002a). Although these bacterial degradation capabilities in many cases have been correlated with chemotactic behavior for remediation of contaminated sites (Lacal et al., 2013), bacterial chemotaxis toward non-metabolizable pollutants has received less attention.

In contrast to the metabolism-independent behavior described above, some chemotactic responses in bacteria require metabolism of a chemical attractant. For this type of chemotaxis, it has been reported that: (i) there is no attraction towards non-metabolizable analogs of metabolizable attractants, (ii) chemotaxis towards a certain attractant disappears if the metabolism of that attractant is completely inhibited, and only (iii) the presence of an easily metabolizable chemical inhibits chemotaxis to other attractants. Furthermore, a direct correlation has been reported between the efficiency of a chemical as a growth substrate and as a chemoeffector (Alexandre et al., 2000). Chemotaxis towards metabolizable aromatic compounds has been widely documented for different bacteria (Table 6.2), including members of *Pseudomonas* (Grimm et al., 1997; Ortega-Calvo et al., 2003; Zhang et al., 2008; Nisenbaum et al., 2014; Meng et al., 2019), *Sphingomonas* (Pandey et al., 2002a; Samanta et al., 2002), *Ralstonia* (Bhushan et al., 2000; Samanta et al., 2000b; Paul et al., 2006), *Acidovorax* (Rabinovitch-Deere et al., 2012) and *Burkholderia* (Pandey et al., 2012), and common soil bacteria such as *Rhizobium* sp., *Bradyrhizobium* sp. (Parke et al., 1985; Dharmatilake et al., 1992), *Azospirillum* sp. (Lopez-de-Victoria et al., 1993) and *Comamonas* (Rabinovitch-Deere et al., 2012; Ni et al., 2013; Wang et al., 2019), among other species. Both metabolism-dependent and metabolism-independent chemotaxis, have been reported for *Comamonas* species (Ni et al., 2013, 2015; Huang et al., 2016).

Some studies (Clancy et al., 1981; Zhulin et al., 1997; Alexandre et al., 2000) suggested that metabolism-dependent chemotaxis shares the signaling pathway with other behavioral responses collectively known as energy taxis. Energy taxis are broadly defined as a behavioral response to stimuli that affect (Zhulin et al., 1993) electron transport and energy generation, such as air taxis, phototaxis, redox taxis, and taxis to alternative electron acceptors. Changes in the speed of electron transport (or a related parameter) act as the signal that gives rise to this type of behavior. A few energy taxis studies towards HCs and aromatic compounds have been reported (Sarand et al., 2008; Rabinovitch-Deere et al., 2012; Luu et al., 2013; Hughes et al., 2017). Taylor and Zhulin (1998) proposed that when bacteria sense a decrease in intracellular energy, energy taxis override chemotaxis, giving bacteria the opportunity to escape to an environment that supports optimal energy levels. This is a versatile response that can be triggered by any condition that impairs cellular energy.

Pseudomonas putida has been used as a model microorganism (Lacal, 2017) and its chemotaxis towards aromatic HCs is perhaps the most studied linkage between chemotaxis and HC biodegradation (Parales et al., 2000). This organism has been reported to be chemotactic to towards benzene, toluene, and ethylbenzene (Mosqueda et al., 1999; Parales et al., 2000), naphthalene (Tsuda et al., 1990), biphenyl (Grimm et al., 1997), ethylbenzene, isopropyl-benzene, trifluorotoluene, perchloroethylene, dichloroethylene, trichloroethylene (Lacal, 2017), and aromatic acids (Harwood et al., 1984). The chemotaxis toward some of these compounds was found to be induced during growth on toluene, suggesting a correlation between chemotaxis and their transformation as toluene dioxygenase substrates (Lacal, 2017). However, non-toluene-degrading mutants of *P. putida* remained chemotactic to toluene, indicating that toluene was directly detected as the attractant. Interestingly, it was reported that non-metabolizable 3- and 4-chlorobenzoate are attractants to *P. putida* previously grown on benzoate or 4-hydroxybenzoate (Harwood et al., 1990).

However, the ligand spectrum was found to be large; most of these ligands are not catabolic pathway substrates and the relevance of chemotaxis to these compounds is not clear (Lacal, 2017).

The bioremediation of HCs in saline environments can be carried out by technologies using halophilic bacteria. The *Halomonas* genus has been reported to be chemotactic to aromatic compounds such as phenol, naphthalene (Tena-Garitaonaindia et al., 2019), phthalate, phenanthrene, (Corti Monzón et al., 2018; Gasperotti et al., 2018), fluorene, and aromatic acids (Corti et al., 2018).

Chemotaxis towards alkanes has been less frequently reported. Alkane chemotaxis was observed in *Flavimonas oryzihabitans* (Lanfranconi et al., 2003) and *Alcanivorax*, including alkane sensing, chemotaxis, transport, and gene regulation (Wang et al., 2014; Fu et al., 2018), *Acinetobacter* (Fan et al., 2017), and *Pseudomonas* (Nisenbaum et al., 2013; Sampedro et al., 2015; Meng et al., 2017). *Pseudomonas aeruginosa* has been reported to be chemotactic towards *n*-hexadecane, 1-undecene, 1-dodecene, and 1-octene (Nisenbaum et al., 2013, 2014). Given the capability of this strain to degrade chlorophenols (Wolski et al., 2006; Durruty et al., 2011), and HCs such as kerosene (Nisenbaum et al., 2014), further research on this bacteria and its use of for HC remediation technologies is warranted.

TABLE 6.2 Some Bacterial Species Reported as Chemotactic and Hydrocarbon Degraders

Bacteria	Target Hydrocarbons	References
Azospirillum sp.	Benzoate Protocatechuate 4-hydroxybenzoate catechol	Lopez-de-Victoria et al. (1993)
Acinetobacter	n-hexadecane	Singer et al. (1984)
Burkholderia sp.	Nitroaromatics Chloronitroaromatics TNT Toluene	Leungsakul et al. (2005); Pandey et al. (2012)
Burkholderia cepacia	Chlorinated ethenes	Vardar et al. (2005)
Bradyrhizobium sp.	Shikimate Quinate Protocatechuate Vanillate	Parke et al. (1985)

TABLE 6.2 *(Continued)*

Bacteria	Target Hydrocarbons	References
Clostridium sp.	Hexahydro-1,3,5-trinitro-1,3,5-triazine (RDX) Octahydro-1,3,5,7-tetranitro-1,3,5,7-tetrazocine (HMX) 2,4,6,8,10,12-hexanitro-2,4,6,8,10,12-hexaazaisowurtzitane (CL-20)	Bhushan et al. (2004)
Comamonas testosteroni	2-nitrotoluene Benzoate 3-hydroxybenzoate 2,6-dihydroxybenzoate Phenol 2-hydroxybenzoate 4-hydroxybenzoate	Rabinovitch-Deere et al. (2012); Ni et al. (2013)
Flavimonas oryzi-habitans	Gas oil n-hexadecane	Lanfranconi et al. (2003)
Halomonas sp	Phenanthrene Gas oil	Gasperotti et al. (2015); Corti Monzón et al. (2018)
Pseudomonas aeruginosa	n-hexadecane Kerosene 1-undecene 1-dodecane 1-octene	Nisenbaum et al. (2013)
Pseudomonas fluorescens	2-nitrobenzoate	Iwaki et al. (2007)
Pseudomonas putida	Toluene Naphthalene Benzene Ethylbenzene Biphenyl and 4-nitrotoluene (4TN) 3- and 4-chlorobenzoate Trichloroethylene	Wackett et al. (1988); Harwood et al. (1990); Tsuda et al. (1990); Rhys-Williams et al. (1993); Grimm et al. (1997); Marx et al. (1999); Mosqueda et al. (1999); Parales et al. (2000)
Pseudomonas stutzeri	Carbon tetrachloride	Criddle et al. (1990)
Pseudomonas synxantha	*n*-hexadecane	Meng et al. (2017)
Pseudomonas pseudoalcaligenes	Biphenyl Benzoic acid Chlorobenzoic acids	Tremaroli et al. (2010)
Ralstonia sp	Nitroaromatics p-nitrophenol	Pandey et al. (2002b); Paul et al. (2006)
Rhizobium sp.	Shikimate	Parke et al. (1985); Dharmatilake et al. (1992)

6.4 THE ROLE OF CHEMOTAXIS IN HYDROCARBON (HC) POLLUTED ENVIRONMENTS

The major evolutionary reason for chemotactic movement is thought to be enhancement of the capacity to access compounds that serve as carbon and energy sources, or as electron acceptors (Krell et al., 2013). The exposure of bacteria to pollutants frequently induces chemoattraction or chemorepellent reactions. Research suggests that the capacity to degrade a toxic compound has co-evolved in some bacteria with the capacity to chemotactically react to it, and has possibly been selected as an advantageous behavior in bacteria under conditions of nutrient limitation (Pandey et al., 2002a). Pollutant chemoreceptors so far identified are encoded on degradation or resistance plasmids (Krell et al., 2013). In natural environments, microorganisms that are chemotactic towards a compound should be more efficient for bioremediation than non-chemotactic microorganisms, particularly in diffusion limited environments (Samanta et al., 2002) where bioavailability may be a major hurdle to overcome (Pandey et al., 2002a). For bacteria to have access to organic compounds, either the compounds must be dissolved in the aqueous phase or the bacteria must adhere to a water-non-aqueous-phase interface, possibly through biofilm formation. Chemotaxis has been shown to play an important role in biofilm formation in several microorganisms (O'Toole and Kolter 1998) and can also allow increased population density if the chemotactic response is large enough (Lauffenburger et al., 1982). Chemotaxis and motility might be required for bacteria within a developing biofilm to move along the surface, thereby facilitating biofilm growth and spread (Stelmack et al., 1999). This is particularly important for low solubility HCs, and biofilm-forming microorganisms that accumulate near or on an oil-water interface (Xhao et al., 2019) increase bioavailability of dispersed oil droplets.

Positive chemotaxis along contaminant gradients overcomes the mass-transfer limitations that impede the bioremediation process and increases the bioavailability of soil and water contaminants, as it improves the contact between degrader organisms and organic compounds. In saturated systems, chemotaxis may be important for bacterial accession to heterogeneously distributed contaminations (Velasco-Casal et al., 2008). Although chemotaxis requires at least some solubility of the target chemical (Ortega-Calvo et al., 2003), it is theoretically possible that poorly soluble contaminants cause chemotaxis via their gas-phase distribution at very low concentrations when present in unsaturated porous media (Nijland et al., 2010). Also, vapor-phase chemical gradients influence the chemotactic dispersal and the spatiotemporal distribution of contaminant degrading bacteria. Chemotactic dispersal hence can affect the dynamics of microbial ecosystems and the ecology of contaminant biodegradation (Nijland et al., 2010). The best example of chemotaxis-enhanced bioavailability of PAHs relates to the capacity of *Pseudomonas putida* G7 to degrade naphthalene (Lacal, 2017), which was found to exceed the substrate mass transfer within the aqueous phase, thus enhancing the rate of substrate acquisition (Marx et al., 2000). These studies provide evidence that chemotactic bacteria may be more effective for bioremediation applications, especially at sites where contaminants are unevenly distributed or adsorbed to soil particles.

Chemotaxis improves the mixing of bacteria with contaminating sources in regions of low permeability, which increases the consumption of pollutants, promotes an increase in the rate of cell growth and, therefore, accelerates the biodegradation process (Wang et al., 2016). Many studies have demonstrated the importance of chemotaxis towards HCs under various circumstances and, in recent years, more quantitative analyzes have been conducted on directed bacterial movement resulting from chemotactic response, and bacterial propensity to bind chemoattractants to chemotactic receptors (Ford et al., 2007).

The term 'biosorption' can be used to refer to the interactions between the microbial cell surface and hydrophobic materials including petroleum HCs. Meng et al. (2019) introduced the term 'chemotactic biosorption' to study the relationship between biosorption and degradation in the context of chemotaxis and a equation was proposed. In their study, a correlation was found between chemotactic biosorption and biodegradation for hexadecane, tetradecane, phenanthrene, and pyrene.

Chemotaxis is a relevant mobilizing factor for PAH-degrading rhizosphere bacteria targeting moderately hydrophobic PAHs such as phenanthrene, anthracene, and pyrene (Ortega-Calvo et al., 2003). Chemotaxis may be induced by root exudates or by contaminating HCs, which increasing contaminant removal by soil microbial communities. Therefore, one possible way to optimize PAH phytoremediation is the use of plants with chemotactic microbial inoculants.

Currently, the performance of pollutant-degrading microbial populations can be optimized *in vitro* by taking advantage of the significant experimental information that documents the beneficial effect of taxis on bioavailability (Krell et al., 2011, 2013). In the context of bioremediation, chemotaxis can work to improve bacterial residence times in areas of contamination and facilitate the transfer of catabolic genes through mobile bacteria in contaminated environments; these processes may promote the coexistence of biodegradation and chemotaxis genes useful for contaminant removal (Xhao et al., 2019). Although the use of chemotactic bacteria in bioremediation remains, in theory, an efficient and low-cost potential technology for restoring contaminated soils and groundwater, further work is needed prior to reliable full scale application (Wang et al., 2018).

6.5 CONCLUSIONS AND FUTURE RESEARCH DIRECTIONS

Biodegradation of HCs in natural and engineered environments is often not as efficient as expected due to limited bioavailability of these compounds. The research available so far has shown, theoretically, and at lab scale, that the chemotactic movement of bacteria increases bioavailability, which in turn has a beneficial role in bioremediation. However, the underlying mechanisms that control chemotaxis and absorption of HCs by bacterial cells remain largely unexplored. To date, how chemoreceptors perceive a wide variety of unconventional stimuli and how signals are transduced, as well as sensory spectra for the majority of chemotactic bacteria, are still unknown. Although the chemotactic response towards pure compounds has been relatively well studied *in vitro*, responses to xenobiotic mixtures, and responses of complex interacting microbial communities at field scale, remain to be explored. It is important to quantify the role of chemotaxis towards increasing HC bioavailability, given that models applicable to this type of behavior are available. In addition, the identification of bacterial bioreporters and an understanding of chemoreceptor gene transfer in the environment, could be used to develop new bioremediation technologies. Genetic engineering to introduce chemotactic genes into biodegrading bacterial strains is theoretically possible, but would have to overcome regulatory barriers prior to being deployed for *in situ* bioremediation.

Considering the diversity of HCs present in the environment, a richer knowledge of bacterial chemotaxis toward HCs is essential for the development of biosensors for the detection of contaminants useful for the management of remediation of contaminated environments. At present, new tools to explore real-time *in situ* chemotactic responses to predict HC bioavailability and bioremediation efficiency are needed.

KEYWORDS

- **bacteria**
- **bioavailability**
- **biodegradation**
- **chemotaxis**
- **detection**
- **hydrocarbons**

REFERENCES

Adler, J., (1966). Chemotaxis in bacteria. *Science, 153*(3737), 708–716.

Adler, J., (1969). Chemoreceptors in bacteria. *Science, 166*(3913), 1588–1597.

Ahmed, R. Z., Ahmed, N., & Gadd, G. M., (2010). Isolation of two *Kocuria* species capable of growing on various polycyclic aromatic hydrocarbons. *African J. Biotechnol., 9*(24), 3611–3617.

Alexandre, G., & Zhulin, I. B., (2001). More than one way to sense chemicals. *J. Bacteriol., 183*(16), 4681–4686.

Alexandre, G., Greer, S. E., & Zhulin, I. B., (2000). Energy taxis is the dominant behavior in *Azospirillum brasilense. J. Bacteriol., 182*(21), 6042–6048.

Berg, H. C., (2000). Motile behavior of bacteria. *Phys. Today, 53*(1), 24–29.

Bhushan, B., et al., (2004). Chemotaxis-mediated biodegradation of cyclic nitramine explosives RDX, HMX, and CL-20 by *Clostridium* sp. EDB2. *Biochem. Biophys. Res. Commun., 316*(3), 816–821.

Bhushan, B., Halasz, A., Thiboutot, S., Ampleman, G., & Hawari, J., (2000). Chemotaxis and biodegradation of 3-methyl- 4-nitrophenol by *Ralstonia* sp. SJ98. *Biochem. Biophys. Res. Commun.,* 275(*1*), 129–133.

Bi, S., & Sourjik, V., (2018). Stimulus sensing and signal processing in bacterial chemotaxis. *Curr. Opin. Microbiol., 45*, 22–29.

Bray, D., Levin, M. D., & Morton-Firth, C. J., (1998). Receptor clustering as a cellular mechanism to control sensitivity. *Nature, 393*(6680), 85–88.

Cai, Q., Li, Z., Ouyang, Q., Luo, C., & Gordon, V. D., (2016). Singly flagellated *Pseudomonas aeruginosa* chemotaxes efficiently by unbiased motor regulation. *MBio, 7*(2), e00013–16.

Celani, A., & Vergassola, M., (2010). Bacterial strategies for chemotaxis response. *Proc. Natl. Acad. Sci. U. S. A., 107*(4), 1391–1396.

Clancy, M., Madill, K. A., & Wood, J. M., (1981). Genetic and biochemical requirements for chemotaxis to L-proline in *Escherichia coli. J. Bacteriol., 146*(3), 902–906.

Corti, M. G., Nisenbaum, M., Herrera, S. M. K., & Murialdo, S. E., (2018). New findings on aromatic compounds' degradation and their metabolic pathways, the biosurfactant production and motility of the halophilic bacterium *Halomonas* sp. KHS3. *Curr. Microbiol., 75*(8), 1108–1118.

Criddle, C. S., DeWitt, J. T., Grbić-Galić, D., & McCarty, P. L., (1990). Transformation of carbon tetrachloride by *Pseudomonas* sp. strain KC under denitrification conditions. *Appl. Environ. Microbiol., 56*(11), 3240 LP–3246.

Davis, M. L., Mounteer, L. C., Stevens, L. K., Miller, C. D., & Zhou, A., (2011). 2D motility tracking of *Pseudomonas putida* KT2440 in growth phases using video microscopy. *J. Biosci. Bioeng., 111*(5), 605–611.

Dharmatilake, A. J., & Bauer, W. D., (1992). Chemotaxis of *Rhizobium meliloti* towards nodulation gene-inducing compounds from alfalfa roots. *Appl. Environ. Microbiol., 58*(4), 1153–1158.

Durruty, I., Okada, E., González, J. F., & Murialdo, S. E., (2011). Degradation of chlorophenol mixtures in a fed-batch system by two soil bacteria. *Water SA, 37*(4), 547–552.

Englert, D. L., et al., (2009). *Chemotaxis, 571*, 1–23.

Fan, X., Zhu, Y., Gu, P., Li, Y., Xiao, G., Song, D., Wang, Y., et al., (2017). Bacterial community compositions of propylene oxide saponification wastewater treatment plants. *RSC Adv., 7*, 22347–22352.

Ford, R. M., & Harvey, R. W., (2007). Role of chemotaxis in the transport of bacteria through saturated porous media. *Adv. Water Resour., 30*(6, 7), 1608–1617.

Fu, X., Lai, Q., Dong, C., Wang, W., & Shao, Z., (2018). Complete genome sequence of *Alcanivorax xenomutans* P40, an alkane-degrading bacterium isolated from deep seawater. *Mar. Genomics., 38*, 1–4.

Gasperotti, A. F., Revuelta, M. V., Studdert, C. A., & Seitz, M. K. H., (2018). Identification of two different chemosensory pathways in representatives of the genus *Halomonas*. *BMC Genomics., 19*, 266.

Gasperotti, A. F., Studdert, C. A., Revale, S., & Seitz, M. K. H., (2015). Draft genome sequence of *Halomonas* sp. KHS3, a polyaromatic hydrocarbon-chemotactic strain. *Genome Announc., 3*(2), e00020–15.

Gestwicki, J. E., & Kiessling, L. L., (2002). Inter-receptor communication through arrays of bacterial chemoreceptors. *Nature, 415*(6867), 81–84.

Gordillo, F., Chávez, F. P., & Jerez, C. A., (2007). Motility and chemotaxis of *Pseudomonas* sp. B4 towards polychlorobiphenyls and chlorobenzoates. *FEMS Microbiol. Ecol., 60*(2), 322–328.

Grimm, A. C., & Harwood, C. S., (1997). Chemotaxis of *Pseudomonas* spp. to the polyaromatic hydrocarbon naphthalene. *Appl. Environ. Microbiol., 63*(10), 4111–4115.

Hanzel, J., Harms, H., & Wick, L. Y., (2010). Bacterial chemotaxis along vapor-phase gradients of naphthalene. *Environ. Sci. Technol., 44*(24), 9304–9310.

Harwood, C. S., Parales, R. E., & Dispensa, M., (1990). Chemotaxis of *Pseudomonas putida* toward chlorinated benzoates. *Appl. Environ. Microbiol., 56*(5), 1501–1503.

Harwood, C. S., Rivelli, M., & Ornston, L. N., (1984). Aromatic acids are chemoattractants for *Pseudomonas putida*. *J. Bacteriol., 160*(2), 622–628.

Huang, Z., Ni, B., Jiang, C., Wu, Y., He, Y., Parales, R. E., & Liu, S., (2016). Direct sensing and signal transduction during bacterial chemotaxis toward aromatic compounds in *Comamonas testosteroni*. *Mol. Microbiol., 101*(2), 224–237.

Hughes, J. G., Zhang, X., Parales, J. V., Ditty, J. L., & Parales, R. E., (2017). *Pseudomonas putida* F1 uses energy taxis to sense hydroxycinnamic acids. *Microbiology, 163*(10), 1490–1501.

Iwaki, H., Muraki, T., Ishihara, S., Hasegawa, Y., Rankin, K. N., Sulea, T., Boyd, J., & Lau, P. C. K., (2007). Characterization of a pseudomonad 2-nitrobenzoate nitroreductase and its catabolic pathway-associated 2-hydroxylaminobenzoate mutase and a chemoreceptor involved in 2-nitrobenzoate chemotaxis. *J. Bacteriol., 189*(9), 3502–3514.

Jeong, H., Lee, S. H., Kim, J. M., Kim, H. E., Kim, Y. G., Yoo, J. Y., Chang, W. S., & Lee, C. S., (2010). Microfluidic monitoring of *Pseudomonas aeruginosa* chemotaxis under the continuous chemical gradient. *Biosens. Bioelectron., 26*(2), 351–356.

Jeong, H., Lee, S., & Lee, C., (2013a). Pump-less static microfluidic device for analysis of chemotaxis of *Pseudomonas aeruginosa* using wetting and capillary action. *Biosens. Bioelectron., 47*, 278–284.

Jin, T., & Hereld, D., (2009). *Chemotaxis: Methods and Protocols*. Springer Protoc. Human Press, Rockville, MD, USA.

Kato, J., Kim, H., Takiguchi, N., Kuroda, A., & Ohtake, H., (2008). *Pseudomonas aeruginosa* as a model microorganism for investigation of chemotactic behaviors in ecosystem. *J. Biosci. Bioeng., 106*(1), 1–7.

Krell, T., et al., (2011). Diversity at its best: Bacterial taxis. *Environ. Microbiol., 13*(5), 1115–1124.

Krell, T., Lacal, J., Reyes-Darias, J. A., Jimenez-Sanchez, C., Sungthong, R., & Ortega-Calvo, J. J., (2013). Bioavailability of pollutants and chemotaxis. *Curr. Opin. Biotechnol., 24*(3), 451–456.

Lacal, J., (2017). The potential of hydrocarbon chemotaxis to increase bioavailability and biodegradation efficiency. In: *Cellular Ecophysiology of Microbe: Hydrocarbon and Lipid Interactions, Handbook of Hydrocarbon and Lipid Microbiology* (pp. 1–14). Springer, Cham.

Lacal, J., Alfonso, C., Liu, X., Parales, R. E., Morel, B., Conejero-Lara, F., Rivas, G., et al., (2010). Identification of a chemoreceptor for tricarboxylic acid cycle intermediates: Differential chemotactic response towards receptor ligands. *J. Biol. Chem., 285*(30), 23126–23136.

Lacal, J., Muñoz-Martínez, F., Reyes-Darías, J. A., Duque, E., Matilla, M., Segura, A., Ortega-Calvo, J. J., et al., (2011). Bacterial chemotaxis towards aromatic hydrocarbons in *Pseudomonas*. *Environ. Microbiol., 13*(7), 1733–1744.

Lacal, J., Reyes-Darias, J. A., García-Fontana, C., Ramos, J. L., & Krell, T., (2013). Tactic responses to pollutants and their potential to increase biodegradation efficiency. *J. Appl. Microbiol., 114*(4), 923–933.

Lanfranconi, M. P., Alvarez, H. M., & Studdert, C. A., (2003). A strain isolated from gas oil-contaminated soil displays chemotaxis towards gas oil and hexadecane. *Environ. Microbiol., 5*(10), 1002–1008.

Law, A. M. J., & Aitken, M. D., (2003). Bacterial chemotaxis to naphthalene desorbing from a nonaqueous liquid. *Appl. Environ. Microbiol., 69*(10), 5968 LP–5973.

Leahy, J. G., & Colwell, R. R., (1990). Microbial degradation of hydrocarbons in the environment. *Microbiol. Rev., 54*(3), 305–315.

Leungsakul, T., Keenan, B. G., Smets, B. F., & Wood, T. K., (2005). TNT and nitroaromatic compounds are chemoattractants for *Burkholderia cepacia* R34 and *Burkholderia* sp. strain DNT. *Appl. Microbiol. Biotechnol., 69*(3), 321–325.

Liu, X., & Parales, R. E., (2009). Bacterial chemotaxis to atrazine and related s-triazines. *Appl. Environ. Microbiol., 75*(17), 5481–5488.

Lopez-de-Victoria, G., & Lovell, C. R., (1993). Chemotaxis of *Azospirillum* species to aromatic compounds. *Appl. Environ. Microbiol., 59*(9), 2951–2955.

Luu, R. A., Schneider, B. J., Ho, C. C., Nesteryuk, V., Ngwesse, S. E., Liu, X., Parales, J. V., et al., (2013). Taxis of *Pseudomonas putida* F1 toward phenylacetic acid is mediated by the energy taxis receptor Aer2. *Appl. Environ. Microbiol., 79*(7), 2416 LP–2423.

Marcos, Fu, H. C., Powers, T. R., & Stocker, R., (2012). Bacterial rheotaxis. *Proc. Natl. Acad. Sci., 109*(13), 4780 LP–4785.

Marshall, A. G., & Rodgers, R. P., (2004). Petroleomics: the next grand challenge for chemical analysis. *Acc. Chem. Res., 37*(1), 53–59.

Marx, R. B., & Aitken, M. D., (2000). Bacterial chemotaxis enhances naphthalene degradation in a heterogeneous aqueous system. *Environ. Sci. Technol., 34*(16), 3379–3383.

Marx, R., & Aitken, M., (1999). Quantification of chemotaxis to naphtalene by *Pseudomonas putida* G7. *Appl. Environ. Microbiol.,* 65(*7*), 2847–2852.

McGenity, T. J., Folwell, B. D., Mckew, B. A., & Sanni, G. O., (2012). Marine crude-oil biodegradation: A central role for interspecies interactions. *Aquat. Biosyst., 8*, 10.

Meijering, E., Dzyubachyk, O., & Smal, I., (2012). Methods for cell and particle tracking. In: Conn, P. M., (ed.), *Methods Enzymol. Imaging Spectrosc. Anal. Living Cells* (Vol. 504, pp. 183–200). Elsevier Inc.

Meng, L., Li, H., Bao, M., & Sun, P., (2017). Metabolic pathway for a new strain *Pseudomonas synxantha* LSH-7′: From chemotaxis to uptake of n-hexadecane. *Sci. Rep., 7*, 39068.

Meng, L., Li, W., Bao, M., & Sun, P., (2019). Great correlation: Biodegradation and chemotactic adsorption of *Pseudomonas synxantha* LSH-7' for oil-contaminated seawater bioremediation. *Water Res., 153*, 160–168.

Mosqueda, G., Ramos-González, M. I., & Ramos, J. L., (1999). Toluene metabolism by the solvent-tolerant *Pseudomonas putida* DOT-T1 strain, and its role in solvent impermeabilization. *Gene, 232*(1), 69–76.

Murialdo, S. E., Sendra, G. H., Passoni, L. I., Arizaga, R., & González, J. F., (2009). Analysis of bacterial chemotactic response using dynamic laser speckle. *J. Biomed. Opt., 14*(6), 064015.

Ni, B., et al., (2015). A novel chemoreceptor MCP2983 from *Comamonas testosteroni* specifically binds to cis-aconitate and triggers chemotaxis towards diverse organic compounds. *Appl. Microbiol. Biotechnol., 99*(6), 2773–2781.

Ni, B., Huang, Z., Wu, Y., Fan, Z., Jiang, C., & Liu, S., (2013). *Comamonas testosteroni* uses a chemoreceptor for tricarboxylic acid cycle intermediates to trigger chemotactic responses towards aromatic compounds. *Mol. Microbiol., 90*(*4*), 813–823.

Nijland, R., & Burgess, J. G., (2010). Bacterial olfaction. *Biotechnol. J., 5*(*9*), 974–977.

Nisenbaum, M., Bouchet, A., Guzmán, M. N., & González, J. F., (2014). Dynamic laser speckle and fuzzy mathematical morphology applied to studies of chemotaxis towards hydrocarbons. *Int. J. Environ. Heal., 7*(1), 58–69.

Nisenbaum, M., Maldonado, E., Martínez, A. J., González, J. F., Passoni, L. I., & Murialdo, S. E., (2016). Video processing analysis for the determination and evaluation of the chemotactic response in bacterial populations. *J. Microbiol. Methods., 127*, 146–153.

Nisenbaum, M., Sendra, G. H., Cerdá, G. G. A., Scagliola, M., González, J. F., & Murialdo, S. E., (2013). Hydrocarbon biodegradation and dynamic laser speckle for detecting chemotactic responses at low bacterial concentration. *J. Environ. Sci., 25*(3), 613–625.

O'Toole, G. A., & Kolter, R., (1998). Flagellar and twitching motility are necessary for *Pseudomonas aeruginosa* biofilm development. *Mol. Microbiol., 30*(2), 295–304.

Okada, E., Nisenbaum, M., Martínez, A. J., & Murialdo, S. E., (2017). Chemotaxis detection towards chlorophenols using video processing analysis. *J. Microbiol. Methods., 142*, 15–19.

Ortega-Calvo, J. J., Marchenko, A. I., Vorobyov, A. V., & Borovick, R. V., (2003). Chemotaxis in polycyclic aromatic hydrocarbon-degrading bacteria isolated from coal-tar- and oil-polluted rhizospheres. *FEMS Microbiol. Ecol., 44*, 373–381.

Pandey, G., & Jain, R. K., (2002a). Bacterial chemotaxis toward environmental pollutants: Role in bioremediation. *Appl. Environ. Microbiol., 68*(12), 5789–5795.

Pandey, G., Chauhan, A., Samanta, S. K., & Jain, R. K., (2002b). Chemotaxis of a *Ralstonia* sp. SJ98 toward co-metabolizable nitroaromatic compounds. *Biochem. Biophys. Res. Commun., 299*(3), 404–409.

Pandey, J., Sharma, N. K., Khan, F., Ghosh, A., Oakeshott, J. G., Jain, R. K., & Pandey, G., (2012). Chemotaxis of *Burkholderia* sp. Strain SJ98 towards chloronitroaromatic compounds that it can metabolise. *BMC Microbiol., 12*(1), 19.

Parales, R. E., & Harwood, C. S., (2002). Bacterial chemotaxis to pollutants and plant-derived aromatic molecules. *Curr. Opin. Microbiol., 5*(3), 266–273.

Parales, R. E., (2004). Nitrobenzoates and aminobenzoates are chemoattractants for pseudomonas strains. *Appl. Environ. Microbiol., 70*(1), 285–292.

Parales, R. E., Ditty, J. L., & Harwood, C. S., (2000). *Toluene Degrading Bacteria are Chemotactic Towards the Environmental, 66*(9), 4098–4104.

Parke, D., Rivelli, M., & Ornston, L. N., (1985). Chemotaxis to aromatic and hydroaromatic acids: Comparison of *Bradyrhizobium japonicum* and *Rhizobium trifolii. J. Bacteriol., 163*(2), 417–422.

Parkinson, J. S., (2007). A 'bucket of light' for viewing bacterial colonies in soft agar. *Methods Enzymol., 423*(07), 432–435.

Paul, D., Singh, R., & Jain, R. K., (2006). Chemotaxis of *Ralstonia* sp. SJ98 towards p -nitrophenol in soil. *Environ. Microbiol., 8*(10), 1797–1804.

Pham, H. T., & Parkinson, J. S., (2011). Phenol sensing by *Escherichia coli* chemoreceptors: A nonclassical mechanism. *J. Bacteriol., 193*(23), 6597–6604.

Poole, P. S., Sinclair, D. R., & Armitage, J. P., (1988). Real-time computer tracking of free-swimming and tethered rotating cells. *Anal. Biochem., 175*(1), 52–58.

Rabinovitch-Deere, C. A., & Parales, R. E., (2012). Three Types of taxis used in the response of *Acidovorax* sp. Strain JS42 to 2-nitrotoluene. *Appl. Environ. Microbiol., 78*(7), 2306 LP–2315.

Rhys-Williams, W., Taylor, S. C., & Williams, P. A., (1993). A novel pathway for the catabolism of 4-nitrotoluene by *Pseudomonas*. *Microbiology., 139*(9), 1967–1972.

Samanta, S. K., & Jain, R. K., (2000a). Evidence for plasmid-mediated chemotaxis of *Pseudomonas putida* towards naphthalene and salicylate. *Can. J. Microbiol., 46*(1), 1–6.

Samanta, S. K., Bhushan, B., Chauhan, A., & Jain, R. K., (2000b). Chemotaxis of a *Ralstonia* sp. SJ98 toward different nitroaromatic compounds and their degradation. *Biochem. Biophys. Res. Commun., 269*(1), 117–123.

Samanta, S. K., Singh, O. V., & Jain, R. K., (2002). Polycyclic aromatic hydrocarbons: environmental pollution and bioremediation. *Trends Biotechnol., 20*(6), 243–248.

Sampedro, I., et al., (2015). *Pseudomonas* chemotaxis. *FEMS Microbiol. Rev.*

Sarand, I., Österberg, S., Holmqvist, S., Holmfeldt, P., Skärfstad, E., Parales, R. E., & Shingler, V., (2008). Metabolism-dependent taxis towards (methyl)phenols is coupled through the most abundant of three polar localized Aer-like proteins of *Pseudomonas putida. Environ. Microbiol., 10*(5), 1320–1334.

Shitashiro, M., Kato, J., Fukumura, T., Kuroda, A., Ikeda, T., Takiguchi, N., & Ohtake, H., (2003). Evaluation of bacterial aerotaxis for its potential use in detecting the toxicity of chemicals to microorganisms. *J. Biotechnol., 101*(1), 11–18.

Si, G., Yang, W., Bi, S., Luo, C., & Ouyang, Q., (2012). A parallel diffusion-based microfluidic device for bacterial chemotaxis analysis. *Lab Chip, 12*(7), 1389–1394.

Silverman, M., & Simon, M., (1974). Flagellar rotation and the mechanism of bacterial motility. *Nature, 249*(5452), 73, 74.

Singer, M. E., & Finnerty, W. R., (1984). Microbial metabolism of straight-chain and branched alkanes. In: *Petroleum Microbiology* (pp. 1–59). Macmillan, New York.

Son, K., Brumley, D. R., & Stocker, R., (2015). Live from under the lens: Exploring microbial motility with dynamic imaging and microfluidics. *Nat. Rev. Microbiology., 13*, 761–775.

Stelmack, P. L., Gray, M. R., & Pickard, M. A., (1999). Bacterial adhesion to soil contaminants in the presence of surfactants. *Appl. Environ. Microbiol., 65*(1), 163–168.

Stocker, R., Seymour, J. R., Samadani, A., Hunt, D. E., & Polz, M. F., (2008). Rapid chemotactic response enables marine bacteria to exploit ephemeral microscale nutrient patches. *Proc. Natl. Acad. Sci., 105*(11), 4209 LP–4214.

Strobel, K. L., McGowan, S., Bauer, R. D., Griebler, C., Liu, J., & Ford, R. M., (2011). Chemotaxis increases vertical migration and apparent transverse dispersion of bacteria in a bench-scale microcosm. *Biotechnol. Bioeng., 108*(9), 2070–2077.

Taute, K. M., Gude, S., Tans, S. J., & Shimizu, T. S., (2015). High-throughput 3D tracking of bacteria on a standard phase-contrast microscope. *Nat. Commun., 6*, 1–9.

Taylor, B. L., & Zhulin, I. B., (1998). *Micro Review in Search of Higher Energy: Metabolism-Dependent Behavior in Bacteria, 28*, 683–690.

Tena-Garitaonaindia, M., Llamas, I., Toral, L., & Sampedro, I., (2019). Chemotaxis of halophilic bacterium *Halomonas anticariensis* FP35 towards the environmental pollutants phenol and naphthalene. *Sci. Total Environ., 669*, 631–636.

Thar, R., Blackburn, N., & Kühl, M., (2000). A new system for three-dimensional tracking of motile microorganisms. *Appl. Environ. Microbiol., 66*(5), 2238–2242.

Timmis, K. N., (2010). *Handbook of Hydrocarbon and Lipid Microbiology*. Springer, Berlin, Heidelberg.

Transportation Research Board and National Research Council, (2003). *Oil in the Sea III: Inputs, Fates, and Effects*. The National Academies Press, Washington, DC.

Tremaroli, V., Vacchi, S. C., Fedi, S., Ceri, H., Zannoni, D., & Turner, R. J., (2010). Tolerance of *Pseudomonas pseudoalcaligenes* KF707 to metals, polychlorobiphenyls and chlorobenzoates: Effects on chemotaxis-, biofilm- and planktonic-grown cells. *FEMS Microbiol. Ecol., 74*(2), 291–301.

Tsuda, M., & Iino, T., (1990). Naphthalene degrading genes on plasmid NAH7 are on a defective transposon. *Mol. Gen. Genet. MGG., 223*(1), 33–39.

Valderrama-Gómez, M. Á., Schomer, R. A., Savageau, M. A., & Parales, R. E., (2020). *TaxisPy: A Python-Based Software for the Quantitative Analysis of Bacterial Chemotaxis*. bioRxiv.

Vardar, G., Barbieri, P., & Wood, T. K., (2005). Chemotaxis of *Pseudomonas stutzeri* OX1 and *Burkholderia cepacia* G4 toward chlorinated ethenes. *Appl. Microbiol. Biotechnol., 66*(6), 696–701.

Vater, S. M., Weiße, S., Maleschlijski, S., Lotz, C., Koschitzki, F., Schwartz, T., Obst, U., & Rosenhahn, A., (2014). Swimming behavior of *Pseudomonas aeruginosa* studied by holographic 3D tracking. *PLoS One, 9*(1), e87765.

Velasco-Casal, P., Wick, L., & Ortega-Calvo, J. J., (2008). Chemoeffectors decrease the deposition of chemotactic bacteria during transport in porous media. *Environ. Sci. Technol., 42*(4), 1131–1137.

Wackett, L. P., & Gibson, D. T., (1988). Degradation of trichloroethylene by toluene dioxygenase in whole-cell studies with *Pseudomonas putida* F1. *Appl. Environ. Microbiol., 54*(7), 1703–1708.

Wang, W., & Shao, Z., (2014). The long-chain alkane metabolism network of *Alcanivorax dieselolei*. *Nat. Commun., 5*(1), 5755.

Wang, X., Gao, B., Gu, S., Zhong, W., Kihaule, K. S., & Ford, R. M., (2018). Preliminary study on dimensionless expression of bacterial chemotaxis in simulated contaminated system. *IOP Conf. Ser. Earth Environ. Sci., 178*, 012009.

Wang, Y. H., Huang, Z., & Liu, S. J., (2019). Chemotaxis towards aromatic compounds: Insights from *Comamonas testosteroni*. *Int. J. Mol. Sci., 20*(11).

Wang, Y., Wang, Q., Li, M., Yang, Y., He, W., Yan, G., & Guo, S., (2016). An alternative anaerobic treatment process for treatment of heavy oil refinery wastewater containing polar organics. *Biochem. Eng. J., 105*, 44–51.

Wolski, E. A., Murialdo, S. E., & González, J. F., (2006). Effect of pH and inoculum size on pentachlorophenol degradation by *Pseudomonas* sp. *Water SA., 32*(1), 93–98.

Wu, J., Wu, J., Wu, X., & Lin, F., (2013a). Recent developments in microfluidics-based chemotaxis studies. *Lab Chip, 13*(13), 2484–2499.

Wu, M., Chen, L., Tian, Y., Ding, Y., & Dick, W. A., (2013b). Degradation of polycyclic aromatic hydrocarbons by microbial consortia enriched from three soils using two different culture media. *Environ. Pollut., 178*, 152–158.

Xhao, X., & Ford, R. M., (2019). The role of bacterial chemotaxis in improving bioremediation efficiency. In: *2019 AIChE Annu. Meet.*

Yu, H. S., & Alam, M., (1997). An agarose-in-plug method to study chemotaxis in the Archaeon *Halobacterium salinarum*. *FEMS Microbiol. Lett., 156*, 265–269.

Zang, X. Q., Li, Z. Y., Zhang, X. Y., Jiang, L., Ren, N. Q., & Sun, K., (2017). Advance in bacteria chemotaxis on microfluidic devices. *Chinese J. Anal. Chem., 45*(11), 1734–1744.

Zhang, J., Xin, Y., Liu, H., Wang, S., & Zhou, N., (2008). Metabolism-independent chemotaxis of *Pseudomonas* sp. strain WBC-3 toward aromatic compounds. *J. Environ. Sci., 20*(10), 1238–1242.

Zhulin, I. B., & Armitage, J. P., (1993). Motility, chemokinesis, and methylation-independent chemotaxis in *Azospirillum brasilense*. *J. Bacteriol., 175*(4), 952–958.

Zhulin, I. B., Rowsell, E. H., Johnson, M. S., & Taylor, B. L., (1997). Glycerol elicits energy taxis of *Escherichia coli* and *Salmonella typhimurium*. *J. Bacteriol., 179*(10), 3196–3201.

CHAPTER 7

Biosurfactants-Mediated Remediation of Hydrocarbon Pollution

PUNNIYAKOTTI PARTHIPAN,[1,2] PATTANATHU K. S. M. RAHMAN,[3,4] MATTHEW L. SMITH,[3] KADARKARAI MURUGAN,[5] YEN-PENG TING,[6] SUBRAMANIA ANGAIAH,[2] and ARULIAH RAJASEKAR[1]

[1]*Environmental Molecular Microbiology Research Laboratory, Department of Biotechnology, Thiruvalluvar University, Serkkadu, Vellore–632115, Tamil Nadu, India, E-mails: rajasekargood@gmail.com; rajasekargood@tvu.edu.in; rajasekar.aruliah@gmail.com (A. Rajasekar)*

[2]*Electro-Materials Research Lab, Center for Nanoscience and Technology, Pondicherry University, Puducherry–605014, Tamil Nadu, India*

[3]*Pharmacy and Biomolecular Sciences, Liverpool John Moores University, Liverpool, UK*

[4]*School of Biological Sciences, University of Portsmouth, Portsmouth, United Kingdom.*

[5]*Division of Entomology, Department of Zoology, School of Life Sciences, Bharathiar University, Coimbatore–641046, Tamil Nadu, India*

[6]*Department of Chemical and Biomolecular Engineering, National University of Singapore, 4 Engineering Drive 4, Singapore–117585*

7.1 INTRODUCTION

Crude oil is a widely used energy source for most industries with its refined petrochemical products and their hydrocarbon (HC) derivatives, such as petroleum oils, gasoline, asphalt, paraffin oils, lubricants, and fuel oil, playing a central role in chemical processing (Wolicka and Borkowski, 2012). Although crude oil is an intrinsic natural resource, its production and processing are hazardous to the environment (Rocha et al., 2011).

Crude oil is formed by geochemical, chemical, and biological processes acting on organic material, which accumulates over long periods. It is transformed from dregs at modest temperatures up to 50°C and due to defunctionalization and compression, kerogen or immature crude oil is eventually formed. From geochemical observations, crude oil contains biomarkers from plants, animals, and microorganisms whose carbon skeletons were not completely degraded by geochemical processes and therefore remained intact (Surygala, 2001).

Accidental spills frequently occur during exploration, production, transport, refining or storage of crude oil and its products. Crude oil released into the environment is estimated to be approximately 600,000 metric tons per annum (Kvenvolden and Cooper, 2003), with accidental and deliberate human discharge being a significant source of soil and water pollution (Sowani et al., 2019). One example of a massive accidental spill is the Prudhoe Bay oil spill in the Alaskan North slope, which resulted in a huge impact on marine habitats. It was estimated that around 212,000 gallons of crude oil were discharged into the seawater due to the inability to resolve the Deepwater leakage problem, which took months to rectify (Jacobson, 2007). HC contamination leads to ecosystem damage through the accumulation of toxic compounds in animal and plant tissue (Pi et al., 2017; Sowani et al., 2019). The techniques usually used for soil remediation include burying, evaporation, washing, and dispersion, however such technologies are costly and do not lead to a complete breakdown of the contaminants (Das and Chandran, 2011). Bioremediation has so far proven to be a sustainable alternative in tackling crude oil contamination and has therefore been the focus of many researchers whose aim is to remediate contaminated environments in a way, which successfully removes HCs while also maintaining an integral environmental benefit. The presence of crude oil and water within an environment generates a specialized niche, which selects for autochthonous microorganisms whose properties prove valuable when considering remediation of the polluted environment (Wolicka and Borkowski, 2012).

In this chapter, we evaluate the role and function of biosurfactants in the biodegradation of hydrophobic pollutants in soil and water environments, particularly in the case of direct uptake of the hydrophobic compounds by microorganisms with a mechanism known as micellization. The importance of HC bioavailability and its impact on biodegradation are discussed. Furthermore, the surface adaptations/modifications of biosurfactants-producing microorganisms toward the utilization of HCs are highlighted.

7.2 BIOREMEDIATION

Industrialization and the vast use of chemicals such as HCs have resulted in the accumulation of contaminants into natural and vulnerable environments, thus representing a serious cause of global concern in this modern era (Sowani et al., 2019). Each year about 1.7 to 8.8 million metric tons of oil are released into natural water sources. About 90% of this oil pollution is associated with accidents due to human failures and activities together with waste disposal (Effendi et al., 2017).

Bioremediation exploits the catabolic machinery of certain microorganisms to treat polluted soil and water (Speight and Arjoon, 2012; Prakash et al., 2020). While bioremediation has been applied for decades, it is still an emerging field that can be improved by scientific and technological developments, in addition to appropriate changes in public policy (Azubuike et al., 2016). At the most basic level, bioremediation processes can be enhanced by improving the availability of materials required for microbial growth and energy production (e.g., providing growth-limiting nutrients and terminal electron acceptors), environmental conditions (e.g., moisture and pH), substrate availability, and microbial dispersal. For substrate bioavailability, the addition of emulsifiers is helpful for low-solubility substrates that tend to partition into non-aqueous environments. Biosurfactants and bioemulsifiers can be used as stimulating agents for bioremediation, thus enhancing its overall efficiency in contaminant removal (Laszlova et al., 2018).

7.3 BIOSURFACTANTS

Some microorganisms produce a range of amphipathic molecules, which accumulate at the interfaces between hydrophobic and hydrophilic phases. They function to decrease surface and interfacial tensions to form emulsions, allowing for the formation of micelles. Biosurfactants differ greatly in their structures and charges, partly a result of the large variations in their origins. These molecules can be classified into two major classes: low-molecular-weight compounds, such as glycolipids and lipopeptides or high-molecular-weight polymers, which include polysaccharides and lipoproteins (Smyth et al., 2010; Parthipan et al., 2018). The most studied biosurfactants are glycolipids with the main examples being rhamnolipids, sophorolipids, trehalolipids, and mannosylerythritol lipids (MELs) (Figure 7.1) (Singh and Tiwary, 2016). Rhamnolipids synthesized by *Pseudomonas* species have been described and potential applications have been reported (Chong and Li, 2017).

Rhamnolipids from *Pseudomonas aeruginosa* are presently commercialized by Jeneil Biosurfactant, USA, primarily as a fungicide in agricultural applications and as an additive to improve bioremediation performance (Hazra et al., 2012).

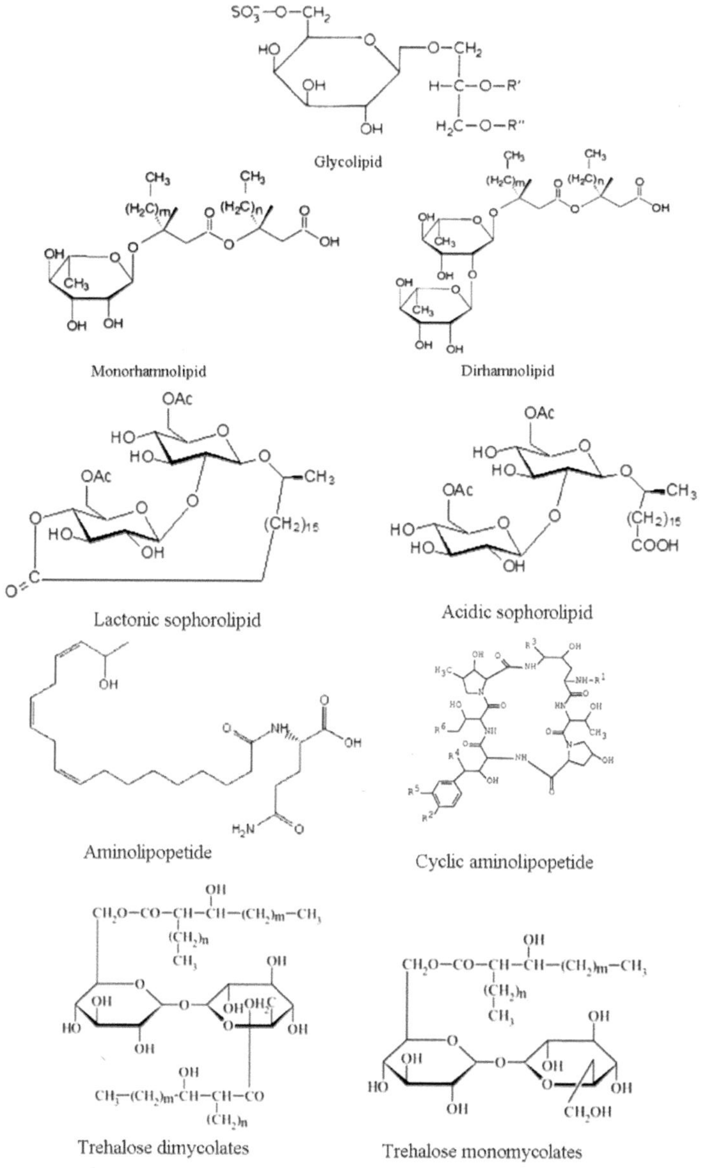

FIGURE 7.1 Chemical structure of different glycolipid biosurfactants.

Microbial adaptation mechanisms in HC-rich environments such as their capability for biosurfactant synthesis and release are crucial, with competence and efficiency being important biosurfactant's characteristics. Competence is based upon critical micelle concentration (CMC) values, whereas efficiency is associated with surface and interfacial tensions. Increasing the concentration of biosurfactants leads to a proportional increase of their ability to reduce surface tension at the substratum interface. The concentration at which the highest value of surface tension reduction is achieved is referred to as the CMC. This CMC value typically ranges from 1 to 2,000 mg/L for many of the well-studied biosurfactants. An efficient biosurfactant can reduce the surface tension of water from 72 to below 30 mN/m and the interfacial tension of n-hexadecane from 40 to 1 mN/m. The molecular weights of biosurfactants range from 500 to 1,500 even though some exceptions exist. Bioemulsifiers are produced by some microbial strains and typically have a high molecular weight. *Pseudomonas* strains grown in the presence of hexadecane produce protein-containing surfactant molecules with molecular weights of about 14,300 (Lang and Wagner, 1987).

7.3.1 ENHANCED CRUDE OIL AND HYDROCARBON (HC) DEGRADATION WITH LOW-MOLECULAR-WEIGHT BIOSURFACTANTS

7.3.1.1 GLYCOLIPIDS

Low-molecular-weight biosurfactants include rhamnolipids, sophorolipids, and trehalolipids, all of which being glycolipids. Glycolipids have been named as such because of the presence of a sugar moiety as the head group, which contributes to their structure and function. One of the most important and widely studied glycolipid biosurfactants are rhamnolipids consisting of one or two rhamnose molecules with up to three hydroxy fatty acids, each having 8–22 carbons (Abdel-Mawgoud et al., 2010). Sophorolipid biosurfactants contain sophorose, a 1,2-disaccharide of glucose, which forms the hydrophilic head of the molecule and is often acetylated with one or two acetyl groups. The hydrophobic portion of a sophorolipid contains a long-chain fatty acid with 16 or 18 carbon atoms (Mekala et al., 2018). Trehalolipids are made up of the disaccharide trehalose, which links at C-6 and C-6′ of mycolic acids (Jahan et al., 2020). Table 7.1 highlights the different types of biosurfactants with their source microorganisms and key achievements in bioremediation.

Kuyukina et al. (2005) studied crude oil desorption and mobilization with glycolipid biosurfactants, which were produced by *Rhodococcus ruber* and enhanced oil removal from soil 1.4–2.3 times higher than that achieved by a synthetic surfactant (tween 60). Biosurfactants were less adsorbed to soil components than synthetic surfactants, thus rapidly penetrating through the soil column and effectively removing 65–82% of crude oil. Cameotra and Singh (2008) isolated two bacterial strains of *P. aeruginosa* and *Rhodococcus* from contaminated sludge samples and used them to degrade oil sludge. They found 90% HC degradation in a laboratory study and the consortium degraded about 91–95% of HC content after 4 weeks of incubation. The addition of crude biosurfactants during the biodegradation process enhanced HC degradation significantly. A similar observation was reported by Nikolopoulou and Alogerakis (2008) who added rhamnolipids along with other nutrients, making a specific impact on the removal of HCs.

Kumari et al. (2012) reported crude oil degradation by two bacterial strains, *Pseudomonas* sp. BP10 and *Rhodococcus* sp. NJ2 isolated from crude oil-contaminated site. Strain BP10 degraded 60.6% of total petroleum HCs, while strain NJ2 metabolized only 49.5% after 30 days of incubation with 2% of crude oil. Zhang et al. (2012) used a *P. aeruginosa* strain for crude oil degradation and found that rhamnolipid biosurfactants produced by this bacterial strain caused 36–46% increase in crude oil removal efficiency within 7 days of incubation.

Recently, Kumari et al. (2018) used five different strains including *Stenotrophomonas maltophilia*, *Ochrobactrum anthropi*, *P. mendocina*, *Microbacterium esteraromaticum* and *P. aeruginosa* to degrade multiple polycyclic aromatic hydrocarbons (PAHs) present in the crude oil. The exposure of the individual bacterial strains to crude oil showed considerable biodegradation capabilities with *M. esteraromaticum* which utilized 81.4% of naphthalene, *P. aeruginosa* utilizing 67.1% of phenanthrene and 61% of benzo(b)fluoranthene and *S. maltophilia* utilizing about 47.9% of fluorene. This degradation efficiency was improved in the presence of a consortium of the five bacterial strains. The addition of rhamnolipid biosurfactants further improved the degradation efficiency by about 10%. The addition of lactonic sophorolipids produced by *Starmerella bombicola* efficiently improved oil dispersion (Shah et al., 2019). In a study by Lee et al. (2018), biosurfactant-producing, indigenous bacteria such as *Bacillus algicola* (003-Phe1), *Rhodococcus soli* (102-Na5), *Isoptericola chiayiensis* (103-Na4) and *Pseudoalteromonas agarivorans* (SDRB-Py1) were used. These strains utilized crude oil as a sole carbon source and the addition of the produced biosurfactants significantly desorbed crude oil in oil-polluted marine sediment. These biosurfactants increased the emulsification of crude oil and significantly assisted crude oil degradation. Similarly, Neifar et al. (2018) reported that an extremely halotolerant bacterial strain *Halomonas desertis* G11 produced glycolipid during crude oil degradation, which can be applied for the remediation of oil spills in extreme high-salt habitats.

Joy et al. (2017) isolated 47 bacterial strains, from which 11 were identified as potential biosurfactant producers with some showing significant HC degradation efficiency. Chandankere et al. (2014) also reported a glycolipid biosurfactants-producing *Bacillus methylotrophicus* USTBa. This bacterial strain enhanced crude oil degradation by about 92% in two weeks with the produced biosurfactants and significantly reduced the surface tension to about 28 mN/m. In most cases, naturally occurring biosurfactants produced by *Bacillus* species are characterized as lipopeptides. The exceptions reported by these authors may be due to the great structural diversity found among biosurfactants, which varies according to both the species and strains within the genus *Bacillus* (Chandankere et al., 2013).

Another recent study by Patowary et al. (2018) described a rhamnolipid biosurfactant produced by *P. aeruginosa* SR17, which showed remarkable HC degradation efficiency. More than 80% of total petroleum HCs were removed by this biosurfactant, proving its degradation efficiency to be far higher than that of sodium dodecyl sulfate (SDS), a synthetic surfactant whose efficiency lies at around 70%. Zenati et al. (2018) used biosurfactants produced by *Marinobacter hydrocarbonoclasticus* SdK644 isolated from a HC-contaminated sediment to enhance the solubilization of crude oil in contaminated seawater. Glycolipids significantly improved crude oil solubilization to an extent, which was higher than that reached with Tween 80, a synthetic surfactant.

Sajna et al. (2015) added glycolipids to a sample of alkane (C_{10}–C_{24}) in the presence of *P. putida* and the degradation efficiency was improved up to 46%. The inclusion of glycolipid biosurfactants at a minimal level leads to significant improvement in the degradation efficiency but we must also consider that at the same time, the increasing addition of biosurfactants, especially above that of the CMC, may lead to growth inhibition, which would be disadvantageous to the degradation process. The greatest degradation rates of HCs, which were achieved, were observed with 2.5 mg/L of biosurfactants. Similarly, Muthukamalam et al. (2017) isolated three bacterial strains of *Corynebacterium aurimucosum*, *Acinetobacter baumannii* and *Microbacterium hydrocarbonoxydans* from crude oil-contaminated soil. Among these three strains, *M. hydrocarbonoxydans* showed efficient crude oil degradation.

7.3.1.2 LIPOPEPTIDES

Surfactin is a cyclic lipoheptapeptide biosurfactant where the peptide moiety is synthesized non-ribosomally by a multienzyme peptide synthase complex. The peptide ring present in surfactin has 7 amino acids and the hydrophobic lipid component is composed of a fatty acid chain. Any structural changes in the fatty acid chain length or changes in the amino acid components of the peptide ring will reflect on the physicochemical properties of the overall product (Jahan et al., 2020).

Ismail et al. (2013) isolated 51 bacterial strains from petroleum-polluted environmental samples, some of which produced biosurfactants/bioemulsifiers by utilizing crude oil as a sole carbon source. A *Bacillus* sp. I-15 was among the screened strains and was found to naturally synthesize lipopeptide biosurfactants. These examples of indigenous biosurfactant-producing strains would make for ideal candidates in the bioremediation of crude oil contaminants. Similarly, Sakthipriya et al. (2015) used a thermophilic bacterial strain of *Bacillus subtilis* for degradation of waxy crude oil, model crude oil and six long-chain paraffins. Biosurfactant production was found to be significantly high in the presence of low-molecular-weight HCs. The results additionally indicated that biosurfactant production was higher in the presence of waxy crude oil than it was in the model crude oil sample due to the presence of short-chain paraffins in the former. The produced biosurfactants exhibited high thermal stability (up to 120°C), thus making them suitable for bioremediation in extreme environments.

Recently, Joy et al. (2017) isolated two bacterial strains of *Ochrobactrum* (GREW1) and *Bacillus* sp. (SB2) which produced lipopeptide biosurfactants. These bacterial strains efficiently degraded crude oil HCs with their degradation efficiency being enhanced in the presence of biosurfactants. Datta et al. (2018) used a *Bacillus subtilis* strain for crude oil degradation and optimized the growth parameters. They found that biosurfactant production during HC degradation leads to effective crude oil degradation with a 91% higher emulsification activity. The role of lipopeptides in crude oil dispersion was described by Rongsayamanont et al. (2017). The authors used *Bacillus* sp. GY19 to produce biosurfactants and found them to lack ecotoxicity when

TABLE 7.1 Biosurfactants and Their Role in Biodegradation of Crude Oil and Hydrocarbons

Biosurfactant	Crude Oil/Hydrocarbons	Source Microorganisms	Key Outcomes/Achievements	References
		Low-Molecular Weight Biosurfactants		
Glycolipid	Crude oil	*Rhodococcus ruber*	65–82% of crude oil degradation was noticed in the presence of biosurfactants	Kuyukina et al. (2005)
Rhamnolipid	Hydrocarbons in oil sludge	*Pseudomonas aeruginosa* SSC2	Biosurfactant addition enhanced the degradation efficiency of sludge hydrocarbons.	Cameotra and Singh (2008)
Rhamnolipid	Petroleum hydrocarbons (C_{19}–C_{34} n-alkanes)	Purchased from Jeneil biosurfactants Co., USA Source microbes (JBR210)	Higher removal (96%) of petroleum hydrocarbons was achieved	Nikolopoulou and Kalogerakis (2008)
Glycolipid	Crude oil	*Pseudomonas* sp. BP10	Higher biosurfactant production capability increased degradation efficiency	Kumari et al. (2012)
Rhamnolipid	Crude oil	*Pseudomonas aeruginosa*	Crude oil degradation was considerably enhanced by biosurfactant production	Zhang et al. (2012)
Glycolipid	Crude oil	*Bacillus methylotrophicus*	Addition of glycolipids increased solubilization of crude oil	Chandankere et al. (2014)
Glycolipid	Alkanes	*Pseudozyma* sp. NII *08165*	Addition of biosurfactants enhanced degradation of alkanes by strain *Pseudomonas putida*	Sajna et al. (2015)
Glycolipids	Crude oil	*Bacillus* sp. (SLDB1)	Glycolipids were confirmed from *Bacillus* sp., with effective surface tension activity	Joy et al. (2017)
Glycolipid	Crude oil	*Corynebacterium aurimucosum*, *Acinetobacter baumannii* and *Microbacterium hydrocarbonoxydans*	Crude oil degradation was enhanced by biosurfactant production along with catabolic enzyme production	Muthukamalam et al. (2017)

TABLE 7.1 *(Continued)*

Biosurfactant	Crude Oil/Hydrocarbons	Source Microorganisms	Key Outcomes/Achievements	References
Monorhamnolipid and dirhamnolipid	Crude oil	*B. algicola*, *R. soli*, *I. chiayiensis*, and *P. sagarivorans*	These indigenous bacteria degrade crude oil hydrocarbons with the aid of biosurfactant production	Lee et al. (2018)
Glycolipid	Crude oil	*Halomonas desertis* G11	Highly stable at halophilic conditions and solubilize crude oil efficiently	Neifar et al. (2018)
Glycolipids	Crude oil	*Marinobacter hydrocarbonoclasticus*	Addition of biosurfactants greatly improved crude oil solubilization much higher than synthetic surfactants	Zenati et al. (2018)
Sophorolipids	Crude oil	*Starmerella bombicola*	Addition of biosurfactants increased oil dispersion	Shah et al. (2019)
Lipopeptide	Crude oil	*Bacillus* sp. I-15	Surface tension was highly reduced which may be applied to enhance crude oil degradation	Ismail et al. (2013)
Surfactin	Waxy crude oil	*Bacillus subtilis*	Biosurfactant produced by this strain is stable at high temperature, pH, and salinity	Sakthipriya et al. (2015)
Lipopeptides	Crude oil	*Ochrobactrum* sp. (GREW1) and *Bacillus* sp. (SB2)	Strains GREW1 and SB2 showed significant surface tension reduction values of 31.14 mN/m and 28.16 mN/m, respectively	Joy et al. (2017)
Lipopeptide	Crude oil	*Bacillus* sp. GY19	Biosurfactant increased oil dispersion in marine environment without harming marine species	Rongsayamanont et al. (2017)
Surfactin	Crude oil	*Bacillus subtilis*	Higher surface tension and emulsification activity greatly enhanced crude oil degradation.	Datta et al. (2018)

TABLE 7.1 *(Continued)*

Biosurfactant	Crude Oil/Hydrocarbons	Source Microorganisms	Key Outcomes/Achievements	References
Lipopeptide	Aliphatic and aromatic hydrocarbons in crude oil	*Agrobacterium fabrum* SLAJ731	Enhanced crude oil degradation	Sharma et al. (2019)
Surfactin, fengycin, and Lichenysin	Crude oil	*Bacillus amyloliquefaciens*	Surfactin enhanced removal of aromatic hydrocarbons	Wang et al. (2020)
High-Molecular Weight Biosurfactants				
Bioemulsifier	Crude oil	*Geobacillus pallidus*	Enhanced degradation of crude oil	Zheng et al. (2011)
Glycoprotein	Crude oil	*Aeribacillus pallidus* SL-1	Played crucial role the hydrocarbon degradation	Tao et al. (2020)

presented to marine organisms and additionally dispersed spilled crude oil in the marine environment by increasing its solubility and therefore its bioavailability.

Sharma et al. (2019) used an indigenous bacterial strain, *Agrobacterium fabrum* SLAJ731, for crude oil degradation. This bacterial strain produced lipopeptides with high CMC values and efficient surface tension reduction. The ability for this strain to synthesize biosurfactants enhanced its capacity for oil degradation and resulted in the breakdown of both aliphatic (47%) and aromatic HCs (70%). Wang et al. (2020) reported that surfactin addition increased the biodegradation rate of alkyl aromatic HCs by 1.5–87.2%, and significantly enhanced the biodegradation of alkylphenanthrenes, alkyl-dibenzothiophenes, and methyltriaromatic steroids.

7.3.2 ENHANCED CRUDE OIL AND HYDROCARBON (HC) DEGRADATION WITH HIGH-MOLECULAR WEIGHT BIOSURFACTANTS

Bioemulsifiers are considered as high-molecular-weight surfactants, they exhibit diverse properties and play an important role in solubilizing hydrophobic contaminants (Camacho-Chab et al., 2013). Emulsan is an example of a high-molecular-weight biosurfactant, which is synthesized by *Acinetobacter calcoaceticus* RAG-1 and *Acinetobacter calcoaceticus* BD4 and consists of anionic lipopolysaccharide and protein components (Desai and Banat, 1997, Ron, and Rosenberg, 2001). Similarly, biodispersan, an extracellular non-dialyzable dispersing agent produced by *Acinetobacter calcoaceticus* A2, is also made up of anionic heteropolysaccharide with 4 reducing sugars (Desai and Banat, 1997). Alasan produced by *Acinetobacter radioresistens* KA-53 consists of anionic alanine-containing heteropolysaccharide-protein (Ron and Rosenberg, 2001).

Zheng et al. (2011) found that two bacterial strains, *Geobacillus pallidus* XS2 and XS3, naturally produce bioemulsifiers in response to HCs, resulting in the degradation of 85% of crude oil. Both bioemulsifiers consisted of carbohydrates alongside additional lipids and proteins. Recently, Tao et al. (2020) studied crude oil degradation with the thermophilic bacterial strain *Aeribacillus pallidus* SL-1 at 60°C. Strain SL-1 had effective degradation on short-chain n-alkanes (< C17) and aromatic HCs from crude oil. Degradation efficiency was significantly enhanced by glycoprotein biosurfactants produced by this bacterial strain, once again highlighting the significance of biosurfactants throughout the degradation process.

7.4 BIOAVAILABILITY OF POLLUTANTS

Many physiochemical variables can make an impact on the bioremediation process and this is dependent on several aspects including characteristics of the site, environmental factors (temperature, pH, nutrient availability, electron donors/acceptors, the nature and concentration of the pollutant), presence of degrading enzymes and the bioavailability of the pollutant (Souza et al., 2014). To date, research on biodegradation has focused on the impact of environmental factors and the biodegradation capabilities of microorganisms. However, while intrinsic degradation potential and appropriate environmental factors are necessary, it is crucial to consider the consequences of pollutant's bioavailability as well. This is particularly important for subsurface environments, where bioavailability may frequently control the incidence and rate of biodegradation (Mao et al., 2015).

Bioavailability can be defined as the number of pollutant molecules available for utilization by microorganisms (Megharaj et al., 2011). Biodegradation rate is proportional to the bioavailability of the pollutants, partly because microorganisms require energy for expression of the genes involved in the synthesis of degradation enzymes. If the apparent pollutant concentration is low, stimulation will not take place as it may result in unnecessary allocation of resources, while at high concentration, the process will be enhanced as the use of energy here can be justified by the cell (Sarubbo et al., 2015). Indigenous soil microorganisms are characteristically slow-developing organisms habitually exposed to oligotrophic conditions (Roszak and Colwell, 1987). The existence of a pollutant can have an important impact on the metabolic potential of such microorganisms. Three cases can be envisioned that would originate from the inconsistent bioavailability of the pollutant. In the first case, degradation will not take place if the amount of bioavailable pollutants is inadequate to satisfy the energy expenses to support biodegradation. In the second case, if the bioavailability is very low, microorganisms can degrade pollutants in a way that is inadequate for the development of new biomass. In the third case, if there is an adequate level of bioavailable pollutant, biodegradation in a mounting stage will be encouraged. Likely, the third case will allow the most favorable rates of bioremediation. Of course, the ability of microorganisms to affect or enhance the availability of some nutrients present in their environment is a very important process. That is what the biosurfactants-producing microorganism effectively do, releasing surface-active molecules to enhance bioavailability and accessibility of many nutrients (Maier, 2000).

Microorganisms developed numerous mechanisms to boost the bioavailability of hydrophobic pollutants. One approach is the synthesis and excretion of biosurfactants. A second approach deals with improvement of cell adhesion to hydrophobic surfaces. In the case of a pollutant with a low aqueous solubility, this permits the degrading microorganisms to attach to the hydrophobic substrate and gain direct access to it (Bezza and Chirwa, 2016). Bezza and Chirwa (2017a) reported that biosurfactants increased the bioavailability of PAHs, which have been recognized as significant pollutants due to their carcinogenic nature. Microcosms supplemented with lipopeptides enhanced the degradation efficiency of 5- and 6-ring PAHs up to 79%. In another study, the authors studied the biodegradation of pyrene by a consortium composed of *P. viridiflava* and *P. nitroreducens* in the presence of lipopeptide biosurfactants produced by *Paenibacillus dendritiformis* CN5 and found that at lipopeptides concentrations of 600 and 300 mg L^{-1}, the biodegradation of pyrene was about 83.5% and 67%, respectively, which was much higher than that achieved in the absence of biosurfactant (only 16%) (Bezza and Chirwa, 2017b). Recently, Parthipan et al. (2017) reported that the HC-degrading *Bacillus subtilis* A1 produced a high amount of lipopeptides, which enhanced crude oil degradation in a short time (7 days). In addition to biosurfactants, different degradative enzymes such as alkane hydroxylase, alcohol dehydrogenase and laccase played a key role in improving the degradation efficiency.

It has been noticed that HC-utilizing microorganisms have an affinity for hydrophobic surfaces such as HC droplets (Patowary et al., 2017). Certainly, numerous microbes stick to surfaces and there are possibilities for both specific and nonspecific mechanisms for compliance with hydrophobic substrates. Hua et al. (2003) studied the impact of biosurfactants on HC degradation by the yeast *Candida Antarctica.* Interestingly, they found that biosurfactants alter the surface properties of microorganisms and promote the biodegradation of n-alkanes. Biosurfactants are involved in HC degradation by increasing the bioavailability of HCs through micellar solubilization, and they interact with the cell surface, modifying its hydrophobicity, which leads to increased association of microbial cells with HCs. Usually, both mechanisms take place but the dominance of one or the other depends on the strain (Bezza et al., 2016).

Rocha et al. (2011) reported that biosurfactants addition extensively promoted the degradation of alkanes, particularly the recalcitrant iso-alkanes. They can enhance the bioavailability of hydrophobic molecules by solubilization, a crucial step in the bioremediation process (Padmapriya et al., 2013). Nevertheless, there are numerous studies reporting that in spite of augmented solubility, biosurfactants can lead to decrease in the biodegradation rate (Zhao et al., 2011). Three factors could be recognized in this instance. The first factor deals with the presence of the pollutant within a biosurfactant micelle, hindering the cells' ability to reach the molecule and therefore reducing its overall bioavailability (Guha and Jaffe, 1996). The next deals with the direct impact of biosurfactants on cell surface characteristics, in some cases, these may interfere with the attachment to hydrophobic substrates, resulting in reduced interaction between the cell and the pollutant itself (Bezza et al., 2016). The third factor deals with the genus-specificity of biosurfactants and it has been revealed that a given biosurfactant might augment the activity of the synthesizing microbe, while inhibiting the action of other microorganisms (Diaz De Rienzo et al., 2015).

7.5 INFLUENCE OF MICELLIZATION ON ACCESS OF HYDROCARBONS (HCS)

When the concentration of a biosurfactant reaches the CMC, the monomers start to form aggregates such as micelles, where the hydrophobic tails are directed to the inside and the hydrophilic heads are oriented to the micelle's surface, making an interface with the aqueous medium (Soberon-Chavez and Maier, 2010; Souza et al., 2014). Above a definite level of the CMC, contaminants commence to form into micellar aggregates. These aggregates are formed because of numerous weak chemical interactions such as Van der Waals and hydrogen bridges. Micelle development leads to decrease in surface and interfacial tension due to reduced force of repulsion between immiscible liquid phases (Aparna et al., 2012). The molecule mobilization mechanism, which provokes the reduction of surface tension, occurs at biosurfactants concentrations below the CMC, while solubilization starts above the CMC due to the development of micelles, which increase oil bioavailability (Urum and Pekdemir, 2004). The emulsification begins when there is surfactant accretion that forms a solution including minute droplets of oil pendant within aqueous medium. The emulsification power of biosurfactants increases as the molecular weight gets higher (Franzetti et al., 2010).

When micelles appear in aqueous solution with hydrophobic pollutants, the hydrophobic ends of the surfactant molecules solubilize the hydrophobic pollutant inside the micelle, while the hydrophilic ends remain firmly associated with the external aqueous medium. Biosurfactants, therefore, have the potential to impede the natural degradation of the pollutants, a consequence of the micellar aggregates shielding the pollutants from the degrading microorganisms (Tondo et al., 2010).

Makkar and Rockne (2003) have evaluated the function of synthetic surfactants and biosurfactants in the degradation of PAHs. In several instances, synthetic surfactants have been found to inhibit PAHs degradation and this was attributed to surfactant degradation, surfactant toxicity or presence of PAHs inside surfactant micelles. Inhibition of micelle formation by biosurfactants may promote bioavailability of biosurfactant-associated PAHs by direct transfer. It was found that emulsan from *Acinetobacter* sp. enhanced the degradation of waterlogged HCs by producing a hydrophilic shell near the oil droplets (Allison, 2000). Shin et al. (2005), however, reported that pseudosolubilization of phenanthrene by rhamnolipids does not increase its accessibility to degrading microorganisms such as *P. putida*.

7.6 ACCESSIBILITY OF HYDROPHOBIC POLLUTANTS IN SOIL

The reduced bioavailability of contaminants in soil is due to long-term pollution often referred to as aging, a result of the chemical oxidation process and less diffusion of the pollutants into small pores, thus incorporating them into the organic material. The addition of biosurfactants during the bioremediation practice helps to overcome bioavailability issues, which may otherwise be a product of aging (Pacwa-Plociniczak et al., 2011). Certainly, the microbial cells might also have mechanisms for uptake of the pollutants. Understanding principles of pollutants uptake by the degrading microorganisms can improve our ability to manipulate them in efficient remediation processes.

When biosurfactant molecules exist in a water-soil mixed system, they may adsorb onto the surface of soil particles, where interactions may occur (Figure 7.2). Usually, the head groups (hydrophilic groups) are able to penetrate the aqueous part and the tail groups (lipophilic groups) orient towards, and possibly link with, the hydrophobic pollutants or soil particles in a way which is energetically favorable. Therefore, surfactants at a low concentration essentially accumulate at solid-liquid or liquid-liquid interface as monomers (Mao et al., 2015).

For efficient biosurfactants-mediated remediation of polluted soils, it is important to consider the adsorption behavior of biosurfactants onto polluted soil, the solubilization power of biosurfactant on the target pollutants and biosurfactant's biodegradability and ecotoxicity. When biosurfactants are added to a water-soil system, a particular level of biosurfactant molecules will be adsorbed by soil particles. In cases where more surfactant is adsorbed, less amount will remain to contribute to the solubilization of pollutants. Moreover, the hydrophobicity of the contaminated soil increases as the surfactants are adsorbed onto the soil particles. Consequently, solubilized organic pollutants will be re-adsorbed on the soil surface (Paria, 2008). Hence, the adsorption behavior of biosurfactants onto polluted soil particles is a significant factor in selecting suitable biosurfactants. The molecular structure of biosurfactants represents the governing factor in adsorption behavior.

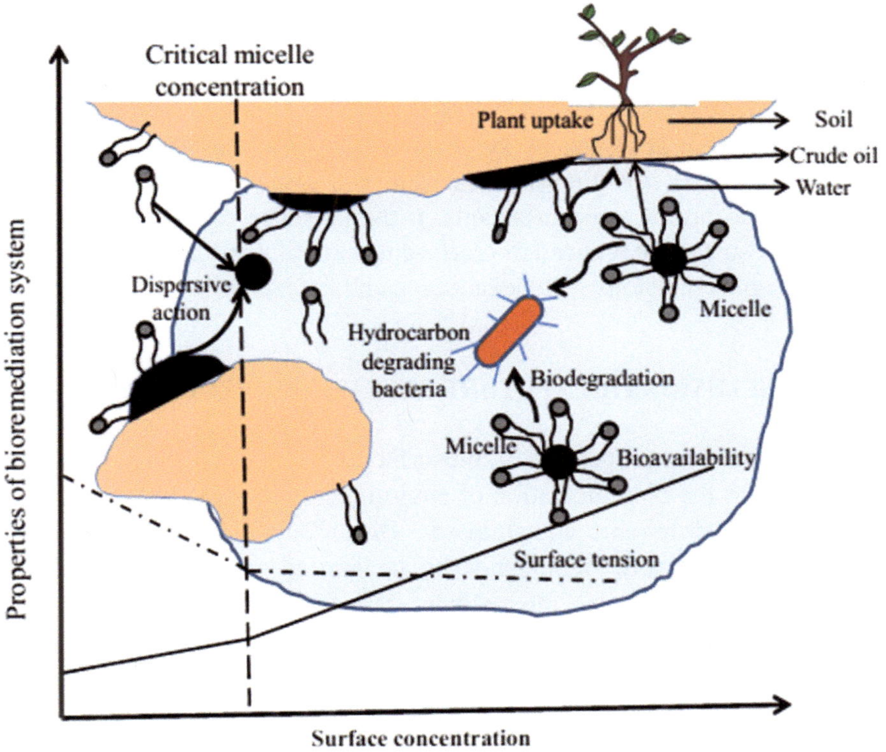

FIGURE 7.2 Schematic illustration of surfactant-enhanced bioremediation of polluted soil.

Pollutants in early stages attach to the soil matrix via reversible sorption. This has been recognized through spiking of soil with pollutants and examining elimination efficiency during washing with surfactants or biodegradation under laboratory conditions. Nevertheless, soil sorption and bioavailability diminish moderately because of the pollutants infiltrating deeper into soil cracks and micropores, which confine the entrance of microorganisms or biosurfactants. Both synthetic surfactants and biosurfactants have occasionally been applied successfully to promote desorption of hydrophobic pollutants from soil, although biosurfactants synthesized by HC-degrading microorganisms are better equipped to carry out the soil remediation process. In addition to the tendency of pollutants to sorb onto soil particles, the surfactant itself tends to sorb strongly to the soil. Subsequently, a higher surfactant concentration is needed to enhance the pseudosolubilization of hydrophobic pollutants nearby soil as contrasted compared to the amount needed for solubilization in aqueous media.

Kavitha et al. (2014) studied solubilization/removal of crude oil from diverse types of soil in an aqueous phase with biosurfactants produced by *Bacillus licheniformis*. The authors reported that biosurfactants achieved a higher oil removal rate (>85%) with a much smaller concentration as compared to that achieved by synthetic surfactants (<60%). Dasari et al. (2014) isolated *P. aeruginosa* strain from petroleum-polluted sites, which was found to produce biosurfactants in the presence of naphthalene. The purified biosurfactant reduced the surface tension of water and was also able to form stable emulsions with hexadecane and kerosene.

7.7 CONCLUSIONS AND FUTURE RESEARCH DIRECTIONS

HC solubilization properties of biosurfactants make them a significant contributor to the bioremediation of environments polluted with petroleum and related hydrophobic contaminants. Biosurfactants play a role in the degradation of organic compounds by facilitating access to hydrophobic substrates via interfacial interactions. Some biosurfactants have been commercialized. Rhamnolipids from *P. aeruginosa* have been a popular choice for industry, likely a result of the wealth of research, which has so far been carried out to optimize their production.

Although very few field studies have so far been conducted in contrast to a greater number of laboratory studies, the reported field studies and many remediation projects show that biosurfactants-based methods can be applied effectively in the bioremediation of different polluted sites. Obviously, additional studies concerning the behavior of biosurfactants in the trajectory and movement of soil pollutants are still required. The need for biosurfactants has increased due to multiple applications in different sectors, such as the remediation of industrial wastewater from oil/gas operations, food processing industry, and other industries. The limitations in the usage of biosurfactants for commercial purposes are associated mainly with their production such as low yield and high costs. Improvements in surfactant chemistry, biotechnology, and remediation technologies are therefore essential to promote the application of biosurfactants in the future.

ACKNOWLEDGMENTS

P. Parthipan gratefully acknowledges the Science and Engineering Research Board (SERB), Department of Science and Technology (DST), for providing research fellowship under the National Postdoctoral Fellowship (PDF/2017/001134).

KEYWORDS

- **bioavailability**
- **biodegradation**
- **bioremediation**
- **critical micelle concentration**
- **petroleum hydrocarbons**
- **rhamnolipids**
- **surface tension**

REFERENCES

Abdel-Mawgoud, A. M., Lepine, F., & Deziel, E., (2010). Rhamnolipids: Diversity of structures, microbial origins and roles. *Appl. Microbiol. Biotechnol., 86*(5), 1323–1336.

Allison, D., (2000). *Community Structure and Co-operation in Biofilms*. Cambridge: Cambridge University Press.

Aparna, A., Srinikethan, G., & Smitha, H., (2012). Production and characterization of biosurfactant produced by a novel *Pseudomonas* sp. 2B. *Colloids Surf. B Biointer., 95*, 23–29.

Azubuike, C. C., Chikere, C. B., & Okpokwasili, G. C., (2016). Bioremediation techniques-classification based on site of application: Principles, advantages, limitations and prospects. *World J. Microbiol. Biotechnol., 32*, 180.

Banat, I. M., Makkar, R. S., & Cameotra, S. S., (2000). Potential commercial applications of microbial surfactants. *Appl. Environ. Microbiol., 53*, 495–508.

Bezza, F. A., & Chirwa, E. M. N., (2017a). The role of lipopeptide biosurfactant on microbial remediation of aged polycyclic aromatic hydrocarbons (PAHs)-contaminated soil. *Chem. Eng. J., 309,* 563–576.

Bezza, F. A., & Chirwa, E. M. N., (2017b). Pyrene biodegradation enhancement potential of lipopeptide biosurfactant produced by *Paenibacillus dendritiformis* CN5 strain. *J. Hazard. Mater., 321,* 218–227.

Bezza, F. A., & Chirwa, E. M., (2016). Biosurfactant-enhanced bioremediation of aged polycyclic aromatic hydrocarbons (PAHs) in creosote contaminated soil. *Chemosphere, 144,* 635–644.

Camacho-Chab, J. C., Guezennec, J., Chan-Bacab, M. J., Rios-Leal, E., Sinquin, C., Salazar, R. M., et al., (2013). Emulsifying activity and stability of a non-toxic bioemulsifier synthesized by *Microbacterium* sp. MC3B-10. *Int. J. Mol. Sci., 14*(9), 18959–18972.

Cameotra, S. S., & Singh, P., (2008). Bioremediation of oil sludge using crude biosurfactants. *Int. Biodeterior. Biodegradation, 62,* 274–280.

Chandankere, R., Yao, J., Cai, M., Masakorala, K., Jain, A. K., & Choi, M. M. F., (2014). Properties and characterization of biosurfactant in crude oil biodegradation by bacterium *Bacillus methylotrophicus* USTBa. *Fuel, 122,* 140–148.

Chandankere, R., Yao, J., Choi, M. M. F., Masakorala, K., & Chan, Y., (2013). An efficient biosurfactant-producing and crude oil -emulsifying bacterium *Bacillus methylotrophicus* USTBa isolated from petroleum reservoir. *Biochem. Eng. J., 74,* 46–53.

Chong, H., & Li, Q., (2017). Microbial production of rhamnolipids: Opportunities, challenges, and strategies. *Microb. Cell Fact., 16,* 137.

Curtis, T. P., Sloan, W. T., & Scannell, J. W., (2002). Estimating prokaryotic diversity and its limits. *Proc. Natl. Acad. Sci. USA, 99,* 10494–10499.

Das, N., & Chandran, P., (2011). Microbial degradation of petroleum hydrocarbon contaminants: *An overview Biotech. Res. Int.,* 941810.

Das, P., & Ma, L. Z., (2013). Pyocyanin pigment assisting biosurfactant-mediated hydrocarbon emulsification. *Int. Biodeterior. Biodegradation, 85,* 278–283.

Dasari, S., Venkata, S. K. C., Wudayagiri, R., & Valluru, L., (2014). Biosurfactant-mediated biodegradation of polycyclic aromatic hydrocarbons—naphthalene. *Bioremediat. J., 18,* 258–265.

Datta, P., Tiwari, P., & Pandey, L. M., (2018). Isolation and characterization of biosurfactant producing and oil-degrading *Bacillus subtilis* MG495086 from formation water of Assam oil reservoir and its suitability for enhanced oil recovery. *Bioresour. Technol., 270,* 439–448.

Desai, J. D., & Banat, I. M., (1997). Microbial production of surfactants and their commercial potential. *Microbiol. Mol. Biol. Rev., 61*(1), 47–64.

Effendi, H., Munawaroh, A., & Ayu, I. P., (2017). Crude oil spilled water treatment with *Vetiveria zizanioides* in floating wetland. *Egypt. J. Aquat. Res., 43,* 185–193.

Franzetti, A., Gandolfi, I., Bestetti, G., Smyth, T. J., & Banat, I. M., (2010). Production and applications of trehalose lipid biosurfactants. *Eur. J. Lipid Sci. Technol., 112,* 617–627.

Guha, S., & Jaffe, P. R., (1996). Bioavailability of hydrophobic compounds partitioned into the micellar phase of nonionic surfactants. *Environ. Sci. Technol., 30,* 1382–1391.

Hazra, C., Kundu, D., & Chaudhari, A., (2012). Biosurfactant-assisted bioaugmentation in bioremediation. In: Satyanarayana, T., Johri, B. N., & Prakash, A., (eds.), *Microorganisms in Environmental Management: Microbes and Environment* (1st edn., pp. 631–664). Springer, New York, NY, USA.

Hua, Z., Chen, J., Lun, S., & Wang, X., (2003). Influence of biosurfactants produced by *Candida Antarctica* on surface properties of microorganism and biodegradation of n-alkanes. *Water Res., 37*, 4143–4150.

Ismail, W., Al-Rowaihi, I. S., Al-Humam, A. A., Hamza, R. Y., El Nayal, A. M., & Bououdina, M., (2013). Characterization of a lipopeptide biosurfactant produced by a crude-oil-emulsifying *Bacillus* sp. I-15. *Int. Biodeterior. Biodegradation, 84*, 168–178.

Jahan, R., Bodratti, A. M., Tsianou, M., & Alexandridis, P., (2020). Biosurfactants, natural alternatives to synthetic surfactants: Physicochemical properties and applications. *Adv. Colloid Interface Sci., 275*, 102061.

Joy, S., Rahman, P. K. S. M., & Sharma, S., (2017). Biosurfactant production and concomitant hydrocarbon degradation potentials of bacteria isolated from extreme and hydrocarbon contaminated environments. *Chem. Eng. J., 317*, 232–241.

Kavitha, V., Mandal, A. B., & Gnanamani, A., (2014). Microbial biosurfactant mediated removal and/or solubilization of crude oil contamination from soil and aqueous phase: An approach with *Bacillus licheniformis* MTCC 5514. *Int. Biodeterior. Biodegradation, 94*, 24–30.

Kumari, B., Singh, S. N., & Singh, D. P., (2012). Characterization of two biosurfactant producing strains in crude oil degradation. *Process Biochem., 47*, 2463–2471.

Kumari, S., Regar, K. R., & Manickam, N., (2018). Improved polycyclic aromatic hydrocarbon degradation in a crude oil by individual and a consortium of bacteria, *Bioresour. Technol., 254*, 174–179.

Kuyukina, M. S., Ivshina, I. B., Makarov, S. O., Litvinenko, L. V., Cunningham, C. J., & Philp, J. C., (2005). Effect of biosurfactants on crude oil desorption and mobilization in a soil system. *Environ. Int., 31*, 155–161.

Kvenvolden, K. A., & Cooper, C. K., (2003). Natural seepage of crude oil into the marine environment. *Geo-Mar. Lett., 23*, 140–146.

Lang, S., & Wagner, F., (1987). Structure and properties of biosurfactants. In: Kosaric, N., Cairns, W. L., & Gray, N. C. C., (eds.), *Biosurfactants and Biotechnology* (pp. 21–45). Marcel Dekker, New York.

Laszlova, K., Dudasova, H., Olejnikova, P., Horvathova, G., Velicka, Z., Horvathova, H., & Dercova, K., (2018). The application of biosurfactants in bioremediation of the aged sediment contaminated with polychlorinated biphenyls. *Water Air Soil Pollut., 229*, 219.

Lee, D. W., Lee, H., Kwon, B., Khim, J. S., Yim, U. H., Kim, B. S., & Kim, J., (2018). Biosurfactant-assisted bioremediation of crude oil by indigenous bacteria isolated from Taean beach sediment. *Environ. Pollut., 241*, 254–264.

Maier, R. M., (2000). Bioavailability and its importance to bioremediation. In: Valdes, J. J., (ed.), *Bioremediation* (pp. 59–78). Springer, Dordrecht.

Makkar, R. S., & Rockne, K. J., (2003). Comparison of synthetic surfactants and biosurfactants in enhancing biodegradation of polycyclic aromatic hydrocarbons. *Environ. Toxicol. Chem., 22*, 2280–2292.

Mao, X., Jiang, R., Xiao, W., & Yu, J., (2015). Use of surfactants for the remediation of contaminated soils: A review. *J. Hazard. Mater., 285*, 419–435.

Mayri, A., De Rienzo, D., Stevenson, P., Marchant, R., & Banat, I. M., (2016). Antibacterial properties of biosurfactants against selected Gram-positive and -negative bacteria. *FEMS Microbiol. Lett., 363*(2), fnv224.

Megharaj, M., Ramakrishnan, B., Venkateswarlu, K., Sethunathan, N., & Naidu, R., (2011). Bioremediation approaches for organic pollutants: A critical perspective. *Environ. Int., 37*, 1362–1375.

Mekala, S., Peters, K. C., Singer, K. D., & Gross, R. A., (2018). Biosurfactant-functionalized porphyrin chromophore that forms J-aggregates. *Org. Biomol. Chem., 16*(39), 7178–7190.

Mnif, I., Rihab, S., & Dhouha, G., (2017). Application of bacterial biosurfactants for enhanced removal and biodegradation of diesel oil in soil using a newly isolated consortium. *Process Saf. Environ. Prot., 109*, 72–81.

Muthukamalam, S., Sivagangavathi, S., Dhrishya, D., & Rani, S. S., (2017). Characterization of dioxygenases and biosurfactants produced by crude oil-degrading soil bacteria. *Braz. J. Microbiol., 48*, 637–647.

Neifar, M., Chouchane, H., Najjari, A., Hidri, D. E., Mahjoubi, M., Ghedira, K., Naili, F., et al., (2018). Genome analysis provides insights into crude oil degradation and biosurfactant production by extremely halotolerant *Halomonas desertis* G11 isolated from Chott El-Djerid Salt-Lake in Tunisian desert. *Genomics*. https://doi.org/10.1016/j.ygeno.2018.12.003.

Nikolopoulou, M., & Kalogerakis, N., (2008). Enhanced bioremediation of crude oil utilizing lipophilic fertilizers combined with biosurfactants and molasses. *Mar. Pollut. Bull., 56*, 1855–1861.

Pacwa-Płociniczak, M., Płaza, G. A., Piotrowska-Seget, Z., & Cameotra, S. S., (2011). Environmental applications of biosurfactants: recent advances. *Int. J. Mol. Sci., 12*, 633–654.

Padmapriya, B., Suganthi, S., & Anishya, R. S., (2013). Screening, optimization and production of biosurfactants by *Candida* species isolated from oil-polluted soils. *Am. Eurasian J. Agric. Environ. Sci., 13*(2), 227–233.

Paria, S., (2008). Surfactant-enhanced remediation of organic contaminated soil and water. *Adv. Colloid Interface Sci., 138*, 24–58.

Parthipan, P., Preetham, E., Machuca, L. L., Rahman, P. K. S. M., Murugan, K., & Rajasekar, A., (2017). Biosurfactant and degradative enzymes mediated crude oil degradation by bacterium *Bacillus subtilis* A1. *Front. Microbiol., 8,* 193. doi: 10.3389/fmicb.2017.00193.

Parthipan, P., Sabarinathan, D., Angaiah, S., & Rajasekar, A., (2018). Glycolipid biosurfactant as an eco-friendly microbial inhibitor for the corrosion of carbon steel in vulnerable corrosive bacterial strains. *J Mol. Liq., 261*, 473–479.

Patowary, K., Patowary, R., Kalita, M. C., & Deka, S., (2017). Characterization of biosurfactant produced during degradation of hydrocarbons using crude oil as sole source of carbon. *Front. Microbiol., 8*, 279. doi: 10.3389/fmicb.2017.00279.

Patowary, R., Patowary, K., Kalita, M. C., & Deka, S., (2018). Application of biosurfactant for enhancement of bioremediation process of crude oil-contaminated soil. *Int. Biodeterior. Biodegradation.* https://doi.org/10.1016/j.ibiod.2018.01.004.

Paul, D., Pandey, G., Pandey, J., & Jain, R. K., (2005). Accessing microbial diversity for bioremediation and environmental restoration. *Trends Biotechnol., 23*, 135–142.

Pi, Y., Bao, M., Liu, Y., Lu, T., & He, R., (2017). The contribution of chemical dispersants and biosurfactants on crude oil biodegradation by *Pseudomonas* sp. LSH-7. *J. Clean. Prod., 153*, 74–82.

Prakash, A. A., Prabhu, N. S., Rajasekar, A., Parthipan, P., AlSalhi, M. S., Devanesan, S., & Govarthanan, M., (2020). Bio-electrokinetic remediation of crude oil contaminated soil enhanced by bacterial biosurfactant. *J. Hazard. Mater.,* 124061. https://doi.org/10.1016/j.jhazmat.2020.124061.

Rocha, C. A., Pedregosa, A. M., & Laborda, F., (2011). Biosurfactant-mediated biodegradation of straight and methyl-branched alkanes by *Pseudomonas aeruginosa* ATCC 55925. *AMB Exp., 1*, 9.

Ron, E. Z., & Rosenberg, E., (2001). Natural roles of biosurfactants. *Environ. Microbiol., 3*(4), 229–236.

Rongsayamanont, W., Soonglerdsongpha, S., Khondee, N., Pinyakong, O., Tongcumpou, C., Sabatini, D. A., & Luepromchai, E., (2017). Formulation of crude oil spill dispersants based on the HLD concept and using a lipopeptide biosurfactant. *J. Hazard. Mater., 334*, 168–177.

Roszak, D. B., & Colwell, R., (1987). Survival strategies of bacteria in the natural environment. *Microbiol. Rev., 51*, 365–379.

Sajna, K. V., Sukumaran, R. K., Gottumukkala, L. D., & Pandey, A., (2015). Crude oil biodegradation aided by biosurfactants from *Pseudozyma* sp. NII 08165 or its culture broth. *Bioresour. Technol., 191*, 133–139.

Sakthipriya, N., Doble, M., & Sangwai, J. S., (2015). Action of biosurfactant producing thermophilic *Bacillus subtilis* on waxy crude oil and long-chain paraffins. *Int. Biodeterior. Biodegradation., 105*, 168–177.

Sarubbo, L., Luna, J., & Rufino, R., (2015). Application of a biosurfactant produced in low-cost substrates in the removal of hydrophobic contaminants. *Chem. Eng. Trans., 43*, 295–300.

Shah, M. U. H., Moniruzzaman, M., Sivapragasam, M., Talukder, M. M. R., Yusup, S. B., & Goto, M., (2019). A binary mixture of a biosurfactant and an ionic liquid surfactant as a green dispersant for oil spill remediation. *J. Mol. Liq., 280*, 111–119.

Sharma, S., Verma, R., & Pandey, L. M., (2019). Crude oil degradation and biosurfactant production abilities of isolated *Agrobacterium fabrum* SLAJ731. *Biocatal. Agric. Biotechnol., 21*, 101322.

Shin, K. H., Ahn, Y., & Kim, K. W., (2005). Toxic effect of biosurfactant addition on the biodegradation of phenanthrene. *Environ. Toxicol. Chem., 24*, 2768–2774.

Singh, P., & Tiwary, B. N., (2016). Isolation and characterization of glycolipid biosurfactant produced by a *Pseudomonas otitidis* strain isolated from Chirimiri coal mines, India. *Bioresour. Bioprocess., 3*, 42.

Soberon-Chavez, G., & Maier, R. M., (2010). Biosurfactants: A general overview. In: Soberon-Chavez, G., (ed.), *Biosurfactants: From Genes to Applications* (pp. 1–11). Springer, Munster, Germany.

Souza, E. C., Vessoni-Penna, T. C., & De Souza, O. R. P., (2014). Biosurfactant-enhanced hydrocarbon bioremediation: An overview. *Int. Biodeterior. Biodegradation, 89*, 88–94.

Speight, J. G., & Arjoon, K. K., (2012). *Bioremediation of Petroleum and Petroleum Products* (pp. 1–38). John Wiley & Sons.

Surygała, J., (2001). *Crude Oil and Environment (in Polish).* Oficyna Wydawnicza Politechniki Wrocławskiej. Wrocław.

Tao, W., Lin, J., Wang, W., Huang, H., & Li, S., (2020). Biodegradation of aliphatic and polycyclic aromatic hydrocarbons by the thermophilic bioemulsifier-producing *Aeribacillus pallidus* strain SL-1. *Ecotoxicol. Environ. Saf., 189*, 109994.

Urum, K., & Pekdemir, T., (2004). Evaluation of biosurfactants for crude oil contaminated soil washing. *Chemosphere, 57*, 1139–1150.

Wang, X., Cai, T., Wen, W., Ai, J., Ai, J., Zhang, Z., Zhu, L., & George, S. C., (2020). Surfactin for enhanced removal of aromatic hydrocarbons during biodegradation of crude oil. *Fuel,* 267, 117272.

Watanabe, K., (2001). Microorganisms relevant to bioremediation. *Curr. Opin. Biotechnol., 12*, 237–241.

Wolicka, D., & Borkowski, A., (2012). Microorganisms and crude oil. In: Romero-Zeron, L., (ed.), *Introduction to Enhanced Oil Recovery (EOR) Processes and Bioremediation of Oil-Contaminated Sites* (pp. 113–142). InTech.

Zenati, B., Chebbi, A., Badis, A., Eddouaouda, K., Boutoumi, H., Hattab, M. E., Hentati, D., et al., (2018). A non-toxic microbial surfactant from *Marinobacter hydrocarbonoclasticus* SdK644 for crude oil solubilization enhancement. *Ecotoxicol. Environ. Saf., 154*, 100–107.

Zhang, X., Xu, D., Zhu, C., Lundaa, T., & Scherr, K. E., (2012). Isolation and identification of biosurfactant producing and crude oil-degrading *Pseudomonas aeruginosa* strain. *Chem. Eng. J., 209*, 138–146.

Zhao, Z., Selvam, A., & Wong, J. W. C., (2011). Synergistic effect of thermophilic temperature and biosurfactant produced by *Acinetobacter calcoaceticus* BU03 on the biodegradation of phenanthrene in bioslurry system. *J. Hazard. Mater., 190*, 345–350.

Zheng, C., He, J., Wang, Y., Wang, M., & Huang, Z., (2011). Hydrocarbon degradation and bioemulsifier production by thermophilic *Geobacillus pallidus* strains. *Bioresour. Technol., 102*, 9155–9161.

CHAPTER 8

Microbial Enhanced Oil Recovery

GEETHA S. J.,[1] YAHYA AL-WAHAIBI,[2] SAIF AL-BAHRY,[1,3] ABDULKADIR ELSHAFIE,[1] ALI AL-BEMANI,[2] and SANKET J. JOSHI[3,4]

[1]*Department of Biology, College of Science, Sultan Qaboos University, Oman*

[2]*Petroleum and Chemical Engineering Department, College of Engineering, Sultan Qaboos University, Oman*

[3]*Oil and Gas Research Center, Sultan Qaboos University, Oman*

[4]*Central Analytical and Applied Research Unit, College of Science, Sultan Qaboos University, Oman*

8.1 HISTORY AND BACKGROUND

Controlled use of fire by early humans was a turning point in the cultural evolution of humans as a species. Fire provided a source of heat, ability to work in the dark, protection from predatory animals, a method to cook food, to make tools, and expand cultural and ceremonial practices. Plant-based wood was most likely used first as fuel, and alternative fuels such as animal dung, plant-based oils and animal fats were adopted later. Sea voyages and hunting of sea animals, lead to harvesting of whale-oils and paved the way for technologies to handle liquid fuels. Although it was widely used during the 19th and early 20th centuries, the discovery of crude oil revolutionized all sectors of human endeavor, including transport, household activities, the energy sector, and other industries. Advances in technologies led to discoveries of terrestrial and deep-sea oil fields, increased crude oil production, and resulted in large-scale separation processes to distill products such as petroleum, diesel, kerosene, and gas; this was a boon for the petroleum engineering sector and related industries. The petroleum industry plays a powerful role in global geopolitics, and now that the peak of resource exploitation may be approaching, it has also catalyzed and increased awareness of the challenges of non-renewable energy sources, and the need for development and deployment of renewable alternatives. While important, the shift to renewable technologies will take many more years to fully integrate into society in a cost-competitive manner. Thus, crude oil and other fossil fuels will continue to play a role in human endeavors for years to come. As such, it is inevitable that the easiest oil reserves to exploit, those harboring light oils, will eventually be fully consumed, demanding better technologies for recovering heavy oil residuals from currently spent wells. One such technology, proposed decades ago and now enjoying renewed interest, is the application of microorganisms and their products downhole to enhance heavy oil recovery once traditional chemical methods are no longer effective. These technologies will be of particular interest in parts of the world dominated by heavy oil reserves, such as Canada, Mexico, Middle East, and Venezuela. The history, current state of the art, and future of microbial enhanced oil recovery (MEOR), are described here.

8.2 CRUDE OIL AND ENHANCED OIL RECOVERY (EOR)

Crude oil is a heterogeneous and complex mixture of hydrocarbons (HCs), resins, heavy metals, polycyclic aromatic HCs, water, salts, and minerals. Crude oils may be broadly classified as sweet-sour, light, heavy, conventional or unconventional, but a better approach is to use descriptive measures such density and gravity, as proposed by the American Petroleum Institute (the API index is based on relative oil densities). Using the API scale, crude oils are classified as light ($API \geq 31$), medium ($22 \leq API < 31$), heavy ($10 \leq API < 22$), and extra heavy ($API \leq 10$). Light oils are considered conventional, and heavy oil, extra-heavy oil, oil or tar sand, shale, and bitumen are considered unconventional (Al-Sayegh et al., 2017; Joshi et al., 2019). Conventional light oils are preferred due to ease of recovery, transportation, and refinability, resulting in higher profits and higher percentages of light petroleum fuels. Easy to recover oil sources are becoming depleted, leaving unconventional resources to account for a major share of world oil reserves. High-viscosity reserves pose challenges upstream during recovery, mid-stream during transportation to refineries and ports, and downstream during refining (Al-Sayegh et al., 2017; Joshi et al., 2019).

Increased demand for fuel and petrochemical derivatives, such as plastics and fine chemicals (Figure 8.1), will continue to build global crude demand. As forecasted by the International Energy Agency, the petrochemical sector will account for almost half of global oil demand growth by 2050, equivalent to almost 7 million barrels per day (Hickin, 2019).

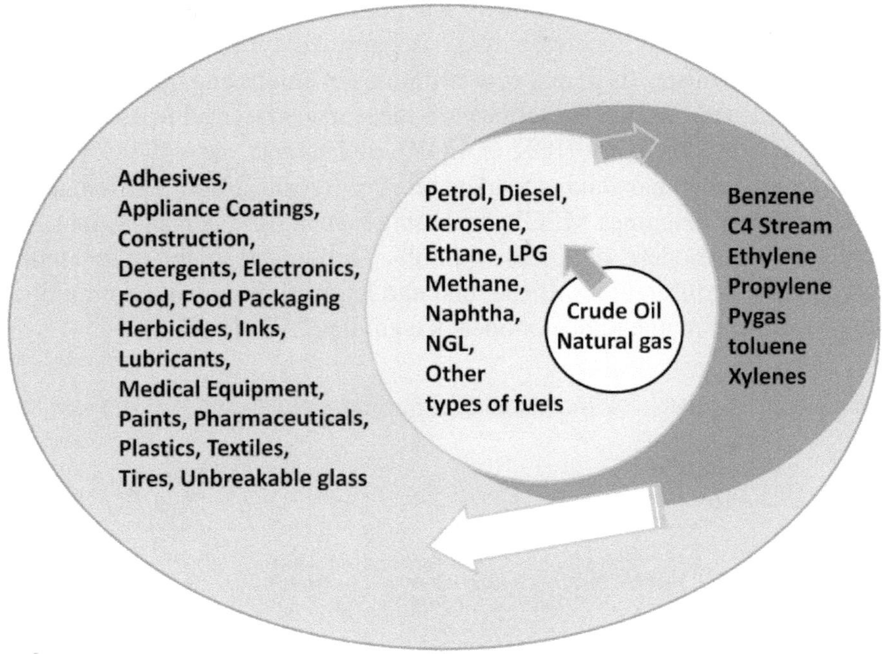

FIGURE 8.1 Refined petrochemicals derived from crude oil and natural gas (Hickin, 2019).

Upstream processes in the petroleum industry include locating and exploration of oil fields, informed by geological analysis of drill cores and water-oil samples, drilling, primary, secondary, and tertiary oil recovery (Figure 8.2), and transport of oil to refineries by rail, truck, pipeline, and ports. Primary production relies on natural pressure release to bring crude oil to well heads for a limited time, and once the natural pressure dissipates, secondary production follows, where reservoir pressure is increased artificially either with beam pumps (artificial lifts) or by injecting water (with optional chemical additives). Oil-water immiscibility and subsurface fractures limits secondary recovery, necessitating deployment of tertiary recovery options. These include chemical, thermal, and microbial technologies. Chemical technologies are widely employed and involve surfactant (anionic or cationic), polymer (partially hydrolyzed polyacrylamides), alkali-surfactant-polymer combination, emulsifier, and gas (CO_2) injections, to name a few. Thermal technologies mainly exploit steam flooding (with or without chemicals) or *in situ* combustion. Microbial methods consist of injecting microbes, microbial nutrients or microbial products such as biosurfactants, biopolymers, gases, biosolvents or biomass. The selection of recovery methods depends on criteria such as: subsurface porosity and permeability, geology, and type of rock (sandstone or carbonates, clays), viscosity, and type of crude oil, and oil field water chemistry (type and concentrations of anions and monovalent and divalent cations). Generally, tertiary oil recovery is referred to as improved or enhanced oil recovery (IOR or EOR), and targets the ~45–60% of oil remaining after primary and secondary recovery (Joshi, 2008). The remainder of this chapter examines MEOR, including mechanistic considerations, the application of modern molecular microbial ecology techniques for understanding the proliferation of beneficial and harmful microbial communities down well, and mathematical models for guiding MEOR efforts.

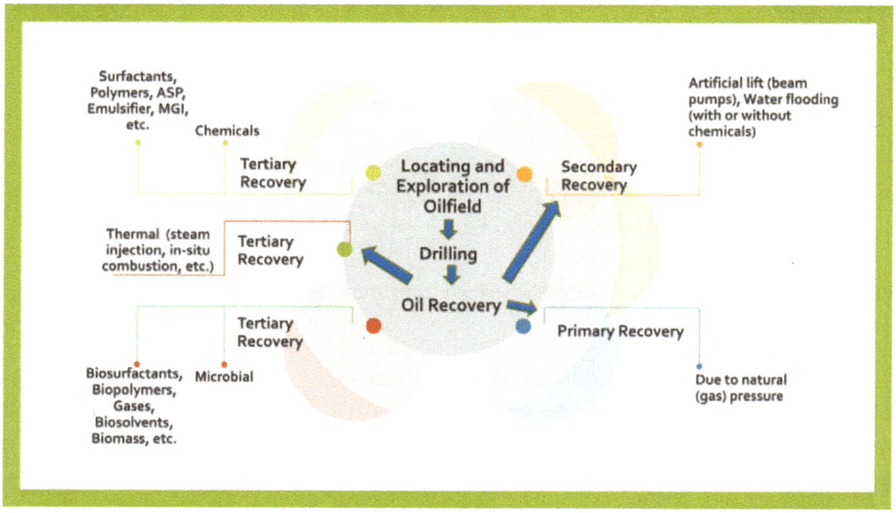

FIGURE 8.2 Crude oil recovery phases and technologies employed for enhanced oil recovery (Joshi, 2008).

8.3 MICROBIAL ENHANCED OIL RECOVERY (MEOR)

Globally, chemically driven EOR technologies are widely utilized due to overall ease of application, availability of reservoir-compatible chemicals, and cost. Recent awareness about the toxicity and recalcitrance of recovery chemicals and their degradation products has led to efforts to develop more environmentally compatible alternatives. Microbial EOR (MEOR) processes are one such alternative, and are gaining attention due to improved biodegradability, ease of use, and decreasing cost of application (Liu and Wei, 2017). First suggested in 1926, a few laboratory studies on MEOR were carried out through 1940, and in 1954 the first MEOR field test was reported in the USA (Brown, 2010). Initially viewed by industry as a scientific curiosity, the knowledge and perception of MEOR as a viable technology has improved, and increasing academic and industry research is being carried out at both laboratory and field-scale trials around the world (Banat, 2010; Al-Sulaimani et al., 2011; Voordouw, 2011; Al-Maaini et al., 2014; Al-Hinai et al., 2014; Al-Sayegh et al., 2017; Geetha et al., 2018).

At its core, it is microbial products that play the most important roles in MEOR, products such as biosurfactants, biopolymers, biogas, bioacids, biosolvents, and microbial biomass (Table 8.1). Biosurfactants, such as rhamnolipids, sophorolipids, surfactin, and lichenysins, are reported to enhance oil recovery by reducing surface and interfacial tension; altering reservoir wettability; emulsifying oil in water; and altering reservoir permeability. Biopolymers such as xanthan, schizophyllan, and gellan enhance oil recovery by increasing injection water viscosity, altering reservoir permeability, and changing water mobility.

TABLE 8.1 Microbial Products and Their Roles in MEOR (McInerney et al., 2005; Lazar et al., 2007; Joshi, 2008; Putra and Hakiki, 2019; Gaol et al., 2019)

Microbial Product	Role in MEOR
Biosurfactants	Surface tension and interfacial tension reduction, wettability alteration, emulsification, relative permeability alteration.
Biopolymers	Increase in injection water viscosity, permeability alteration, water mobility control.
Biogases (CO_2, CH_4, N_2, H_2)	Reservoir repressurization, reduction in oil viscosity, solubilization of carbonates.
Biosolvents	Reduction in oil viscosity, interfacial tension reduction, emulsification.
Bioacids	Dissolution of rock and minerals, relative permeability alteration, wettability alteration.
Microbial biomass (microbial permeability profile modification)	Permeability alteration, selective plugging of high permeability thief zones, wettability alteration, heavy oil upgrading by degradation, emulsification of oil by enzymes and other bioproducts.

MEOR technologies are applied in either *in situ* or *ex-situ* mode. For *in situ* MEOR, either nutrients (such as molasses), or nutrients and exogenous microbes, are injected to stimulate the production of metabolites such as biosurfactants, biopolymers, solvents, acids, and biomass. Applied in huff and puff (one well injection) mode, or in four or five-point injection (one injector well and four producers) mode, *in situ* MEOR is generally more economical and more widely adopted than *ex-situ* MEOR, but challenges such as uncertainty of success rate, long incubation times, undesirable indigenous microbial activities (e.g., sulfate reduction) must be faced. *Ex-situ* MEOR involves producing microbial metabolites outside of the reservoir prior to injection. The advantage of *ex situ* MEOR is ease of application using existing techniques for chemically-enhanced recovery, and the opportunity to lab test prior to field application taking into account parameters such as microbial product stability, adsorption, and effects on wettability. The biggest challenge for *ex situ* MEOR is cost-biosurfactants and biopolymers remain expensive to produce and obtain at scale. Other issues include high adsorption to reservoir rock and degradation by indigenous microbes, all of which has translated into very few reports of field application (Geetha et al., 2018), even with successful lab-scale MEOR studies available in the literature (Dastgheib et al., 2008; Al-Hattali et al., 2013; Joshi and Desai, 2013; Al-Wahaibi et al., 2016; Joshi et al., 2016a, b; Jha et al., 2016; Elshafie et al., 2017; Shibulal et al., 2018; Al-Ghailani et al., 2018). On the whole, reports of field-scale MEOR application have been positive: Safdel et al. (2017) compiled 47 MEOR field trial cases from 21 different countries; Maudgalya et al. (2007) compiled 407 reported MEOR field trials in sandstone and carbonate reservoirs, with some reports being of particular interest (Bhupathiraju et al., 1993; Lazar et al., 2007; Brown, 2010; Le et al., 2015; Patel et al., 2015; Liu and Wei, 2017; Gao, 2018).

8.4 OMICS TOOLS FOR IDENTIFICATION OF BENEFICIAL AND HARMFUL MICROBIAL FLORA IN OIL RESERVOIRS

Microorganisms are environmentally omnipresent, and oil reservoirs are no exception to alterations catalyzed by the metabolic activities of microbial extremophiles which can survive under conditions of high salinity, wide ranges of temperatures and pH, oxygen limitation, and high levels of toxic organic and inorganic matter. It is imperative to characterize and understand the beneficial (e.g., MEOR, biodegradation) and harmful (e.g., souring, plugging) activities of microbial communities in oil reservoirs. Recent advances in analytical techniques, multi-omic technologies for molecular microbial ecology, bioinformatics, and computer technologies, are expanding our perceptions of key MEOR processes (Van Hamme et al., 2003; Voordouw, 2011; Ismael et al., 2017).

Traditional culture-dependent, and modern culture-independent, microbial community ecology methods applied to onshore and offshore oil reservoirs, with or without EOR application, are expanding our understanding of oil field microbial community structure and metabolic potential (Mueller and Nielsen, 1996; Orphan et al., 2000, 2003; Chai et al., 2015; Zhang et al., 2010, 2012; Kim et al., 2018). Culture-dependent methods can only characterize a fraction of microorganisms in environmental matrices (Al-Bahry et al., 2013), and currently it is thought that the majority of indigenous oil reservoir microorganisms are non-culturable (Tüccar et al., 2019). Despite this, combinations of culture-dependent and independent methods are yielding insights into oil field microbial communities (Al-Bahry et al., 2013) in China, Norway, Oman, Canada, USA, Thailand, and Nigeria (Table 8.2). Much of this work has focused on produced water and crude oil due to the technical difficulties involved with collecting uncontaminated reservoir rocks and sediments. However, the majority of oil field microbes are attached to the subsurface strata, and data obtained from core samples should be more relevant to MEOR than those obtained from produced water or oil (Tüccar et al., 2019). Given that indigenous microbes respond to external nutrient addition during drilling, EOR operations, production, and transport, it is critical that sampling techniques for obtaining relevant samples from all subsurface strata be developed for microbiological characterization.

Oilfield microbial communities have been characterized using culture-dependent techniques, now outdated culture-independent techniques (e.g., denaturing gradient gel electrophoresis (DGGE)), and newer high-throughput sequencing techniques which can also capture metagenomes and metatranscriptomes (Al-Bahry et al., 2013; Zhang et al., 2014; Hu et al., 2016; Vigneron et al., 2017; Kim et al., 2018; Tüccar et al., 2019). Growing datasets are supporting the hypothesis that microbes are thriving under the harsh environmental conditions in oil reservoirs and that the addition of chemicals and nutrients during treatment of oil wells for EOR and MEOR field applications leads to alterations in indigenous microbial community makeup and activity. Voordouw et al. (1993, 1996) reported a protocol for quantitative analysis of microbial communities using 'reverse sample genome probing,' cloning and sequencing of PCR-amplified 16S rRNA genes, allowing them reproducible fingerprinting of the microbial community in oil field production waters from different sites, in contact with the production waters. They detected a variety of gram-negative, sulfate-reducing bacteria (SRB) and a number

TABLE 8.2 Global Oil Field Microbial Community and Diversity Studies

Oil Field	Region/Country	Reservoir Conditions	Most Abundant Microbes	Reference
Sinopec shengli oil field	China	Temperature: 55–91°C; salinity: 3,000–20,000 mg/l; permeability: 207×10^{-3}–$6,900 \times 10^{-3}$ μm^2	Archaebacteria and thermophilic bacteria (*Thermus, Thermincola, Thermanaeromonas*), hyperthermophilic archaea (*Geoglobus*), halophilic bacteria	Lin et al. (2014)
Daqing oil field	Northeast China	Sandstone reservoir; 44.6°C; average permeability 0.414 μm^2; polymer flooding	*Pseudomonas, Acinetobacter, Thauera, Azovibrio, Arcobacter, Helicobacter, Desulfitobacterium*, and *Clostridium*	Gao et al. (2018)
Bahja oil field	Oman	Sandstone; salinity 71–78,000 ppm; permeability and porosity in the range of 100–2,000 and 22–28; bottom-hole temperature of 61–64°C	*Thermotoga* spp., *Halomonas* spp., *Petrotoga* spp., *Clostridium* spp., *Geobacter* spp., *Desulfotomaculum* spp., *Rhodococcus* spp., *Diaphorobacter* spp., *Thermotoga* spp., and *Actinobacterium* spp.	Al-Bahry et al. (2013)
Different heavy oil wells	Southern Oman	–	*Bacillus* sp., *Paenibacillus, Brevibacillus, Planococcus*, and *Flavobacterium*	Shibulal et al. (2018)
Northwest of the HeNan oilfield	China	Crude oil pour point of 45°C and wax content of 46.72%; formation pressure is 1977.8 psi, and the temperature approx. 65°C	*Bacillus* sp. *and C. ultunense*, *Acinetobacter* sp., *Pseudomonas* sp.	He et al. (2016)
The Dan and Halfdan oil fields	North sea, ~200 km west off Southern Jutland (Denmark)	Formations is situated about 2,000 m below the surface; low-permeability chalks; temperature 80C	*Firmicutes* (Bacteria), *Thermococcales* (Archaea), heterotrophic nitrate reducers (*Deferribacterales*), archaeal SRP (*Archaeoglobus*), thermophilic genera (*Desulfacinum* and *Desulfotomaculum*)	Gittel et al. (2009)
Chengdong petroleum reservoir, Shengli oilfield	Yellow River Delta of Shandong province adjacent to the Bohai Bay of China	Temperature 55–65°C; different water contents (66.8–98.0%); Oil viscosity (230.8–2866 mPa s)	Archaeal communities (*Archaeoglobus, Methanothermobacter, Methermicoccus, Thermococcus, Methanosaeta*)	Liang et al. (2018)

TABLE 8.2 *(Continued)*

Oil Field	Region/Country	Reservoir Conditions	Most Abundant Microbes	Reference
Liaohe oil field (J68-50 well group)	China	–	*Comamonas*, *Brevundimonas*, *Azospirillum*, *Achromobacter*, *Pseudomonas*, and *Hyphomonas*	Su et al. (2018)
Bamianhe oilfield	Dongying, east China's Shandong Province	Temperature 60°C; water salinity 35,000 mg/L; the average buried depth is 950–1210 m; an average porosity of approximately 33.7% and permeability of approximately 0.458 μm^2; oil viscosity 4896 mPa·s	Thermophilic genera (*Thermanaeromonas*, *Thermolithobacter*, *Thermacetogenium*, *Hydrogenophilus*, *Tepidiphilus*, *Petrobacter*, *Bacillus*, and *Deferribacter)*, unclassified bacteria, and halotolerant bacterial species *(Paenibacillaceae and Marinobacter)*	Lujun et al. (2018)
Jilin oilfield	Northeast of China	Temperature 40°C; total dissolved solids 7.2 g/l; pH 8.2	*Sulfurospirillum*, *Oceanobacter*, *Pseudomonas*	Song et al. (2018)
Five different high-temperature oil reservoirs near Segno, Texas, and Crossfield, Alberta	USA and Canada	Temperature ~75–85°C	Hydrogenotrophic methanogens (*Methanosaetaceae*, *Methanobacterium*, and *Methanoculleus*), bacteria belonging to Clostridiaceae and Thermotogaceae	Kim et al. (2018)
Olla oil field in LaSalle Parish, Louisiana	USA	CO_2-EOR flood	Methanogenic archaea (*Methanothermococcus* spp., *Methanohalophilus halophilus*, *Methanolobus* sp.), bacteria (*Thermovirgaceae*, *Alicyclobacillus* sp., *Deferribacter* sp. and *Acinetobacter lwoffii*)	Shelton et al. (2018)
Mae soon reservoir of Fang oil field	Chiang Mai, Northern Thailand	Formation temperature 63–72°C; grain density 2.63–2.64 g/cm^3; permeability 0.1–193 mD; porosity 11.20–23%	*Deinococcus-Thermus*, *Betaproteobacteria*, *Thermus*, *Acinetobacter*, and *Pseudomonas*	Phetcharat et al. (2018)

TABLE 8.2 *(Continued)*

Oil Field	Region/Country	Reservoir Conditions	Most Abundant Microbes	Reference
Oil reservoir	Northeast China	Temperature 46–52°C; salinity 1,583–1,934 mg/l	*Ochrobactrum*, *Acinetobacter*, *Lactococcus*, *Pseudomonas*, *Prauseria*, and *Bacillus*	Liu et al. (2019)
Obigbo North facility	Nigeria	Temperature 25–45°C; pH 6.85–8.32; NaCl 157–179 mM	*Marinobacter*, *Azovibrio*, *Thauera*, *Methanolobus*, *Methanolobus*, and *Methanobacterium*	Okoro et al. (2018)
Shen84 oil reservoir, Liaohe oil field	China	Temperature 43°C; permeability 200–2000 mD;	*Pseudomonas* and *Thauera* members from Proteobacteria phylum	Wang et al. (2019)
Jin 35 block of Liaohe sandstone reservoir	Liaoning province of China	Temperature 45–51°C	*Lysinibacillus*, *Acinetobacter junnii*, *Sulfurihydrogenibium*	Wang et al. (2019)
Shengli oil field	Delta of the yellow river, China	Sandstone; temperature 63°C; porosity 30%; water permeability 247–556×10^{-3}μm^2	Proteobacteria, firmicutes, *Thermotogae*, Synergistetes, *Actinobacteria*, *Bacteroidetes*, *Deferribacteres*	Song et al. (2019)

of anaerobic, fermentative, or acetogenic bacteria. The newly discovered sulfide oxidizer, reported by them, could provide a vital link in the oil field sulfur cycle (microbial sulfate or sulfur reduction). An et al. (2013) compared 160 microbial community compositions from different HC resource environments and sequenced metagenomes to characterize their metabolic potential. They observed several aerobic taxa and genes not commonly reported from such environments, these observations may influence strategies to recover crude oil from these environments by *in situ* microbial processes. Singh et al. (2014) compiled and reported oil field microbial community structures (and changes in response to the chemicals introduced during various oil field operations), using metagenomics techniques, and their role on the natural microbiological processes and the impacts on oil recovery techniques.

While currently limited by the technological challenges discussed above, and both industry and national intellectual property barriers, these data will contribute to the development of beneficial MEOR technologies if properly exploited with modern computational modeling.

8.5 MICROFLUIDICS, MICROMODELS, MATHEMATICAL MODELING AND SIMULATIONS FOR MEOR

Generally, microfluidics, micromodels, mathematical, and numerical models, and reservoir simulators are used to perform, predict, and understand the processes driving EOR prior to field application. These models are informed by laboratory-based core flood experiments (or simulations), field trials, and operational field experience, and are important tools for optimizing operational performance (Haq et al., 2019; Jeong et al., 2019a).

8.5.1 MICROFLUIDICS AND MICROMODELS

Although the term 'oil reservoir' implies large subsurface pools of liquid petroleum, in reality crude oil resides in micro and nano-scale pores in rock. Direct visualization of fluid flow dynamics is critical for understanding and manipulating complex fluid (crude oil, water, chemicals, and bioproducts) flow in porous oil reservoir media during EOR. However, porous media such as core-plugs are opaque, which makes direct visualization challenging to impossible. The typical conventional reservoir fluid analyses (e.g., to check effect of pressure-volume-temperature (PVT) on EOR and formation damage) used in the petroleum industry require high capital and labor investment, and long run times. As an alternative, micromodels and microfluidic chip systems (also known as lab-on-a-chip) may be used. These are transparent interconnected porous networks that enable optical of the complex fluid phase dynamics (viscosity, wetting behavior, surface, and interfacial tension, and capillary effects) occurring at the pore scale at reservoir temperatures, pressures, and geometries (Sinton, 2014; Bao et al., 2017; Anbari et al., 2018). Depending on the fabrication method, 2D and 3D micromodels can be prepared from glass, silicon, and polymer materials such as polydimethylsiloxane and polymethyl methacrylate (Anbari et al., 2018) allowing for: faster processing, smaller experimental footprint, control over temperature, pressure, and chemical concentrations, and lower sample volumes (Sinton, 2014; Bao et al., 2017). Microfluidic systems have been used to analyze oil recovery performance impacts of chemical and biological surfactants and polymers, perhaps performing better than traditional sand-pack methods, with higher repeatability when pore structure and wettability is controlled (Chandler et al., 2019).

Gaol et al., (2019) investigated MEOR mechanisms using microfluidic technology, quantifying the contribution of plugging and fluid mobility changes to incremental oil recovery. They used a novel single-phase and two-phase flooding experimental setup enabling the investigation of MEOR in micromodels under elevated pressure and temperature under anaerobic and sterile conditions. The study was conducted for ten days with static incubation, and bacterial cells and *in situ* gas production were visualized and quantified by using an image processing algorithm; both differential and absolute pressures were measured throughout the single-phase experiment. Their results provided insights into the pore-scale mechanisms of MEOR processes in porous media under defined reservoir conditions. Wang et al. (2013) studied the microscopic interrelation between bacterial cells and crude oil in a transparent gas-permeable chip with an internal microchannel. The biodegradation of crude oil by *Dietzia* under aerobic conditions was characterized revealing the production of biosurfactants, reducing the interfacial tension and thus increasing the dissolution of the crude oil. Rahayyem et al. (2019) reported enzyme-enhanced oil recovery (EEOR) using enzyme (Apollo GreenZyme®) and a commercial surfactant (J13131) in polydimethylsiloxane (PDMS) microfluidic devices. They reported up to 92% oil recovery using GreenZyme in a sandstone PDMS microfluidic device.

8.5.2 MATHEMATICAL MODELING AND SIMULATIONS

Even though MEOR technology is recognized as an economical technology, it has not been widely used in the oil industry due to the inconsistency of its successes. This must be addressed, and accurate performance predictions under various operating conditions must be achieved for MEOR technology to be successfully applied at scale. This requires mathematical and numerical modeling, challenging as accurately capturing the complexities of microbial community dynamics under reservoir conditions requires a granular understanding of microbial growth and metabolism, physiology, gene expression, speciation, drift, and dispersal (Jeong et al., 2019). The associated biochemical reactions and their kinetics must be captured in models in order to drive rational approaches to selective plugging for enhancing oil mobility without compromising oil quality and oilfield safety.

To date, modified Monod kinetics have been applied in reservoir simulations (using software such as CMG STARS, CMG CMOST, Eclipse) to describe the biochemical mechanisms underlying MEOR (Table 8.3). Most of those simulations considered wettability and permeability alterations, interfacial tension reduction in the presence of biosurfactants and biopolymers, the plugging effects of microbial biomass, and dissolution activities of bioacids and biosolvents. However, extensive experimental work, mathematical modeling, and simulation studies are still required for better understanding of MEOR mechanisms, before deploying field-scale applications.

8.6 CONCLUSIONS AND FUTURE RESEARCH DIRECTIONS

With turbulent global economic markets, growing populations, and decreasing conventional oil supplies, MEOR has the potential to improve recovery of light and heavy crude oils. Gaps between understanding microbial growth and metabolism, and controlling these processes for the benefit of the oil industry, stand in the way of large-scale adoption of MEOR. Omics technologies, microfluidics, and mathematical simulations have recently yielded valuable datasets with the potential to begin filling these gaps, but much work remains before MEOR can be reliably implemented as a major oil industry practice. If intellectual property barriers can be reduced, and large multi-national studies can be deployed for building large, quality-controlled public datasets, there is no doubt that MEOR technologies would mature at a much faster rate for the benefit of all.

TABLE 8.3 Models and Simulations for MEOR Lab or Field Scale Applications

Type of Model	Software/Equations Used	Type of Medium	Microorganism and/or Product	Factors/Components Considered	Outcome	References
Temperature-dependent coupled biokinetic and thermal model	CMG STARS (2017) and CMG CMOST (2017)	Homogenous Reservoir (cold-water MEOR)	*Bacillus subtilis* and Surfactin	Nutrient concentration, injection rate, and temperature	Optimal injection design improved oil recovery up to 53%	Hong et al. (2019)
One-dimensional two-phase, five-component mathematical model	Numerical simulator (microbial transport simulator, MTS) numerical simulator was established in the Fortran	Porous medium (core flood)	Biopolymer	Oil, water, microorganisms, nutrients, and metabolites	Biopolymer-producing microorganisms improved 7–8% of oil recovery	Wang et al. (2018)
General 3D flow and transport model An axiomatic continuum modeling approach	The axiomatic continuum modeling approach, for numerical modeling a finite element method and COMSOL Multiphysics®	Under laboratory conditions at core scale	Microbial biomass and biosurfactant	Core, oil, microorganisms, and metabolites	The experimental results were accurately predicted by the simulations, and the model can be extended and applied to other scenarios	Diaz-Viera et al. (2019)
Surfactant model	Modified Eclipse reservoir simulator	Under laboratory conditions at core scale	Biosurfactant and solvent	JF-2 bio-surfactant and butanol	Simulated oil recovery curve matched with the experimental curve	Haq et al. (2019)
One-dimensional isothermal model	Monod kinetics	Simulation	Biosurfactant	Oil, water, bacteria, substrate, and metabolite (surfactant) s	An incremental recovery of 38–44%	Nielsen et al. (2010)

TABLE 8.3 *(Continued)*

Type of Model	Software/Equations Used	Type of Medium	Microorganism and/or Product	Factors/Components Considered	Outcome	References
Multiple-ion exchange transport model with multiphase fluid flow and multi-species reactive transport models	Computer codes were developed in C language to solve the complete set of governing equations	Berea sandstone rock system	*Pseudomonas putida* MTCC 2467	Injection and formation Water, oil, microbes, and nutrients	Provides criteria for the selection of suitable reservoir, microbes, and injection water salinity to attain higher oil recovery	Sivasankar and Kumar (2018)
A mathematical model for microbial enhanced oil recovery considering the double-bacterial competition mechanism	One-dimensional rectangular grid to simulate the reservoir and a one-injection/one-production model is adopted for the well pattern.	Simulation	Biosurfactants and biopolymers	Oil, water, microbe 1, microbe 2, nutrients, metabolite 1 and metabolite 2	By using surfactant and polymer model 12.3% incremental oil recovery was observed	Tianyuan et al. (2019)
Non-standard model	Continuity equations for water and oil, a balanced equation for the oil-water interface and advective-dispersive transport equations for bacteria, nutrients, and surfactants	Simulation	Biosurfactants	Interfacial area (IFA) between the oil and water, bacteria, nutrients, surfactants, and two-phase flow	Highlighted the importance of including IFA and chemotaxis in simulation of MEOR	Landa-Marbán et al. (2017)

TABLE 8.3 *(Continued)*

Type of Model	Software/Equations Used	Type of Medium	Microorganism and/or Product	Factors/Components Considered	Outcome	References
3D model	UTCHEM simulator	Simulation	Biosurfactants	Temperature, salinity, and pH, biosurfactant adsorption, maximum growth rate, and nutrient concentration	Observed 10–15% incremental oil recovery using in situ produced biosurfactant	Hosseininoosheri et al. (2016)
Systematic modeling approach	Stoichiometric equation and kinetic reactions using batch model simulation	Simulation and sand pack model	biopolymer dextran-*Leuconostoc mesenteroides*	*In situ* selective plugging by biopolymer	With the optimized MEOR process 36.7% improvement of the oil recovery	Jeong et al. (2019b)

KEYWORDS

- **crude oil**
- **enhanced oil recovery**
- **mathematical modeling and simulations**
- **microbial byproducts**
- **microbial enhanced oil recovery**
- **microfluidics and micromodels**
- **omics**

REFERENCES

Al-Bahry, S. N., Elshafie, A. E., Al-Wahaibi, Y. M., Al-Bemani, A. S., Joshi, S. J., Al-Maaini, R. A., Al-Alawi, W. J., et al., (2013). Microbial consortia in Oman oil fields: A possible use in enhanced oil recovery. *J. Microbiol. Biotechnol., 23*(1), 106–117.

Al-Ghailani, T., Al-Wahaibi, Y. M., Joshi, S. J., Al-Bahry, S. N., Elshafie, A. E., & Al-Bemani, A. S., (2018). Alkaline-biosurfactant-biopolymer process and its potential for enhancing oil recovery in Omani oil field. In: *SPE EOR Conference at Oil and Gas West Asia*. Society of Petroleum Engineers.

Al-Hattali, R., Al-Sulaimani, H., Al-Wahaibi, Y., Al-Bahry, S., Elshafie, A., Al-Bemani, A., & Joshi, S. J., (2013). Fractured carbonate reservoirs sweep efficiency improvement using microbial biomass. *J. Petrol. Sci. Eng., 112*, 178–184.

Al-Hinai, B. S., Al-Bahry, S. N., Elshafie, A. E., Al-Wahaibi, Y. M., Al-Bemani, A. S., & Joshi, S. J., (2014). Potential of microbial biotechnology in heavy oil enhanced recovery. In: *Biotechnology and Conservation of Species from Arid Regions* (Vol. 2, pp. 495–506). Nova Science Publishers, Inc.

Al-Maaini, R. A., Al-Bahry, S. N., Elshafie, A. E., Al-Wahaibi, Y. M., Al-Bimani, A. S., Joshi, S. J., & Al-Alawi, W. J., (2014). Indigenous bacterial consortia from oil wells and their role in in-Situ microbial enhanced oil recovery. In: *Biotechnology and Conservation of Species from Arid Regions* (Vol. 2, pp. 455–470). Nova Science Publishers, Inc.

Al-Sayegh, A., Al-Wahaibi, Y., Al-Bahry, S., Elshafie, A., Al-Bemani, A., & Joshi, S., (2017). Enhanced oil recovery using biotransformation technique on heavy crude oil. *Int. J. Geomate., 13*, 75–79.

Al-Sulaimani, H., Joshi, S., Al-Wahaibi, Y., Al-Bahry, S., Elshafie, A., & Al-Bemani, A., (2011). Microbial biotechnology for enhancing oil recovery: Current developments and future prospects. *Biotechnol. Bioinf. Bioeng., 1*(2), 147–158.

Al-Wahaibi, Y., Al-Hadrami, H., Al-Bahry, S., Elshafie, A., Al-Bemani, A., & Joshi, S., (2016). Injection of biosurfactant and chemical surfactant following hot water injection to enhance heavy oil recovery. *Petrol. Sci., 13*(1), 100–109.

An, D., Caffrey, S. M., Soh, J., Agrawal, A., Brown, D., Budwill, K., Dong, X., Dunfield, P. F., Foght, J., Gieg, L. M., Hallam, S. J., et al., (2013). Metagenomics of hydrocarbon resource environments indicates aerobic taxa and genes to be unexpectedly common. *Environ. Sci. Technol., 47*(18), 10708–10717.

Anbari, A., Chien, H. T., Datta, S. S., Deng, W., Weitz, D. A., & Fan, J., (2018). Microfluidic model porous media: Fabrication and applications. *Small, 14*(18), 1703575.

Banat, I. M., (1995). Biosurfactants production and use in microbial enhanced oil recovery and pollution remediation: A review. *Bioresour. Technol., 51*, 1–12.

Bao, B., Riordon, J., Mostowfi, F., & Sinton, D., (2017). Microfluidic and nanofluidic phase behavior characterization for industrial CO_2, oil and gas. *Lab on a Chip, 17*(16), 2740–2759.

Bhupathiraju, V. K., McInerney, M. J., & Knapp, R. M., (1993). Pretest studies for a microbially enhanced oil recovery field pilot in a hypersaline oil reservoir. *Geomicrobiol. J., 11*(1), 19–34.

Brown, L. R., (2010). Microbial enhanced oil recovery (MEOR). *Curr. Opin. Microbiol., 13*(3), 316–320.

Chai, L., Zhang, F., She, Y., Banat, I., & Hou, D., (2015). Impact of a microbial-enhanced oil recovery field trial on microbial communities in a low-temperature heavy oil reservoir. *Nature Env. Pol. Technol., 14*(3), 455–462.

Chandler, B. D., Kim, J., & Willmott, E., (2019). *Utilizing Microfluidics as a Rapid Screening Technology for Enhanced Oil Recovery.* U.S. Patent 0218897.

Dastgheib, S. M. M., Amoozegar, M. A., Elahi, E., Asad, S., & Banat, I. M., (2008). Bioemulsifier production by a halothermophilic *Bacillus* strain with potential applications in microbially enhanced oil recovery. *Biotechnol. Lett., 30*(2), 263–270.

Diaz-Viera, M. A., Ortiz-Tapia, A., Hernandez-Perez, J. R., Castorena-Cortes, G., Roldan-Carrillo, T., & Olguin-Lora, P., (2019). A flow and transport model for simulation of microbial enhanced oil recovery processes at core scale and laboratory conditions. *Int. J. Numer. Anal. Mod., 16*(1).

Elshafie, A., Joshi, S. J., Al-Wahaibi, Y. M., Al-Bahry, S. N., Al-Bemani, A. S., Al-Hashmi, A., & Al-Mandhari, M. S., (2017). Isolation and characterization of biopolymer producing Omani *Aureobasidium pullulans* strains and its potential applications in microbial enhanced oil recovery. In: *SPE Oil and Gas India Conference and Exhibition.* Society of Petroleum Engineers.

Gao, C., (2018). Experiences of microbial enhanced oil recovery in Chinese oil fields. *J. Petrol. Sci. Eng., 166*, 55–62.

Gao, P., Li, G., Le, J., Liu, X., Liu, F., & Ma, T., (2018). Succession of microbial communities and changes of incremental oil in a post-polymer flooded reservoir with nutrient stimulation. *Appl. Microbiol. Biotechnol., 102*(4), 2007–2017.

Gaol, C., Wegner, J., Ganzer, L., Dopffel, N., Koegler, F., Borovina, A., & Alkan, H., (2019). Investigation of pore-scale mechanisms of microbial enhanced oil recovery MEOR using microfluidics application. SPE-195553-MS. In: *SPE Europe Featured at 81st EAGE Conference and Exhibition.* Society of Petroleum Engineers.

Geetha, S. J., Banat, I. M., & Joshi, S. J., (2018). Biosurfactants: Production and potential applications in microbial enhanced oil recovery (MEOR). *Biocatal. Agric. Biotechnol., 14*, 23–32.

Gittel, A., Sørensen, K. B., Skovhus, T. L., Ingvorsen, K., & Schramm, A., (2009). Prokaryotic community structure and sulfate reducer activity in water from high-temperature oil reservoirs with and without nitrate treatment. *Appl. Environ. Microbiol., 75*(22), 7086–7096.

Haq, B., Liu, J., Liu, K., Malaki, E., & Al-Shehri, D., (2019). Modification of Eclipse simulator for microbial enhanced oil recovery. *J. Petrol. Explor. Prod. Technol., 9*, 2247–2261.

He, Y., Zhang, F., Banat, I. M., Pu, C., Dong, H., Shu, F., et al., (2016). Deposit reduction in a high pour point oil reservoir due to the activity of indigenous bacterial communities. *Int. Biodeter. Biodegr., 110*, 87–98.

Hickin, P., (2019). Insight: The middle east's petrochemicals push signals oil's future. *OPAL, 9*, 40–44.

Hong, E., Jeong, M. S., Kim, T. H., Lee, J. H., Cho, J. H., & Lee, K. S., (2019b). Development of coupled biokinetic and thermal model to optimize cold-water microbial enhanced oil recovery (MEOR) in homogenous reservoir. *Sustainability, 11*(6), 1652.

Hosseininoosheri, P., Lashgari, H. R., & Sepehrnoori, K., (2016). A novel method to model and characterize in situ biosurfactant production in microbial enhanced oil recovery. *Fuel, 183*, 501–511.

Hu, P., Tom, L., Singh, A., Thomas, B. C., Baker, B. J., Piceno, Y. M., et al., (2016). Genome-resolved metagenomics analysis reveals roles for candidate phyla and other microbial community members in biogeochemical transformations in oil reservoirs. *mBio., 7*, e01669–15.

Ismail, W. A., Van, H. J. D., Kilbane, J. J., & Gu, J. D., (2017). Petroleum microbial biotechnology: Challenges and prospects. *Front. Microbiol., 8*, 833.

Jeong, M. S., Lee, J. H., & Lee, K. S., (2019a). Critical review on the numerical modeling of in situ microbial enhanced oil recovery processes. *Biochem. Eng. J.*, 107294.

Jeong, M. S., Noh, D. H., Hong, E., Lee, K. S., & Kwon, T. H., (2019b). Systematic modeling approach to selective plugging using In-situ bacterial biopolymer production and its potential for microbial-enhanced oil recovery. *Geomicrobiol. J., 36*(5), 468–481.

Jha, S. S., Joshi, S. J., & Geetha, S. J., (2016). Lipopeptide production by *Bacillus subtilis* R1 and its possible applications. *Braz. J. Microbiol., 47*(4), 955–964.

Joshi, S. J., & Desai, A. J., (2013). Bench-scale production of biosurfactants and their potential in ex situ MEOR application. *Soil Sediment Contam., 22*(6), 701–715.

Joshi, S. J., (2008). *Isolation and Characterization of Biosurfactant Producing Microorganisms and their Possible Role in Microbial Enhanced Oil Recovery (MEOR)*. PhD Dissertation, MS University of Baroda, Vadodara, India.

Joshi, S. J., Al-Wahaibi, Y. M., Al-Bahry, S. N., Elshafie, A. E., Al-Bemani, A. S., Al-Bahri, A., & Al-Mandhari, M. S., (2016a). Production, characterization, and application of *Bacillus licheniformis* W16 biosurfactant in enhancing oil recovery. *Front. Microbiol., 7*, 1853.

Joshi, S. J., Al-Wahaibi, Y. M., Al-Bahry, S., Elshafie, A., Al-Bemani, A. S., Al-Hashmi, A., Samuel, P., et al., (2016b). Production and application of schizophyllan in microbial enhanced heavy oil recovery. In: *SPE EOR Conference at Oil and Gas West Asia*. Society of Petroleum Engineers.

Joshi, S. J., Al-Wahaibi, Y., & Al-Bahry, S., (2019). Biotransformation of heavy crude oil and biodegradation of oil pollution by arid zone bacterial strains. In: *Microbial Metabolism of Xenobiotic Compounds* (pp. 103–122). Springer, Singapore.

Kim, D. D., O'Farrell, C., Toth, C. R., Montoya, O., Gieg, L. M., Kwon, T. H., & Yoon, S., (2018). Microbial community analyses of produced waters from high-temperature oil reservoirs reveal unexpected similarity between geographically distant oil reservoirs. *Microbial Biotechnol., 11*(4), 788–796.

Landa-Marbán, D., Radu, F. A., & Nordbotten, J. M., (2017). Modeling and simulation of microbial enhanced oil recovery including interfacial area. *Transport Porous Med., 120*(2), 395–413.

Lazar, I., Petrisor, I. G., & Yen, T. F., (2007). Microbial enhanced oil recovery (MEOR). *Petrol. Sci. Technol., 25*(11), 1353–1366.

Le, J. J., Wu, X. L., Wang, R., Zhang, J. Y., Bai, L. L., & Hou, Z. W., (2015). Progress in pilot testing of microbial-enhanced oil recovery in the Daqing oilfield of north China. *Int. Biodeterior. Biodegradation, 97*, 188–194.

Liang, B., Zhang, K., Wang, L. Y., Liu, J. F., Yang, S. Z., Gu, J. D., & Mu, B. Z., (2018). Different diversity and distribution of archaeal community in the aqueous and oil phases of production fluid from high-temperature petroleum reservoirs. *Front. Microbiol., 9*, 841.

Lin, J., Hao, B., Cao, G., Wang, J., Feng, Y., Tan, X., & Wang, W., (2014). A study on the microbial community structure in oil reservoirs developed by water flooding. *J. Petrol. Sci. Eng., 122*, 354–359.

Liu, J., Wu, J., Lin, J., Zhao, J., Xu, T., Yang, Q., Zhao, J., et al., (2019). Changes in the microbial community diversity of oil exploitation. *Genes, 10*(8), 556.

Liu, K., & Wei, X., (2017). Oil recovery: Experiences and economics of microbially enhanced oil recovery (MEOR). In: Lee, S., (ed.), *Consequences of Microbial Interactions with Hydrocarbons, Oils, and Lipids: Production of Fuels and Chemicals, Handbook of Hydrocarbon and Lipid Microbiology* (pp. 1–18). Springer, Cham.

Lujun, C., Fan, Z., Yuehui, S., Banat, I. M., & Xianqing, L., (2018). Characteristics of bacterial communities in high-temperature heavy oil reservoir. *J. Eng. Sci. Technol. Rev., 11*(2), 153–159.

Maudgalya, S., Knapp, R. M., & McInerney, M., (2007). Microbially enhanced oil recovery technologies: A review of the past, present and future. In: *Production and Operations Symposium*. Society of Petroleum Engineers.

McInerney, M. J., Nagle, D. P., & Knapp, R. M., (2005). Microbially enhanced oil recovery: Past, present, and future. In: *Petroleum Microbiology*. ASM Press, Washington, DC.

Mueller, R. F., & Nielsen, P. H., (1996). Characterization of thermophilic consortia from two souring oil reservoirs. *Appl. Environ. Microbiol., 62*, 3083–3087.

Nielsen, S. M., Shapiro, A. A., Michelsen, M. L., & Stenby, E. H. D., (2010). Simulations for microbial enhanced oil recovery with metabolite partitioning. *Transport Porous Med., 85*(3), 785–802.

Okoro, C. C., & Amund, O. O., (2018). Microbial community structure of a low sulfate oil producing facility indicate dominance of oil-degrading/nitrate reducing bacteria and methanogens. *Petrol. Sci. Technol., 36*(4), 293–301.

Orphan, V. J., Goffredi, S. K., Delong, E. F., & Boles, J. R., (2003). Geochemical influence on diversity and microbial processes in high temperature oil reservoirs. *Geomicrobiol. J., 20*, 295–311.

Orphan, V. J., Taylor, L. T., Hafenbradl, D., & Delong, E. F., (2000). Culture-dependent and culture-independent characterization of microbial assemblages associated with high-temperature petroleum reservoirs. *Appl. Environ. Microbiol., 66*, 700–711.

Patel, J., Borgohain, S., Kumar, M., Rangarajan, V., Somasundaran, P., & Sen, R., (2015). Recent developments in microbial enhanced oil recovery. *Renew. Sust. Energ. Rev., 52*, 1539–1558.

Phetcharat, T., Dawkrajai, P., Chitov, T., Wongpornchai, P., Saenton, S., Mhuantong, W., Kanokratana, P., et al., (2018). Effect of inorganic nutrients on bacterial community composition in oil-bearing sandstones from the subsurface strata of an onshore oil reservoir and its potential use in microbial enhanced oil recovery. *PloS One, 13*(11), e0198050.

Putra, W., & Hakiki, F., (2019). Microbial enhanced oil recovery: Interfacial tension and biosurfactant-bacteria growth. *J. Petrol. Explor. Prod. Technol.*, 1–22. https://doi.org/10.1007/s13202-019-0635-8.

Rahayyem, M., Mostaghimi, P., Alzahid, Y. A., Halim, A., Evangelista, L., & Armstrong, R. T., (2019). Enzyme-enhanced oil recovery EEOR: A microfluidics approach. SPE-195116-MS. *SPE Middle East Oil and Gas Show and Conference*. Society of Petroleum Engineers.

Safdel, M., Anbaz, M. A., Daryasafar, A., & Jamialahmadi, M., (2017). Microbial enhanced oil recovery, a critical review on worldwide implemented field trials in different countries. *Renew. Sust. Energ. Rev., 74*, 159–172.

Shelton, J. L., Andrews, R. S., Akob, D. M., DeVera, C. A., Mumford, A., McCray, J. E., & McIntosh, J. C., (2018). Microbial community composition of a hydrocarbon reservoir 40 years after a CO_2 enhanced oil recovery flood. *FEMS Microbiol. Ecol., 94*(10), fiy153.

Shibulal, B., Al-Bahry, S. N., Al-Wahaibi, Y. M., Elshafie, A. E., Al-Bemani, A. S., & Joshi, S. J., (2018). Analysis of bacterial diversity in different heavy oil wells of a reservoir in south Oman with alkaline pH. *Scientifica, 2018*.

Shibulal, B., Al-Bahry, S., Al-Wahaibi, Y., Elshafie, A., Al-Bemani, A., & Joshi, S., (2018). Microbial-enhanced heavy oil recovery under laboratory conditions by *Bacillus firmus* BG4 and *Bacillus halodurans* BG5 isolated from heavy oil fields. *Colloids Interfaces, 2*(1), 1–18.

Singh, A., Van, H. J. D., Kuhad, R. C., Parmar, N., & Ward, O. P., (2014). Subsurface petroleum microbiology. In: Parmar, N., & Singh, A., (eds.), *Geomicrobiology and Biogeochemistry. Soil Biology* (Vol. 39). Springer, Berlin, Heidelberg.

Sinton, D., (2014). Energy: The microfluidic frontier. *Lab on a Chip, 14*(17), 3127–3134.

Sivasankar, P., & Kumar, G. S., (2018). Modeling the influence of interaction between injection and formation brine salinities on in situ microbial enhanced oil recovery processes by coupling of multiple-ion exchange transport model with multiphase fluid flow and multi-species reactive transport models. *J. Petrol. Sci. Eng., 163*, 435–452.

Song, W. F., Wang, J. W., Yan, Y. C., An, L. Y., Zhang, F., Wang, L., Xu, Y., et al., (2018). Shifts of the indigenous microbial communities from reservoir production water in crude oil-and asphaltene-degrading microcosms. *Int. Biodeter. Biodegr., 132*, 18–29.

Song, Z., Chen, S., Zhao, F., & Zhu, W., (2019). The whole metagenome of injected and produced fluids reveal the heterogenetic characteristics of the microbial community in a water-flooded oil reservoir. *J. Petrol. Sci. Eng., 176*, 1198–1207.

Su, S., Dong, H., Chai, L., Zhang, X., Banat, I. M., Wang, Z., Hou, D., Zhang, F., & She, Y., (2018). Dynamics of a microbial community during an effective boost MEOR trial using high-throughput sequencing. *RSC Adv., 8*(2), 690–697.

Tianyuan, W., Li, Y., Jianlong, X., Lixin, H., Qingfeng, C., & Yuandong, M., (2019). A mathematical model for microbial enhanced oil recovery considering the double-bacterial competition mechanism. *J. Petrol. Sci. Eng., 178*, 336–343.

Tüccar, T., Ilhan-Sungur, E., Abbas, B., & Muyzer, G., (2019). Coexistence of sulfate reducers with the other oil bacterial groups in Diyarbakır oil fields. *Anaerobe, 59*, 19–31.

Van, H. J. D., Singh, A., & Ward, O. P., (2003). Recent advances in petroleum microbiology. *Microbiol. Mol. Biol. Rev., 67*(4), 503–549.

Vigneron, A., Alsop, E. B., Lomans, B. P., Kyrpides, N. C., Head, I. M., & Tsesmetzis, N., (2017). Succession in the petroleum reservoir microbiome through an oil field production lifecycle. *ISME J., 11*, 2141–2154.

Voordouw, G., (2011). Production-related petroleum microbiology: Progress and prospects. *Curr. Opin. Biotechnol., 22*(3), 401–405.

Voordouw, G., Armstrong, S. M., Reimer, M. F., Fouts, B., Telang, A. J., Shen, Y. I. N., & Gevertz, D., (1996). Characterization of 16S rRNA genes from oil field microbial

communities indicates the presence of a variety of sulfate-reducing, fermentative, and sulfide-oxidizing bacteria. *Appl. Environ. Microbiol., 62*(5), 1623–1629.
Voordouw, G., Shen, Y., Harrington, C. S., Telang, A. J., Jack, T. R., & Westlake, D. W., (1993). Quantitative reverse sample genome probing of microbial communities and its application to oil field production waters. *Appl. Environ. Microbiol., 59*(12), 4101–4114.
Wang, L., Tang, Y. Q., Guo, P., Luo, Y. J., Wu, X. L., & Wang, H., (2013). Microbial functioning on crude oil in a gas-permeable single microfluidic channel. *J. Petrol. Sci. Eng., 104*, 38–48.
Wang, T., Yu, L., Xiu, J., Ma, Y., Lin, W., Ma, T., Wang, X., & Wang, L., (2018). A mathematical model for microbial enhanced oil recovery using biopolymer-producing microorganism. *Fuel., 216*, 589–595.
Wang, X., Li, X., Yu, L., Huang, L., Xiu, J., Lin, W., & Zhang, Y., (2019). Characterizing the microbiome in petroleum reservoir flooded by different water sources. *Sci. Total Environ., 653*, 872–885.
Wang, X., Li, X., Yu, L., Li, Y., Huang, L., Lin, W., & Li, D., (2019). Distinctive microbial communities imply the main mechanism in a MEOR trial in high pour-point reservoir. *J. Petrol. Sci. Eng., 175*, 97–107.
Zhang, F., She, Y. H., Banat, I. M., Chai, L. J., Huang, L. Q., Yi, S. J., Wang, Z. L., et al., (2014). Genomovar assignment of *Pseudomonas stutzeri* populations inhabiting produced oil reservoirs. *Microbiologyopen, 3*(4), 446–456.
Zhang, F., She, Y. H., Chai, L. J., Banat, I. M., Zhang, X. T., Shu, F. C., Wang, Z. L., et al., (2012). Microbial diversity in long-term water-flooded oil reservoirs with different in situ temperatures in China. *Sci. Rep., 2*(1), 1–10.
Zhang, F., She, Y. H., Ma, S. S., Hu, J. M., Banat, I. M., & Hou, D. J., (2010). Response of microbial community structure to microbial plugging in a mesothermic petroleum reservoir in China. *Appl. Microbiol. Biotechnol., 88*(6), 1413–1422.

CHAPTER 9

Biocatalytic Upgrading of Opportunity Crudes

M. M. RAMÍREZ-CORREDORES

Idaho National Laboratory, Idaho Falls, ID–83402, USA

9.1 INTRODUCTION

The world's major oil reserves are constituted by heavy oils and bitumen, spread over more than 70 countries. In Canada and Venezuela, technically recoverable reserves of this type are estimated to be 175 and 300 billion barrels (bbl), respectively; larger than that of light oil reserves in Saudi Arabia, estimated at 240 billion bbl (BP International Plc, 2014). The exploitation of viscous heavy oils and of bituminous deposits, such as tar sands, started accelerating a few decades ago, with production rising quickly from the 1980s' through the 2000s,' with recent declines forced by the competition from shale oil.

These heavy oils are characterized by the presence of large macromolecular structures such as the asphaltenes, and by the presence of a high concentration of heteroatomic compounds bearing sulfur, nitrogen, oxygen, and metal atoms. These constituents concentrate towards the higher boiling point fractions (the non-distillable residue).

Low quality oils pose upgrading difficulties forcing producers to offer discounted prices giving rise to the term "Opportunity Crudes," a term that emphasizes the price difference existing between these heavy crudes and those of higher quality. Despite the discounted prices, refiners soon realized that these crudes brought more problems than opportunities, and the postmortem analysis of the state of the art of technology indicates that the real opportunities were for R&D of new upgrading catalysts and processes. Extensive efforts have provided significant advances with limited commercial deployment.

For oils in the higher viscosity range, such as bitumen and extra-heavy oils, viscosity has to be reduced at the front end, for mobilizing the resources from the production sites. Thus, upgrading of opportunity crudes concerns to processes, in which the API gravity[1] of the crude oil is increased or its viscosity is decreased. Several process technologies for reducing viscosity of heavy oil to facilitate transport are commercially available (Heck and Degnan, 1989; Gray, 1994; de Bruijn, 2002; Biasca et al., 2003).

Physical upgrading processes, with no chemical modifications (e.g., deasphalting), as well as chemical conversion of heavier molecules into smaller compounds are commercially available for licensing. Examples of processes based on physical methods include dilution, addition of surface-active agents (e.g., for adsorptive removal) and emulsification. Commercially available chemical processes include thermal and catalytic processing, in the absence or presence of high pressure hydrogen, using conventional refining technologies.

These refining techniques were originally developed for conventional crude oils containing lower concentrations of high molecular weight and metal-containing compounds. Process modifications have focused on coping with issues such as catalyst fouling and deactivation caused by problematic compounds present in residues from heavy, extra heavy oils and bitumen. Mitigation strategies and devices are commercially available, however a holistic technological solution requires further R&D.

Two broad chemical conversion approaches are:

1. **C-Rejection Processes:** It is to eliminate a (large) subset of C-atoms from the molecules to reduce molecular weight, with the production of solid carbon (coke) as a by-product.
2. **H-Addition Processes:** It is to increase the H-content of molecules, decreasing their density.

Examples of the former include visbreaking, hydrovisbreaking, coking, and fluid cracking; and of the latter hydrocracking (HCK), hydroconversion (HDC), and hydroprocessing (HDP) in general.

Naphthenic acids (NAs, the largest class of oxygen-bearing molecules in high acid crudes (HAC)), asphaltenes, and metal-containing compounds negatively affect the performance and operability of these processes. Technologies for the removal of asphaltenes are commercially operative or available for licensing. Some of these technologies have been adapted for

[1] API gravity = (141.5 ÷ Specific Gravity)–131.5.

the simultaneous removal of asphaltenes and metal-containing compounds, as the metal-containing compounds concentrate in the asphaltene fraction. The effectiveness of these technologies is limited, and further upgrading is required prior to downstream processing (e.g., fluidized catalytic cracking (FCC)) or finishing products. Unfortunately, specific technologies for metals and NAs removal are not available commercially, meaning that these compounds are dealt with differently across refineries, depending on their configurations. Usually, refinery streams containing mixtures of metals and NAs are treated via hydrotreatment (HDT) prior to being fed to a process where severe negative effects may occur that could result in unit or refinery shut down.

These commercially available technologies have been integrated in refining operations for more than 50 years and can be considered mature technologies. Most original patents have expired, allowing for new developmental opportunities and operational deployment. Literature on these conventional technologies is abundant throughout the long history of oil refining, for example: books (Heck and Degnan, 1989; Le Page et al., 1990; Gray, 1994; Wauquier, 1995; Speight, 1999, 2010; Self et al., 2000; Trambouze, 2000; Gary and Handwerk, 2001; Hsu and Robinson, 2006; Treccani, 2008; Fahim et al., 2010; Ramirez-Corredores, 2017), reviews, and reports (Orr and Van Sant, 1960; Materials Program, 1980; Ingram, 1996; Office of Technology Assessment Materials Program, 1996; Morel et al., 1997; Solari, 2000; de Bruijn, 2002; Biasca et al., 2003; Chem System, 2003; Flint, 2004; Frost and Sullivan, 2004; Hydrocarbon Publishing Co., 2006; Rahimi and Gentzis, 2006; Robinson, 2006; Ancheyta and Rana, 2007; Clark et al., 2007; Delbianco and Montanari, 2007; Dusseault, 2007; Hydrocarbon Processing, 2008), articles, patents, and brochures (Speight et al., 1985; Thakur and Thomas, 1985; Oelderik et al., 1989; Mohanty et al., 1990; Weissman, 1997; Slavcheva et al., 1999; de Bruijn, 2002; Flint, 2004; Ali and Abbas, 2006; National Energy Technology Laboratory, 2006; Rana et al., 2007; Sawarkar et al., 2007; Zhang et al., 2007; Mullins, 2008; Deshannavar et al., 2010; Cai and Tian, 2011; Angeles et al., 2014; Barbosa et al., 2014; Magomedov et al., 2015; Sahu et al., 2015).

Refinery configuration, oil quality, refinery diet changes and technology options complicate the selection process to decide for the most convenient match. The refinery configuration determines the crude fractionation requirements, based on the feed specifications required by the downstream units (FCC, isomerization, reforming, etc.). Changes in input oil quality through incorporation of opportunity crudes effects the quality of the heavier refined fractions (e.g., atmospheric residue (AR); vacuum gas oil (VGO); and vacuum residue (VR)) pushing the refiner to adjust fractionation to comply with specifications for intermediate streams. For example, the final VGO boiling point will be the lowest boiling point of a contaminant molecule that must be removed prior to downstream processing. Montaghi et al. (Montaghi et al., 2011) highlighted factors to consider when selecting heavy oil processing technologies: the needs for flexibility on feed quality, demonstration of feed type flexibility, conversion level on a once-through basis for a variety of feeds, product quality and its differential to finished products, large scale demonstration and/or commercial experience, reliability, and overall economics.

Biological catalysts and processes have been studied and tested for particular upgrading reactions of model compounds and extracts present in, or from, heavy oils. This chapter reviews the bioupgrading of opportunity crudes and some other initiatives that could serve as motivation and inspiration for defining new upgrading routes.

9.2 OPPORTUNITY CRUDES: BITUMEN, EXTRA-HEAVY, HEAVY OILS AND HIGH ACID CRUDES (HAC)

The quality of a crude oil defines its market value that with geopolitical and market forces are critical variables in the price equation. Therefore, part of the price differential is driven by quality differences between benchmark high quality crudes (light and sweet) and low quality crudes (heavy and sour/acid). Low quality "Opportunity Crudes" or "Advantageous Crudes" are "Unconventional Oils" characterized by discounted selling prices, due to extraordinary production, transportation, and refining needs. The price discount affected by quality is typically referenced to the quality and/or price of benchmark crude Brent. The key quality parameters are sulfur content, total acid number (TAN) and API gravity. Some reports indicated that the price discount/bbl associated with sulfur and acidity might be determined directly from these properties (Rabinow, 2001; Hartmann, 2003; McRae, 2017), as for instance, the estimated discounts of \$0.82/per unit S and \$0.88/ per unit TAN (Bacon and Tordo, 2004). Meanwhile, others indicated a dependence on the market price of the benchmark crude (Tahmassebi, 1986; Bacon and Tordo, 2005). An explanation for this dependence in terms of the API gravity, for instance, is that each API degree costs 18.8¢ when the Brent price is US\$40/bbl but it prices 27.2¢ at US\$60 per Brent barrel. Thus, each degree increase in API was estimated to raise the barrel price by US\$0.007

per dollar of Brent, while a percentage point of sulfur would lower the barrel price by US$0.056 per dollar of Brent. The acidity of Brent is reported to be of TAN = 0.07 and any additional TAN creates a discount estimated at $0.051 per dollar of Brent (Bacon and Tordo, 2005).

Acid and sour crude oils are also considered to be opportunity crudes on the basis of their acidity and sulfur content. Nevertheless, the discounted and differential prices cannot be directly translated into profits without facing the technical challenges brought about by the lower quality.

Since petroleum products are mainly characterized by their boiling point ranges, the distillation curve is probably the most relevant attribute of any crude oil from which to assess physical properties, chemical composition, and overall quality. In refining, a first distillation tower is operated at atmospheric pressure, and the residue obtained in this tower is then subjected to vacuum distillation to deepen the first fractionation. Bench-scale distillations can roughly predict refinery products yield. The mixture of compounds with low boiling points (low molecular weight, MW) are called "light ends." Similarly, those boiling at higher temperatures (high MW) are called "heavy ends." The heavy ends fraction is also known as the "bottom of the barrel" (BotB). (Table 9.1).

TABLE 9.1 Boiling Point Ranges of Crude Oil Streams Produced by Distillation

Stream	Boiling Point Range (°C)	Carbon Number Range
Light ends (refinery gas and liquefied petroleum gas, LPG)	< 40	C_1–C_4
Naphtha, gasoline	40–215	C_5–C_{17}
Light gas oil (LGO), kerosene, diesel	215–342	C_8–C_{18}
Heavy gas oil (HGO), fuel oil (FO)	342–370	C_{12}–C_{20}
Light vacuum gas oil (HVGO), lube oil (LO) and waxes	370–509	$>C_{20}$
Heavy vacuum gas oil (HVGO)	509–550	$>C_{35}$
Vacuum residue (VR), asphalt, and coke	> 550	$>C_{50}$

Viscosity and density, being directly associated to boiling point (Figure 9.1 (Speight, 2001; Jones, 2015)) are markers for oil quality and value. Instead of density, the oil industry commonly uses specific gravity in API degrees. The presence of heavier, high molecular weight hydrocarbons (HCs) increases mean boiling point, typically found in lower API gravities crude oils.

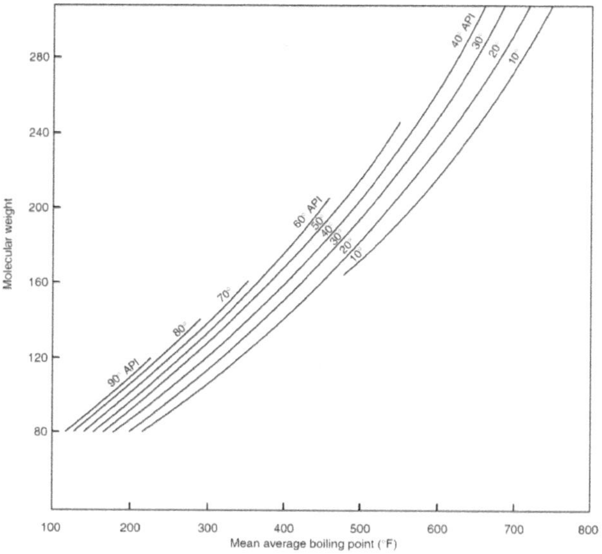

(A)

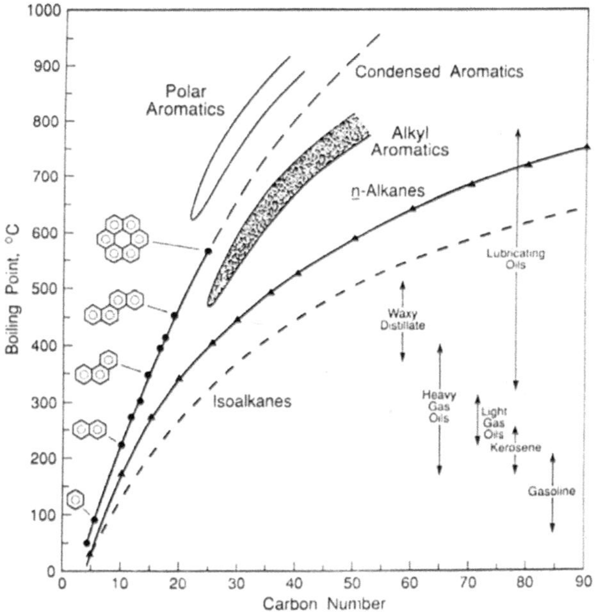

(B)

FIGURE 9.1 Variation of boiling point with (A) molecular weight (MW); (B) carbon number (C#) for various hydrocarbons and petroleum products.

Source: Reproduced from: Jones (2015); with permission from Springer and from Speight (2001) with permission from John Wiley and Sons, respectively.

It is important to stress that lighter polar compounds may have higher boiling points than some heavier non-polar compounds due to intermolecular forces such as hydrogen bonding, a fact that has not dispelled the common notion that boiling point is directly associated with MW. In other words, lack of understanding of fundamental molecular interactions has resulted in the erroneous correlation of "heavy" and "high boiling" with "high molecular weight" (Figure 9.2), as explained by Boduszynski (Boduszynski, 1987).

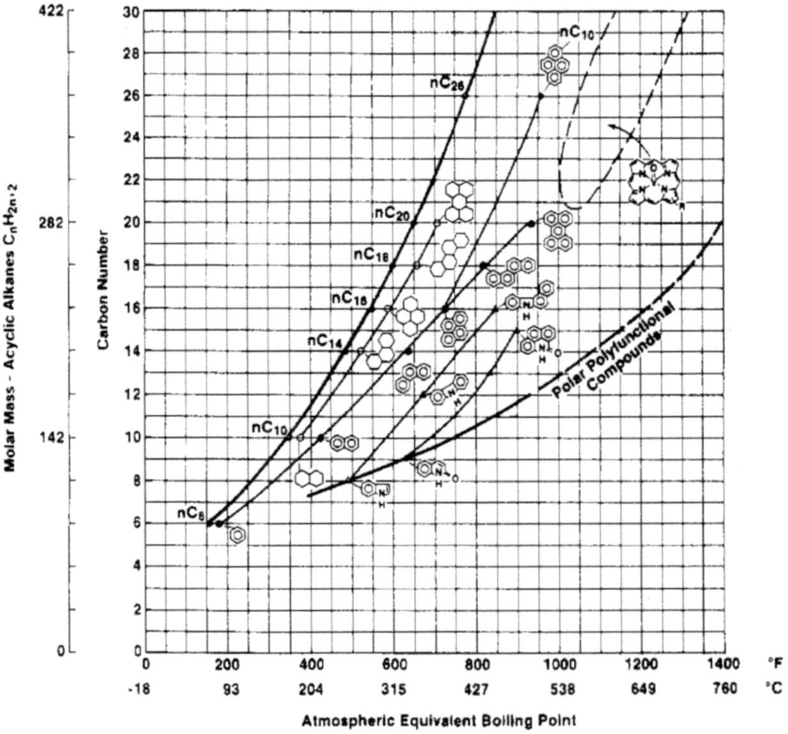

FIGURE 9.2 Molecular structure and boiling point.
Source: Reproduced from: Boduszynski (1987); with permission from ACS Publications.

The boiling point profile as well as the API gravity are used for categorizing crude oil as extra-light, light, medium, heavy, and extra-heavy. For crudes, API gravity ranges are: > 40 for extra-light, 30–40 for light, 20–30 for median, 10–20 heavy and > 10 extra-heavy. Distillation yields are API dependent (Figure 9.3), with the proportion of heavy ends increasing with decreasing API gravity, resulting in a range of crude oil properties (Table 9.2).

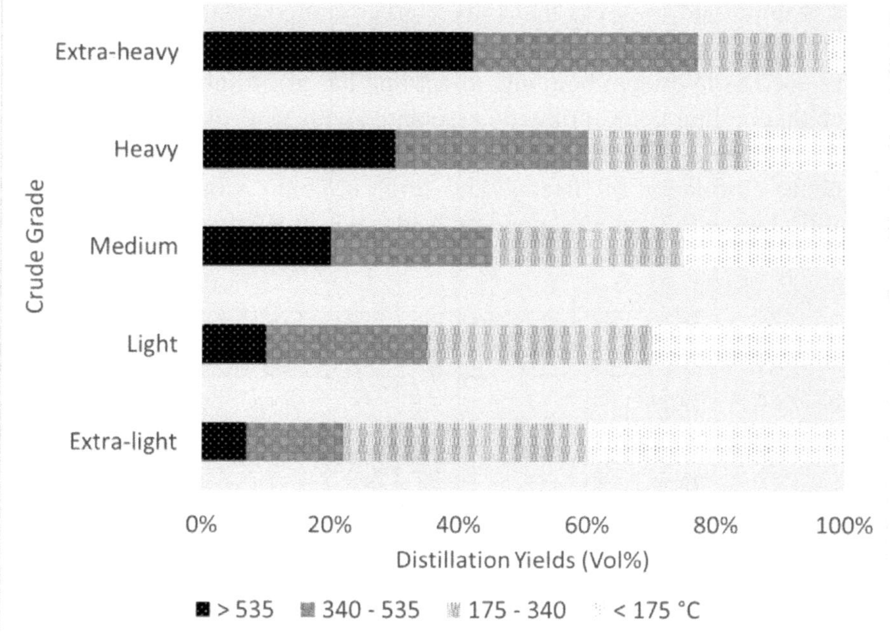

FIGURE 9.3 Examples of distillation yielsds for different crude grades.

TABLE 9.2 Value Ranges for Selected Properties of Different Crude Oil Grades

Property	Extra-Light	Light	Medium	Heavy	Extra-Heavy
API gravity (°)	>39	31.1–39	22.3–31.1	10.0–22.3	<10
H/C ratio	2.2	1.9	1.8	1.5	0.8
Density (g/cc)	<0.83	0.87–0.83	0.92–0.87	1.0–0.92	>1.0
Viscosity (cSt)	<2	2–10	10–100	100–5,000	>5,000
Mean BP (°C)	<350	250–450	300–500	400–600	>500
Molecular Weight	<200	200–275	275–400	400–600	>600
Sulfur (wt.%)	<0.1	0.1–1	0.5–1	1–3	>3
Metals (ppm)	ND	<10	10–100	100–300	>300
Conradson Carbon	<1	1–4	2–12	10–16	>12

The United Nations Institute for Training and Research (UNITAR) and the American Petroleum Institute (API) have established formal quality ranges for light, medium, and heavy oil in terms of API gravity and S-content (Table 9.3).

TABLE 9.3 Value Ranges of Quality Attributes for UNITAR Definition of Oil Classes

	Light Sweet	Light Sour	Medium Sour	Heavy Sweet	Heavy Sour
API gravity (°)	30–40	30–40	26–35	10–25	10–25
Sulfur (wt.%)	< 0.5	0.5–1.5	~ 1.1	< 0.5	1.0–3.0
Metals (ppm)	Low	Low	Low-medium	Low-high	High
Acidity	Low	Low-medium	Low-medium	Low	Medium-high
Nitrogen (ppm)	Varies	Varies	Varies	Varies	Varies

Bitumen is thought to be *in situ* oil degradation products, comprised of refractory heavy molecules not amenable to physical, chemical, or biological molecular weight reduction. While density is employed to differentiate crude oils, viscosity distinguishes crude oil classes from bitumen from *oil sands*. In general, the term oil sand refers to any high permeability, non-cemented sandstone containing immobile viscous oil or bitumen, mixed with clay, water, and other substances. Bitumen is also a tar, hence *tar sands* is a term sometimes used synonymously for oil sands. The density of raw bitumen, with high-S content (in ranges from about 960 to 1,020 kg/m^3), is comparable to that of extra-heavy oils (API < 10°), making the distinction between bitumen and extra-heavy crude oils blurry. However, viscosity distinguishes between these materials, while extra heavy oils topping out at 10^4 mPa-s, bitumen viscosity ranges from around that value to up to 2×10^6 mPa-s. In 1982, UNITAR brought together an expert panel to propose that viscosity and density should be evaluated to classify crude oils and bitumen (Figure 9.4). By considering viscosity first, crude oils are differentiated from bitumen, and subsequently, density is used to distinguish extra-heavy and heavy crude oils. The grades were defined as: *bitumen* having viscosities greater than 10,000 mPa-s, and *crude oils* with viscosities less than or equal to 10,000 mPa-s. Viscosities are reported as gas-free and are referenced to original reservoir temperature. Within the crude oils, API gravities are measured at 15.60°C. *Extra-heavy crude oils* have API gravities of less than 10° (densities greater than 1,000 kg/m^3) and *heavy crude oils* from 20° to 10° (densities from 934 to 1,000 kg/m^3). UNITAR did not provide classification for crude oil with API gravities > 20° or densities less than 934 kg/m^3 (Martinez, 1984).

The need for viscosity reduction has forced the upstream sector of the oil industry to set up and operate upgraders. Different technologies are chosen in *primary upgraders* for achieving the required level of upgrading, which depends on the original viscosity of the crude or bitumen, prior to pumping and transporting bitumen, heavy or extra-heavy oils. Viscosity reduction

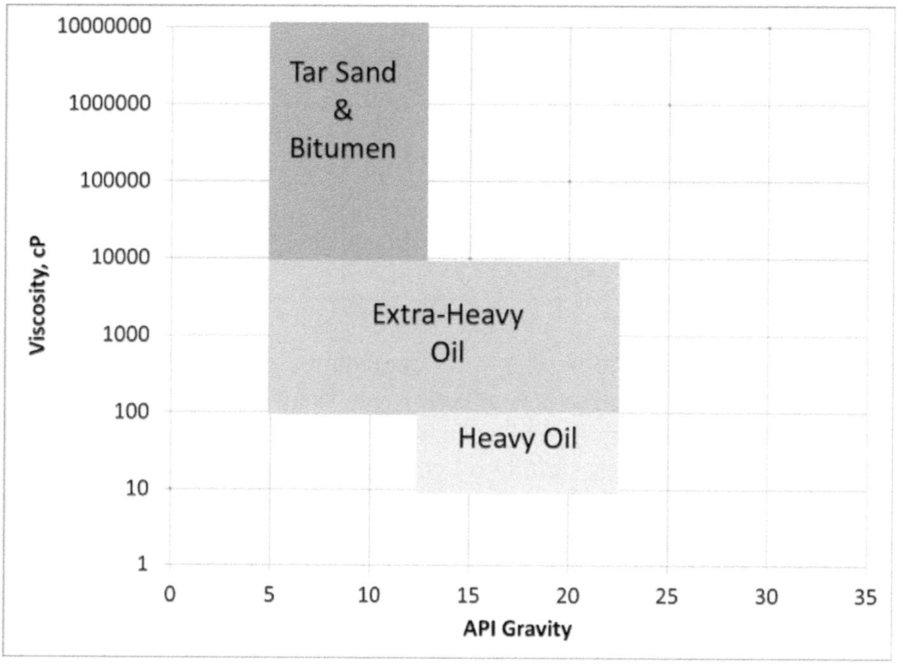

FIGURE 9.4 Viscosity and API gravity among crude oil grades and bitumen.
Source: Reproduced from: Ramirez-Corredores (2017); with permission from Elsevier.

becomes obligatory for complying with pipeline specifications, typically requiring viscosities of less than 350 mPa-s and density below 940 kg/m^3 at temperatures of 7.5–18.5°C. Primary upgrading gives rise to commercial grades of oils derived from tar sands and extra-heavy oils (Figure 9.5). In general, the term synthetic crude or *syncrude* corresponds to viscosity-reduced heavy crude oil suitable for pumping and distribution. *Syncrude's* are classified based on the method used for viscosity reduction:

1. **Diluted Bitumen or Dilbit:** It is a dilution of (~70%) bitumen with a (~30%) diluent, to obtain a density in the range of 650 to 750 kg/m^3. Diluents include either naphtha, natural gas liquids, ultra-light sweet crudes, condensates, or condensates refinery components[2].
2. **Synthetic Crude Oil (SCO):** It is the result of a mild upgrading of bitumen or heavy oil, for viscosity reduction and does not include chemical quality upgrade. Nominally, SCO is residue-free (or at least

[2] Condensates are very light hydrocarbons with API gravity greater than 50° and less than 80°. Condensates may exist in underground formations either in a separate phase from the crude oil or dissolved in the crude oil.

residue has been minimized) increasing its refinery value (improved yield/value), which is higher than that of Dilbit. Densities of SCOs range from 840 to 870 kg/m^3.

3. **Synthetic Bitumen Oil or SynBit:** Being a blend of (~50%) SCO and (~50%) bitumen, has higher refinery value than Dilbit but lower than SCO. SynBits have similar properties to a medium sour crude.
4. **SynDilBits:** These are blends of diluent, hydrotreated light SCO and bitumen. Their properties, behaviors, and refinery values fall between DilBits and SynBits.
5. **Coker Bitumen (CokerBit):** It is a blend of bitumen with coker distillates as diluent. Based on the quality and proportion of diluent, its refinery value may be higher than DilBit.
6. **Medium SCO:** It is a partially upgraded bitumen meeting the quality and properties of a medium sour crude grade.
7. **Sweet SCO:** It is a partially upgraded bitumen meeting the quality and properties of a medium sweet crude.

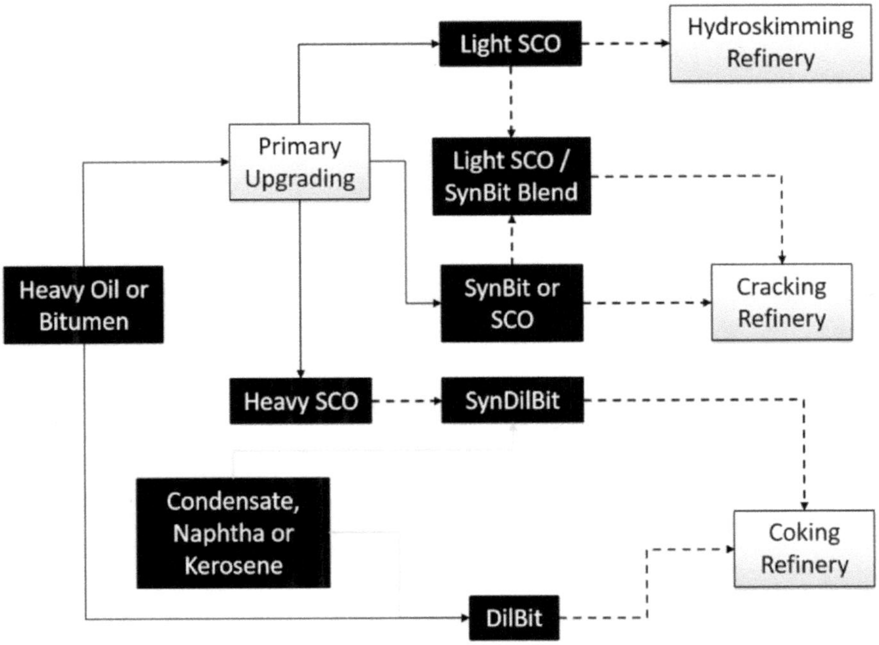

FIGURE 9.5 Commercial grades of oils derived from tar sands and extra-heavy oils and quality fit for differenst refinery complexity.

Source: Reproduced from: Ramirez-Corredores (2017); with permission from Elsevier.

Primary upgrading fulfills two objectives, firstly to meet viscosity and density specifications for pipeline transportation and secondly, to satisfy the quality requirements defined by the purchasing refinery (see Figure 9.5). The availability and logistic accessibility of diluents, oil production technologies, market price, and agreements between producer and refiner, are some of the many variables in the grade production equation. Therefore, quality differences between heavy oil and bitumen grades might be expected among producers, as have been reported by Sayles (Sayles and Routt, 2011) between Canada (bitumen) and Venezuela (Orinoco Belt). A web tool can be accessed to find the basic properties of crude oils of all commercially available grades (Crude Quality Inc., 2015).

Besides the higher viscosity and lower API gravity, heavy crudes are characterized by being difficult to vaporize, having high carbon Conradson/microcarbon residue (CCR/MCR) and asphaltenes-content, and having high levels of heteroatom contaminants (S, N, O, and metals) concentrated in the BotB fractions (Figure 9.6) (Ramirez-Corredores, 2017).

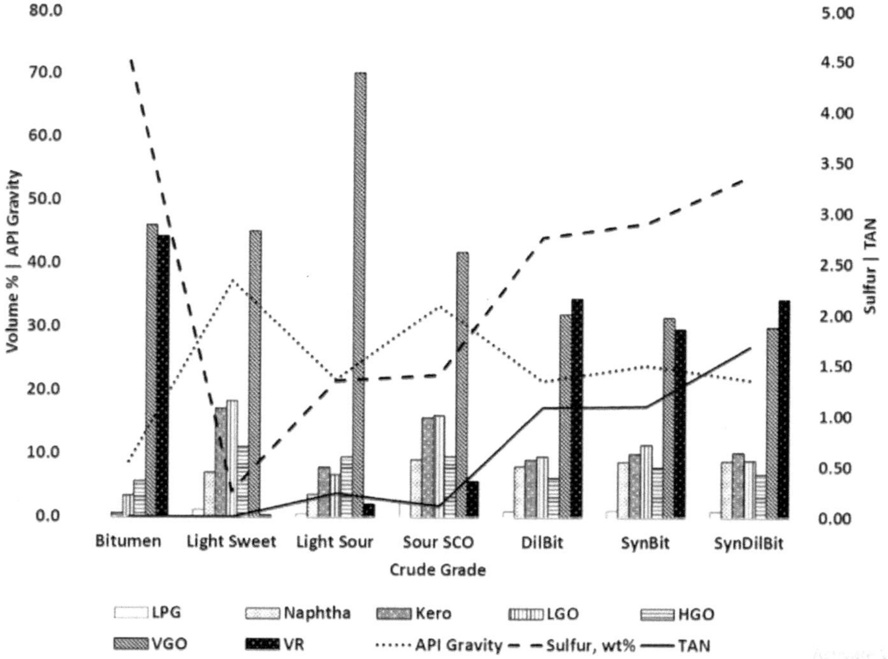

FIGURE 9.6 Quality exemplification of commercial grades of bitumen and derived extra-heavy crude oils.

Source: Reproduced from: Ramirez-Corredores (2017); with permission from Elsevier.

Canada and Venezuela concentrate these heavy oil resources, holding respectively some 1.5 trillion and 2.5 trillion bbl. If these reserves were proven with a 20% recovery rate, these two countries would account for more proven reserves than the conventional reserves of the Middle East. The largest accumulation of oil sands occurs in Canada; with an estimated total potential of 315 billion bbl, these deposits are relatively shallow with 30% porosity at depths of up to 600 m. In 2003, Canada proved more than 175 billion bbl and became the second-largest proven reserves in the world, after Saudi Arabia. The provinces of Alberta and Saskatchewan are the home for 97% of Canada's reserves. Considering their location within a politically stable country with existing infrastructure linked to US markets, the bitumen in Athabasca, Cold Lake and Peace River regions remains a substantial opportunity for long-term development. The Canadian Association of Petroleum Producers (CAPP) reported an expected growth of more than 75%, under moderate growth assumptions, in the period 2008 to 2020 (CAPP, 2014).

Meanwhile, the current situation in Venezuela is quite the opposite, with recent slow development linked to political, social, and economic uprising. Nevertheless, Venezuela follows Canada as the second-largest commercial developer worldwide. The opportunity crude reserves in Venezuela are concentrated in the southeast of the country, north of the Orinoco River in a region known as the *Orinoco oil belt*. Technically these are not bitumen reserves and have been reclassified as extra-heavy oil deposits. These oils exhibit low API gravity (below 8°) but have not been degraded to the same extent as Canadian oil sands. The advantageous subsurface temperatures in the Orinoco belt formation (40–50°C) contribute to the lower viscosity of the oil compared to that of Canadian bitumen, facilitating extraction through horizontal drilling techniques. China, Russia, India, Indonesia, and Ecuador also hold smaller deposits of oil sands and heavy oils (Dusseault, 2007).

A high proportion of compounds with very low reactivity (asphaltenes, resins, and some heteroatomic compounds) and heteroatomic moieties create severe problems for the refining processes of these unconventional oils. Operability problems not only lead to frequent shutdowns, but also are a source of margin losses. Production of stranded streams (low value by-products that have neither application nor market in refineries) and their disposal is one of the sources for margin losses. These low-value products not meeting market specifications, or stranded streams with unacceptable composition characteristics, require safe and secure disposal practices that carry high cost and complex technical challenges. Sometimes, the refiner has to sacrifice high-value products for blending with stranded streams to generate saleable products. The composition of the VR from heavy/extra-heavy oils and bitumen generates a significant stranded stream for most refineries processing these types of crude (Montaghi et al., 2011). Consequently, the objectives of upgrading include not only the conversion of high MW compounds into lower MW compounds in order to reduce density and viscosity, but also the removal of contaminants, such as sulfur, nitrogen, metals (V, Ni, Fe, etc.).

Refineries with simpler configurations (skimming or cracking refineries) struggle with processing heavy crude oils or bitumen, resulting in large volume of stranded VR or low quality fuel oil. Only refineries with higher complexity can handle the low API gravity and the high level of contaminants present in opportunity crudes, but are not free from the penalty of high operating costs.

Another attribute to keep in mind is the fuel value of the HC mixture, which is directly associated with the H/C ratio. The lower H/C ratio (as low as 0.8) in opportunity crude oils is indirectly caused by the presence of highly conjugated or aromatic systems, and of heteroatomic functional groups, which replace hydrogen atoms in the HC skeleton. In order to improve the quality of the BotB fraction, it requires not only removing the heteroatomic moieties, but also increasing the H/C ratio to at least 1.6 (Table 9.4).

TABLE 9.4 Examples of H/C Ratio of Different Materials and Compounds

Material or Compound	H/C Ratio
Bitumen	0.8
Benzene	1.0
Asphaltenes	1.0
Heavy crude	1.5
Shale oil	1.5
Light crude	1.8
Transportation fuels	1.9
LPG	2.0
Hexane	2.3
Butane	2.5
Methane	4.0

In general, asphaltenic macromolecules with high degrees of conjugation or high levels of condensation of aromatic systems are the main components responsible for the low H/C ratio in opportunity crudes. The molecular structure of asphaltene molecules remains unknown though some of their characteristic features have been elucidated. Strausz et al. (1992) and Speight (2004) conducted systematic and comprehensive characterization of asphaltenes. Strausz et al. (1992) proposed the molecular structure shown in Figure 9.7, containing all the identified features, by 1992.

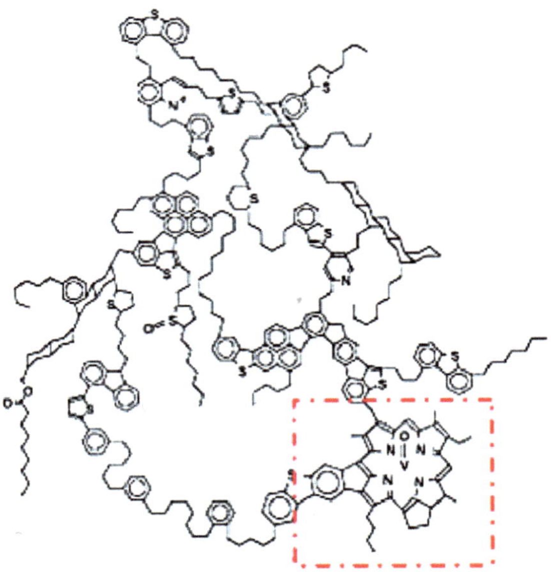

FIGURE 9.7 Molecular structure of asphaltene proposed by Strausz.
Source: Reproduced from: Strausz et al. (1992); with permission from Elsevier.

A supramolecular assembly of molecules has been proposed to explain the associating behavior and architecture of the molecular arrangements of asphaltenes. This model includes the following interactions: aromatic π-π stacking, cooperative binding by H-bonding, acid-base interactions, metal coordination complexes, and interactions between cycloalkyl and alkyl groups to form hydrophobic pockets (Figure 9.8) (Gray et al., 2011). Supramolecular assemblies of this type could be the main reason for the high viscosity of heavy oils and bitumen.

9.3 BIOLOGICALLY MEDIATED PATHWAYS FOR UPGRADING

The limitations offered by, and/or the lack of commercially available technologies for processing crude oils containing large concentrations of heteroatomic contaminants and refractory macromolecular compounds present in opportunity crudes have motivated and inspired R&D activities for decades. However, refiners have been forced to use mitigating measures to cope with the problems brought about by the processing of these crude oils. Both, R&D as well as industrial operations have fertilized the soil for knowledge to seed, grow, and nurture into emerging technologies. The science supporting the development of most of these emerging technologies along with the evolution of existing technologies have been comprehensively reviewed recently (Ramirez-Corredores, 2017).

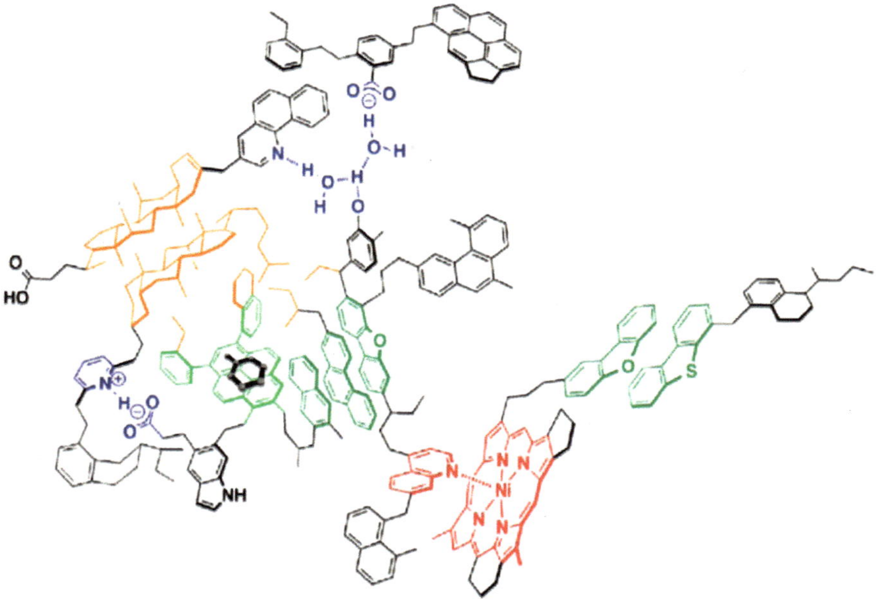

FIGURE 9.8 Gray supramolecular model of asphaltene association.
Source: Reproduced from: Gray et al. (2011); with permission from ACS Publications.

Biological means for improving crude oil and fuel quality were studied intensively from the 1970s' through the first decade of the 21st century, with a focus on the potential use of biological catalysis for oil refining (Ramirez-Corredores and Borole, 2007) and broader applications of biotechnologies throughout the oil industry (Vazquez-Duhalt, 2004). Biological removal of heteroatomic species (sulfur, nitrogen, oxygen, and metal), cleavage of cyclic and aromatic molecules, and biophysical reductions of viscosity and density through incorporation of bio-alcohols or surfactants, have been explored. Biologically mediated changes in the composition or viscosity of heavy crudes, which results in an improvement in the quality and value of oil is referred to as bioupgrading, and may potentially be applied upstream or downstream of conventional refining processes.

Bioupgrading research areas might be divided into the following categories based on the concepts currently applied in the oil industry:

1. Viscosity reduction
2. Biotreatments (biodesulfurization (BDS) or sulfur removal, biodenitrogenation (BDN) or nitrogen removal, biodemetalization (BDM) or metals removal).

9.3.1 VISCOSITY REDUCTION

Biological reduction of viscosity might provide the level of upgrading required to address transportation limitations of heavy oils and partially replace primary upgrading, in some instances. Viscosity bioupgrading may follow three different pathways:

i. Introduction of oxygen into asphaltenic or other heavy molecules (e.g., polycyclic aromatics, resins, etc.), to change molecular structure;
ii. Splitting of the aliphatic chains or aromatic rings in heavy molecules to reduce molecular weight;
iii. Ring saturation or hydrogenation of aromatic rings.

9.3.1.1 UPSTREAM EXPERIENCE

Upstream, microorganisms have been applied in enhanced oil recovery (MEOR) (Safdel et al., 2017), for improving operations and yields and are thought to achieve this through isolated or combined effects of:

i. degrading high molecular weight compounds;
ii. producing biosurfactants;
iii. physically displacing oil adhering to particles of sand in the formation;
iv. plugging the most porous portions of a reservoir, thereby reducing the tendency of water to "finger" through the reservoir; and
v. producing organic acids, which dissolve cementing materials in the formation.

These five modes of action do not operate separately during stimulated oil recovery. The organisms originally considered for R&D were those well adapted to oil well conditions. The presence of microorganisms in oil wells was not anticipated, and their resistance to biocides was much higher than expected (Lazar et al., 2007). Once discovered, oil well microbial communities were collected and subjected to selective enrichment in order to promote the desirable biochemical activities (e.g., lactic acid production). An example extracted from the patent literature is that developed by Sheehy (Sheehy, 1992a, b). Very little is revealed in the patent but it appears that biocides such as acids (e.g., hydrochloric, nitric, hydrofluoric acids), alkalis (e.g., sodium or potassium hydroxides), detergents, osmotic lysis reagents, and formaldehyde were applied in an effort to eliminate part of the extant microbial community. The surviving species were then treated with selected supplementation nutrients (e.g., a non-glucose-containing carbon source and one or more growth limiting non-carbonaceous nutrients) to selectively enrich target species with enhanced hydrophobic characteristics. The final step was to submit the reservoir to nutrient depletion conditions, by allowing the resident microorganisms to exhaust the supplied nutrients through their metabolic processes. Sheehy tried to take advantage of water-flooding operations by promoting certain characteristics of the resident microbial community or enriching the strain with those microorganisms exhibiting the desired characteristics, i.e., selectively controlling the growth of hydrophobic cells and/or with enhanced surface-active capabilities. At the time of filing his patent application, there was not clear understanding on this phenomenon and still there exist knowledge gaps. However, his results indicated that the observed improvements might be associated with an increase in hydrophobic lipids in the microbial cell membrane or with a change in configuration of membrane lipids to increased saturation. The EOR application of this microbial treatment soon faced severe limitations brought about by well volume, topography, geography, temperature, and salinity, as well as biocide potency and solubility, some of which were addressed in more recent patent applications (Sheehy et al., 2014; Marcotte et al., 2018; Sheehy, Govreau, et al., 2018a, b, Sheehy et al., 2018).

It is clear that a variety of metabolites created by different microorganisms would be significantly beneficial for improving the reservoir conditions (porosity, permeability, and wettability), but also could modify the physical properties (e.g., viscosity) of the crude oil (Guo et al., 2015), enabling EOR as well as the production of an upgraded crude oil. Such is the case of the production of biosurfactants, one of the five modes of actions listed above, at the beginning of this subsection. The biosurfactants produced during MEOR operations (Patel et al., 2015; Safdel et al., 2017) can also be used for oil emulsification, which is a common practice in the oil industry for enabling the transportation of heavy crudes and bitumen (Salacier, 2009). Furthermore, biosurfactants may be preferred over chemical surfactants because of their biodegradability (Mohan et al., 2006), low toxicity, and stability in different environments (Reis et al., 2013). The oil industry has been reluctant to implement MEOR on a wide scale. However, there is always the possibility for the industrial production of the biosurfactants elsewhere (Reis et al., 2013) and then applying it into the oil operations, either EOR (as already indirectly proposed by Sheehy (1992a) or for oil upgrading as viscosity reduction agents. The cited references include lists of microorganisms, surface-active metabolites and some of the genetic insights of the application (Reis et al., 2013; Patel et al., 2015; Safdel et al., 2017).

The production of acids defined above as the fifth MEOR mode of action is the other idea that may have potential for oil upgrading. The lactic acid-producing bacteria (Coleman et al., 1992), such as *Lactobacillus* or *Pediococcus* genera have been considered for MEOR based on this capability. Actually, these genera constitute the basis of a stimulating method for oil recovery. A mixture of the selected strain inoculum and the nutrients are injected into the reservoir through the drilled wellbore. Fermentation will take place and the required time for producing an adequate proportion of lactic acid is let to elapse prior to oil recovery from the reservoir. The removal of scale from oilfield equipment, especially those of carbonate or iron represents an additional application for either the bacteria or the produced lactic acid.

The simultaneous BDS and partial oxidation of naphthalene present in gas oil results in an increase in lubricity of the produced gas oil (Enomoto et al., 1998). This effect could be applied for physically displacing oil adhered to particles of sand in a formation. The biocatalyst is derived from a selected group of bacterial strains, with the dual capability for desulfurization of dibenzothiophenes (DBTs) and for the partial oxidation of naphthalene, including for example, *Pseudomonas putida*, *Pseudomonas fluorescens*, *Pseudomonas aeruginosa*, the genus *Bacillus*, the genus *Corynebacterium*, the genus *Rhodococcus*, the genus *Nocardia*, *Streptomyces*, bacteria belonging to the genus *Sphingomonas*. Among these, *Pseudomonas putida*, *Pseudomonas aeruginosa*, bacteria belonging to the genus *Rhodococcus* such as *Rhodococcus erythropolis* have the ability to oxidize naphthalene, and potentially the ability to decompose DBTs (Enomoto et al., 1998). This concept was proven, by growing the bacterial strain in a mineral medium containing glucose and naphthalene for carbon source. The cells were then centrifuged and re-suspended in a phosphate buffer solution. The process was carried out by contacting the catalytic suspension with a diesel feedstock containing about 0.005 wt.% naphthalenes and up to 0.05 wt.% sulfur, at 30°C in a shaker-incubator, for about 24 hours. The acidity of the product gas oil is kept within 0.005–1 wt.% of alkali extractable. The microbial catalyst was capable of desulfurizing up to about 90%, the S-content of the diesel product is adjusted either by hydrotreating the feedstock or by blending with a more severely treated product (containing a lower amount of sulfur). The oxidation reactions were associated with the improvements on the cold flow properties and lubricity of the produced diesel. This example of diesel BDS and aromatics oxidation is probably, the only one that claimed increases in lubricity and improvements on the cold flow properties of the product that can be considered as means for viscosity reduction upgrading.

9.3.1.2 STRUCTURAL SPLITTING OF HETEROATOMIC SYSTEMS

The presence of heteroatomic moieties in the heavy molecules of crude oils are partially responsible for further molecular associations, which lead to viscosity increases. Figure 9.9 includes some examples of heteroatomic compounds relevant to the bioupgrading reactions discussed in this Section. The difficulties for the removal of these heteroatoms increase with the bond order of the chemical bonds linking the heteroatom to the carbon skeleton. The single or aliphatic bonds are easier to break than double, triple or aromatic bonds. The more stable the compound is, the lower its reactivity or the higher its recalcitrance. Both, bond order and reactivity of these compounds affect the action of microbial catalysts and their capabilities towards reducing viscosity via splitting. For instance, heteroatoms bound in aromatic systems might require a more severe oxidative pathway.

As can be seen from Figures 9.7 and 9.8, asphaltene molecules or their molecular assemblies may contain aliphatic sulfur or ester linkages and bridges between two or more polyaromatic rings. If present, these sulfur and oxygen linkages may be broken to render smaller assemblies or condensed structures. Van Hamme et al. (Van Hamme et al., 2004) have shown that sulfide linkages in linear chains between larger aromatic rings can be cleaved by microorganisms under sulfur-limiting conditions, using a low molecular weight model compound. They achieved this by developing an effective screening technique to identify bacteria able to remove sulfur from *bis*-(3-pentafluorophenylpropyl)-sulfide (PFPS), a molecule resistant to terminal attack thanks to the presence of pentafluorinated aromatic rings. *Rhodococcus* sp. strain JVH1 proved to be capable of removing sulfur from the sulfidic linkages in PFPS and other aryl-aryl/aryl-alkyl moieties such as dioctyl sulfide, didodecyl sulfide, dibenzyl sulfide and 1,4-dithiane, but not from DBT or phenyl sulfide. Neither JVH1 nor other strains have been reported to have similar activity against high molecular weight compounds in opportunity crudes.

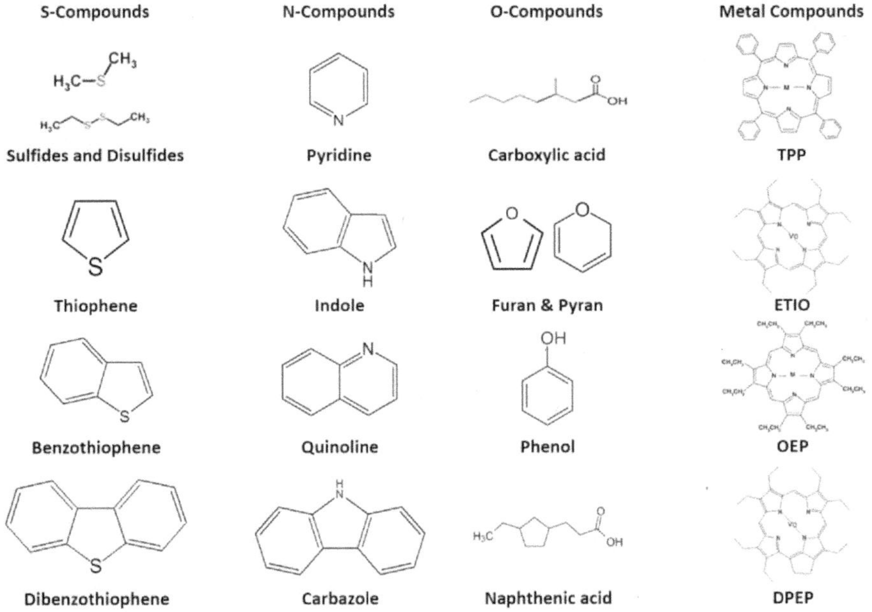

FIGURE 9.9 Some heteroatomic organic compounds relevant to upgrading.

Cytochrome C from Horse heart, from *Saccharomyces cerevisiae*, from *Pseudomonas aeruginosa*, and from *Desulfovibrio vulgaris* have been tested for the conversion of organosulfur molecules such as thiophenes and sulfides to their sulfoxide form, in 10% water-miscible solvent mixture (Vazquez-Duhalt et al., 1993). Polyaromatic sulfur heterocycles can be oxidized in the presence of chloroperoxidase when using a mixed solvent consisting of an aqueous phase containing a buffered water-miscible solvent and a water-immiscible solvent. The conversion of thiophenes and sulfides in an aqueous system with 20% acetonitrile was observed by contacting the solvent mixture containing 0.2% diesel with the enzyme (Ayala et al., 2000).

Environmental Bioscience Corporation (an extinct company) claimed that BDS via the 4S pathway (Figure 9.10) can reduce viscosity of petroleum or petroleum-derived liquids (Sohrabi et al., 2012; Boniek et al., 2014) through contact with *Rhodococcus rhodocrous* bacteria, ATCC No. 53968 (Haney III and Monticello, 1992). The original patent application (Ser. No. 631642) was filed in the US in 1990 and 17 claims of the 23 filed in the patent cooperation treaty (PCT) were allowed in 1996 by the US Patents and Trademark Office. It was postulated that the proposed biocatalytic agent was capable of selectively cleaving carbon-sulfur bonds in sulfur heterocycles without attacking the carbon-carbon bonds, thereby converting those molecules into products which no longer confer high viscosity.

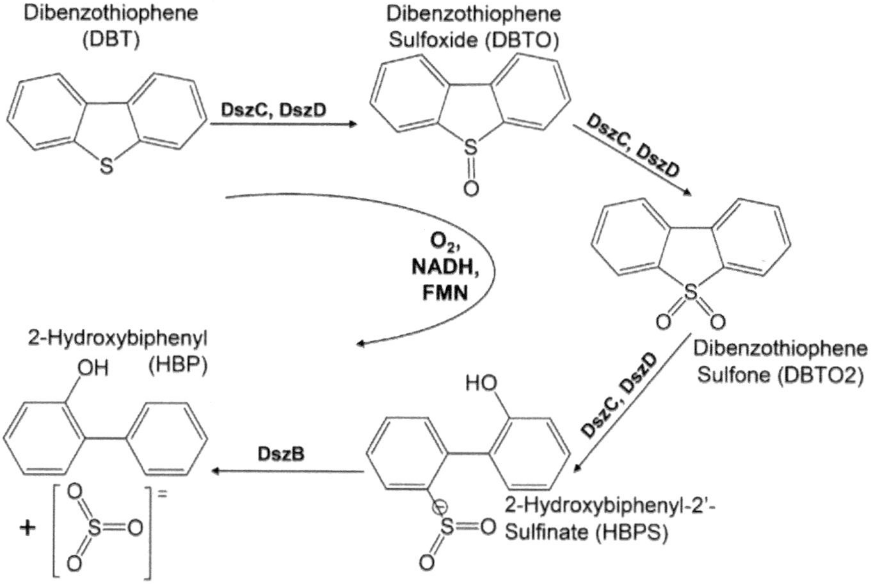

FIGURE 9.10 The 4S pathway for biodesulfurization (BDS).

Source: Reproduced from: Ramirez-Corredores and Borole (2007); with permission from Elsevier.

Although the 4S-pathway works as a BDS reaction for model compounds and surrogate feedstocks (Santos et al., 2007), it is now known that is not particularly effective for sulfur removal from real feedstocks such as intermediate gasoline and diesel streams, and even less effective for crude oil and residues (Monticello, 2000; Srinivas et al., 2013; Mohamed et al., 2015). In these cases, the last step of the 4S pathway appears to become very slow and the pathway mostly oxidizes the S-moiety, specifically converting thiophenic sulfur into sulfones.

The rigid, planar condensed ring structures of S-bearing aromatic heterocycles, such as DBT, tend to form stable, highly organized molecular assemblies through intermolecular interactions (hydrophobic and Van der Waals interactions) in petroleum liquids. If effective, it is postulated that the 4S pathway for the sulfur-bearing heterocycles (as those shown in Figure 9.9) would convert these compounds into molecules having structures with inhibited S-centers (inhibited for intermolecular associations), such as in 2-hydroxybiphenyl (2-HBP) (Figure 9.10). If biocatalysts could efficiently add polar substituents to S-heteroatoms (as shown for compounds DBT sulfoxide, DBTO, and DBT sulfone, $DBTO_2$ in Figure 9.10) in petroleum mixtures, their planar structures would be disrupted and prevent the formation of stable intermolecular associations. The production of free rotation sites via breakage of heterocyclic bonds in ring structures would similarly prevent the formation of stable hydrophobic intermolecular mesostructures. However, the added polar substituents could interact through H-bridge or Van der Waals forces to form other complex arrangements. To date, the practical contribution of a 4S-type pathway or other BDS reactions to viscosity reduction remains elusive and unlikely.

9.3.1.3 PARTIAL OXIDATION OF ASPHALTENES

The presence of nutrient elements such as sulfur, nitrogen, carbon, hydrogen, and oxygen in asphaltene molecules that are essential to the growth of bacteria represents an opportunity for bacterial upgrading and a motivation for overcoming the mass transfer limitations inhibiting microbial activity against these molecules.

Oxidative conversion is a common biocatalytic reaction, in asphaltenes it might result in the introduction of hydroxyl, keto, or carboxyl functionalities in aromatic rings, leading to two major effects on the overall properties of whole crude: destabilization of aromatic stacking, or creation of additional ionic interactions and hydrogen bonds. If the forces responsible for asphaltene stacking could be biologically destabilized, they would segregate into smaller clusters, which would potentially result in viscosity reduction. In addition, decreasing the degree of aromatic condensation by ring-opening and degradation would lead to a reduction in the apparent MW of asphaltenes and resins. Another potential source of viscosity reduction is the surfactant nature oxygenated molecules, which may result in crude-water emulsions.

As with the partial DBT oxidation reactions described above, partial asphaltene oxidation may disrupt the native three-dimensional association of asphaltene molecules while giving rise to a greater number of hydrogen bonds and other attractive forces. The magnitude of the required conversion to render measurable changes in viscosity has not been accomplished yet, as discussed below.

The presence of polycondensed aromatic moieties in asphaltenes, and the postulated mode of π-stacking for their molecular association merit some attention. The splitting reactions of stacked polyaromatic molecules (PAHs) contribute to viscosity reduction. Partial PAH oxidation has been demonstrated using both whole cells (Filatov et al., 2016) and purified enzymes (Borole et al., 2004). Typically, whole-cell biocatalysts lead to complete degradation and cannot be used for partial oxidation unless the organisms are genetically engineered. Whole-cell biocatalytic processes are also limited by diffusion or transport of large molecules across microbial membranes, presenting intractable engineering challenges (Nordblad et al., 2018).

For example, for whole-cell biocatalysts, most published studies on petroleum biodegradation has been focused on the effectiveness of bioremediation following accidental oil release into the environment. In general, most bioremediation efforts effectively removed aliphatic and lightly aromatic fractions but left the more recalcitrant polyaromatic and heavy fractions untouched, most likely due to mass transfer limitations. Biodegradation of heteroatomic compounds in crude oil and its fractions is considered as an unattractive route to upgrading since it is a destruction process that reduces fuel value. However, imaging the loss in heating value when fractions of contaminated fuel containing 5,000 ppm S, 3,000 ppm N and 500 ppm V + Ni are disposed of, and losses associated with the gas yield and coke deposition occurring in a typical hydrotreating operation, controlled biodegradation may remain attractive and justifies additional research.

Two strains, *Bacillus* sp. and *Serratia liquefaciens* were reported to be capable of degrading up to 50% of the aromatic and asphaltenic fraction present in the extracted oil from a polluted soil sample, which was collected from a site in Cactus, Chiapas nearby a petrochemical facility (Rojas-Avelizapa et al., 2002). This study was based on the disappearance of vibrations in the region of wavelength 1,500 and 2,300 cm^{-1} of the infrared spectrum of the asphaltenic fraction. As is typical for claims of this nature, this study did not provide insights into the molecular changes caused by degradation, the reaction mechanisms, or specific details on which compounds or components of the asphaltene fraction were modified or degraded.

Mayan crude oil asphaltenes were used as the sole carbon and energy source to grow *Corynebacterium*, *Bacillus*, *Brevebacterium*, and *Staphylococcus* species (Pineda-Flores et al., 2004). Microbial growth was followed by measuring carbon dioxide evolution from respiration, and it was found that the CO_2 release rates were four times higher than that of asphaltene-free control samples. This finding might be explained by the degradation of the more reactive alkyl substituents, bridges or aliphatic chains hanging from the asphaltenic molecules. Although, these results do not constitute absolute proof of asphaltenes nutrient role, this possibility merits a more systematic approach and better controlled studies.

An analogous study indicated that asphaltene biodegradation was possible only when n-paraffins were present as co-substrate. In this case, C6-precipitated asphaltenes were subjected to biodegradation using a strain isolated from foams sampled from a creek polluted with refinery effluent HCs. The asphaltenes were chemically fractionated into five fractions, which were spectroscopically (UV, IR, and NMR) characterized to identify the changes caused by the microbial treatment. In a 10-days treatment, asphaltene biodegradation of about 46% was observed, which was due to PAHs oxidation and trimming of the aliphatic substituents. Additionally, it was found that less than 25% of the biotransformed asphaltenes were assimilated by the bacterial culture. Nevertheless, neither mechanistic insights nor the role of the paraffin were reported (Rontani et al., 1985). For these reasons, it is believed that mixed cultures are needed for incorporating species capable of producing biosurfactants and others for degradation of the asphaltenic molecules. The aliphatic chains in asphaltenes reacted through the same degradation mechanism as that of saturated HCs, while the degree of condensation of the polyaromatic rings remain unchanged. So far, there does not seem to exist any evidence of disruption of the polyaromatic system nor of any breaking of linkages within the polyaromatic rings, which significantly affect the MW of asphaltenes.

Extracellular enzymes may be thought of as an effective biodegradation approach for high MW molecules, if the level of degradation can be effectively controlled. For example, extracellular enzymes such as lignin peroxidases, manganese peroxidases, and laccases play an important role in lignin degradation in nature (Schoemaker et al., 1989; Hammel, 1996), and these enzymes have been studied for their activity against coal and coal fractions. Wondrack observed an oxidative effect of lignin peroxidase (from *Phanerochaete chrysosporium*) on coal polymers (Wondrack et al., 1989) resulting in the reduction of MW from above 75,000 amu to about 12,000–14,000 amu.

The extracellular enzymes from the wood-decaying fungus, *Nematoloma frowardii* b19, was reported to depolymerize coal (Hofrichter and Fritsche, 1996) based on the decolorization of coal-derived humic acids, which were converted to fulvic acid-like compounds. The depolymerization effect was found to be stimulated by mediators, such as veratryl alcohol, an alcohol known to stimulate lignin degradation. A more extensive review of the work on microbial conversion of coal was published in 2001 (Hofrichter and Fakoussa, 2001) and exemplified more recently by Saha and Sarkar (Saha and Sarkar, 2018). Whether these enzymes can be used for conversion of the asphaltenes present in opportunity crudes remains to be discovered.

In nature, aromatic ring oxidation by enzymes such as monooxygenases and dioxygenases is a first step in the degradation of PAHs. *In vitro* testing of PAH oxidation in the presence of hydrogen peroxide has been conducted using lignin peroxidase (Vazquez-Duhalt et al., 1994), cytochrome P450, cytochrome c (Tinoco and Vazquez-Duhalt, 1998; Fischer and Fetzner, 2000), hemoglobin (Ortiz-Leon et al., 1995; Torres and Vazquez-Duhalt, 2000), myoglobin, horseradish peroxidase (Vazquez-Duhalt et al., 1992), chloroperoxidase (Ayala et al., 2000). The oxidation of PAHs (anthracene, methylanthracenes, acenaphthene, fluoranthene, and pyrene) catalyzed by lignin peroxidase from *Phanerochaete chrysosporium* tends to introduce hydroxyl and keto groups to aromatic rings, and has been reported to be enabled by the use of 10% aqueous-miscible solvent, such as acetonitrile, isopropanol, methanol, *N,N*-dimethylformamide, or tetrahydrofuran. Heterocycles such as carbazole and DBT can also be oxidized, with oxygenation occurring on the heteroatom (Vazquez-Duhalt et al., 1994). The hydrophobicity of horseradish peroxidase was chemically modified for performing in water-immiscible solvents, leading in a 16-fold increase in activity when tested with guaiacol instead of PAHs as a substrate (Vazquez-Duhalt et al., 1992). Three different chemical modifications were studied: deglycosylation with N-glycosidase F (peptide-Na-[N-acetyl-/3-glucosaminyl]), reductive alkylation with benzylaldehyde, and PEG attachment. Guaiacol is a common substrate used to assess the activity of peroxidases. The developed capability for acting in water-immiscible solvents might enable performing reactions in the oil phase with reduced mass transfer limitations. However, for some unknown reasons, research in this area seems to have been discontinued.

Partial oxidation of PAHs by peroxide can be catalyzed by heme-containing hemoglobin. Improved activity and substrate specificity was obtained following PEG and methyl esterification modifications, rendering a more hydrophobic protein. The modified protein exhibited a four-fold increase in catalytic efficiency for pyrene oxidation, compared to that of the unmodified protein. Acenaphthene, anthracene, azulene, benzo(a)pyrene, fluoranthene, fluorene, and phenanthrene could be oxidized, however, no reaction was observed with chrysene and biphenyl (Torres and Vazquez-Duhalt, 2000). The size and number of the PEG moieties attached to the enzyme could be varied to optimize catalytic efficiency at a given PEG/protein mass ratio. The optimized catalytic efficiency was found to occur at a PEG/protein mass ratio of 2.8. Simultaneously, thermal stability was also optimized at that PEG/protein mass ratio. The modified cytochrome c was active and stable within the temperature range of 30 to 120°C and activity maximized between 60 and 80°C. Although the conformation of the native enzyme was modified initially by the chemical treatment, as revealed from the structural analysis, the actual reason for improvement in thermal stability was the formation of the PEG polymer protecting layers on the protein, which inhibits unfolding at higher temperatures. This shell-like structure was reported to act on:

- Stabilization of the heme environment;
- Embedding water molecules between the hydrophilic regions of bound polymer and protein; and
- Enabling interaction between PEG hydrophobic regions and the surface hydrophobic clusters (Garcia-Arellano et al., 2004).

An enhancement on the substrate affinity up to 30-fold were attained by modifying hemoglobin with *p*-nitrophenol and *p*-aminophenol. Additionally, an increase in hydrophobicity of the active site was associated with the increase in solvent concentration, at which the enzyme showed maximum activity compared to the unmodified proteins (Torres et al., 2002). Various studies of enzymes in organic solvents have emphasized the important role that enzyme and solvent hydrophobicity play in catalytic activity (e.g., Ryu et al., 2002; García-Arellano et al., 2004; Steudle et al., 2015; Sheehy et al., 2018a).

The improvements on hydrophobicity and on thermal stability expand the operability and process reliability windows of this type of biocatalysts, giving rise to a greater margin of upgrading opportunities. Bioupgrading would result in increased solubility of asphaltenes in more solvents, and therefore will improve their "pumpability," reactivity, and accessibility to other downstream biocatalysts. A hydrophobicity parameter was developed to explain the behavior of lignin peroxidase and cytochrome c. This parameter was a function of thermodynamic activity and polarity of the solvent and it was used to relate solvent hydrophobicity with its effect on enzymatic activity (PAH oxidation activity). The parameter represents a threshold solvent concentration at which the activity was halved to its initial value hence it is specific of a given enzyme-substrate system (Torres et al., 1996).

Cytochrome c has been reported to oxidize anthracene and pyrene to quinones (Torres et al., 1995), meanwhile pyrene can be partially oxidized to quinones using lignin peroxidase, hemoglobin, as well as cytochrome c, in various aqueous-miscible solvents (Torres et al., 1997). Chemical modification and genetic engineering have led to improvements of the activity, substrate specificity, and stability of cytochrome c biocatalysts. Cytochrome c from *Saccharomyces cerevisiae* was genetically engineered to improve activity by substitution of the glycine residue present at position 82 in the active site with a hydrophobic aromatic residue, phenylalanine. This substitution improved the binding of the PAH substrates to the enzyme active site (Torres et al., 1995), but unfortunately lowered peroxide and solvent stability. A double modification of cytochrome c, by attaching PEG on free amino groups and methyl esters on carboxylic groups, widened the substrate portfolio of this enzyme from 8 PAHs to 17 (Tinoco and Vazquez-Duhalt, 1998).

A study addressing solvent stability was carried out using pinacyanol chloride as substrate. Although this substrate is not a PAH, the results obtained are relevant to biocatalytic partial oxidation. The considered aqueous-miscible solvents were tetrahydrofuran, acetonitrile, isopropanol, methanol, and *N,N*-dimethylformamide. The activity and stability of cytochrome c to catalyze reactions in solutions containing high proportions of water-miscible organic solvents could be associated with the polarity of the solvent, and more particularly with its solvation free energy (the Dimroth-Reichardt parameter). The biocatalyst stability was found to be highly impacted by the presence of peroxide in the reaction media, indicating a need for controlling peroxide concentration. Both, chemical modification and immobilization were able to improve this behavior in systems containing 90% tetrahydrofuran (Vazquez-Duhalt et al., 1993).

The advantages and limitations of cytochrome c as a biocatalyst for PAH oxidation were reviewed by Pickard et al. (Pickard et al., 1999). The main limitation for its application in crude oil upgrading is its inability to perform in organic media, caused by its insufficient active site hydrophobicity. Chemical and genetic modifications have not been successful in overcoming this limitation for performing in organic media. On the other hand, the biocatalyst has received attention for environmental remediation of PAH-contaminated sites. Besides its low cost, it can be advantageously used under extreme conditions such as, pH between 2 and 11, temperatures up to 120°C, and organic solvent concentrations up to 90%. Unfortunately, regardless of the progress achieved in the first decade of the 2000s,' further advancements on process development and solvent engineering principles are needed to realize the improvements and to enable the scaling and implementation of a bioupgrading technology. Regrettably, the area is not receiving as much support as it received in the 1980s' and only a few groups are still working on it.

In general, the key issue limiting the use of enzymes in organic solvents for PAH conversion is the ability of an enzyme to bind a substrate in that environment. Examinations of the thermodynamic limitations governing the activity of a given enzyme in organic media indicate that the parameter dictating the ability of an enzyme to perform a particular conversion is, as expected, the free energy of the reaction. The relative free energy for oxidation of pyrene by cytochrome c in an aqueous-acetonitrile mixture is determined by the substrate-enzyme binding free energy and the free energy of activation. The substrate-enzyme binding free energy changes with solvent hydrophobicity changes. Thus, for a given solvent environment, the rate of the reaction is a function of the binding free energy between the substrate and the enzyme. Consequently, due to the high hydrophobicity of oil as a solvent, the substrates would partition favorably into the oil phase rather than interacting with the enzyme active site. Improvements can be realized by making the active sites either more suitable for substrate binding or more hydrophobic (Borole et al., 2004).

A few reports described the use of enzymes for the biocatalytic upgrading of crude oil. Chloroperoxidase from *Caldariomyces fumago* was capable of converting petroporphyrins and asphaltenes, and conversion was increased by the presence of organic solvent in a mixed solvent system (Fedorak et al., 1993a). Unfortunately, this enzyme requires the addition of chloride together with hydrogen peroxide in order to perform the partial oxidation. While the organosulfur molecules were converted to sulfoxides and sulfones (see also Ref. (Ayala et al., 2000)), the PAHs were chlorinated. Chlorination is an undesirable reaction, due to the environmental implication of chlorinated organic molecules. These changes on the asphaltenes and petroporphyrins decrease their hydrophobicity and so benefit the mass transfer to the reacting aqueous phase. The substrate reactivity for this catalytic system was higher for organosulfur compounds than it was for PAHs, offering a potential opportunity for controlling chlorination (Fedorak et al., 1993a; Mogollon et al., 1998). The proper choice of a solvent system that promotes improved interaction between the enzyme and the substrate may shift the negative binding free energy to overcome enzyme-substrate preferences. It is known that for this enzyme, diesel PAHs[3] conversion in 20% aqueous-miscible solvent in the aqueous phase enabled transfer of substrate from the oil phase to the aqueous (enzyme-containing) phase. In this reacting system, the molecules of interest react in the aqueous phase due to a solvent-enabled favorable partitioning of the substrate.

In all of the reported experiments, catalytic activity declined as solvent concentration was increased in buffered mixtures, reaching a negligible value at a particular concentration of the solvent, which depended on both the HC mix and biocatalyst. This behavior was explained as due to substrate partitioning between the active site of the enzyme and the bulk solvent. In general, for this type of biocatalyst, stability is measured against denaturing factors such as peroxide concentration, solvent composition or temperature increases. Fast inactivation can be caused by the action of peroxide and increases in the proportion of the organics with solvent capabilities, such as lower alkanes and aromatics.

Clearly, biocatalytic applications for crude oil and its fractions or products requires biocatalysts with wide substrate specificity, in order to enable conversion of the broad range of PAHs existing in these streams. However, the desired PAH transformation is such that minimizes value losses, while still representing quality improvements, e.g., preclude aromatic stacking, decreases viscosity, increases reactivity towards bond breaking or towards HDP.

9.3.1.4 AROMATIC RING SATURATION

In comparison to carbon rejection, upgrading via hydrogen addition achieves deeper quality improvement of the product. In oil refining, catalytic hydrogenation processes carried out under cracking conditions (HCK) results in MW reduction. Biological catalysts such as hydrogenases produce protons and electrons from hydrogen gas, which can then be transferred to heteroatomic compounds (mainly, oxygenates or sulfur-containing compounds). Can such enzymes be modified to act upon HCs? These enzymes and microbial catalysts (that can express or excrete these enzymes) have been broadly studied

[3] Aromatics concentration in diesel is controlled due to their negative effect on cetane number and their promoting role in the formation of smoke and particle matter, in the exhaust emissions.

in applications related with the food industry (vegetable oils, fatty acids, fats, lipids, etc.) (Jayanegara et al., 2012; Petersen and Jensen, 2014; Guerreiro et al., 2016) and the bioenergy sector (biodiesel, biomass, green diesel, etc.) (Xu et al., 2008; Loganathan, 2011).

The hydrogenation or hydrogenolysis of aromatic compounds can be achieved if a pretreatment stage of partial oxidation (hydroxylation) is carried out. The reactivity of an aromatic ring towards the reactions with hydrogen is supposedly increased by hydroxylation. The designed functional features of the biocatalyst were incorporated by (i) cloning the genes to express both the dioxygenase and the dehydrogenase that catalyze the formation of a 1,2-dihydroxylated aromatic compound; (ii) hosting in a solvent resistant strain; and (iii) enhancing the activity by using additional enzymes to achieve a desirable level of conversion. The new whole-cell biocatalyst was tested at temperatures lower than 100°C and pressures between 1 and 10 atm, and contacting 100 mg of the aromatic compound with the microbial cells (2 ml of OD600 = 2.0) at pH=7.25, for up to 5 h (Coyle et al., 2000).

Anaerobic subsurface conversion of organic ring compounds, at a depth of at least 500 feet below ground surface yielded a liquid product, characterized by a lower aromaticity and containing paraffinic HCs and organic acids. The reactions were studied in a fermenter operated either concurrently or counter currently, in order to examine the effect of mixing conditions (Compere, 1974).

Actinomycetes such as *Rhodococcus*, *Nocardia*, and *Mycobacterium* have been shown to asymmetrically reduce 2-nitro-l-phenyl-1-propene to produce active 2-nitro-l-phenylpropane, with *Rhodococcus rhodochrous* IFO 3338 exhibiting the best results. These biocatalysts are particularly active for the hydrogenation of nitro-olefins, but not aromatics (Sakai et al., 1985). Would this be an effect of the nitro-substitution? Or could substrate specificity be broadened? Is there a solvent effect limiting the action of the biocatalyst? Although, these questions need to be answered, the resulting hydrogenation might be a promising route for the development of a biocatalyst for hydrogenating upgrading.

In some instances, the heavy compounds present in crude oils are conveniently compared to similar compounds present in coals, such as PAHs and asphaltenes. R&D efforts expand from chemical characterization to reactivity and processing. An example of coal bioprocessing into fuels and fertilizers was reported, using two consortia of bacteria designated KSARC56 and Mic-1, under aerobic and anaerobic conditions, respectively. Ordinarily, the product mixture included methane, lower alcohols, volatile fatty acids, and/or humic acid, while for aromatic feeds the main products were phenols (and phenolics), methane, and carbon dioxide. A slurry containing less than 20%

of the biocatalyst and no more than 50% of feed, at pH of 7.8 is processed either aerobically, for 48 h at 60°C or in a mixture of N_2:CO_2 of about 80:20 with Mic-1. This Mic-1 consortium, ATCC 55801, was derived from *Zootermopsis* sp, three other consortia were considered and for instance Mic-4 was derived from *Nasutitermes* sp (Walia and Srivastava, 1997, 1998).

Hydroliquefaction of highly aromatic solid materials has been widely considered as a valorization route for coal. The hydrogen uptake/biohydrogenation of coal, pretreated coal, asphaltenes, and model (control) compounds such as fumarate and diphenylmethane was studied under anaerobic conditions using *Desulfovibrio desulfuricans* and *Acidianus brierleyi*. During the biotreatment of diphenylmethane catalyzed by *D. desulfuricans*, the hydrogen uptake was 150 μmol g^{-1} h^{-1}. *D. desulfuricans* also hydrogenated 36% of 1 M fumarate to succinate in the presence of hydrogen. In general, the presence of a biocatalyst facilitates hydrogen uptake during treatment. The net hydrogen uptake by coals varied depending upon the coal type and the microorganism. The best microbial catalyst, *D. desulfuricans* showed the greatest hydrogen uptake with untreated coal KCER No. 4677. In this case, the net H-uptake during the biotreatment with *D. desulfuricans* was 1,878 μmol g^{-1}, while in the absence of the biocatalyst, i.e., the coal control experiment it was 245 μmol g^{-1}, and the uptake by cells alone was 127 μmol g^{-1}. The biotreated coal sample was subjected to direct liquefaction, under hydrogen in the presence of tetralin as H-donor solvent. A net increase in liquefaction yield of 5.4% was observed for the (*D. desulfuricans*) biotreated coal KCER No. 4677 compared to the yield observed from the direct liquefaction of the untreated coal sample. The comparison between the two biocatalysts led to the conclusion that the H-uptake was not enough for improving the liquefaction yield, but that biohydrogenation of the coal compounds during biotreatment was needed (Murty et al., 1994).

Although studies of biohydrogenation of unsaturated and aromatic HCs are scarce, there might be some potential in this area for finding new biocatalysts and a heavy oil upgrading application would be developed. The development of a biocatalytic application require studies on the changes in the substrate specificity of the enzymes and of the solvent effects. On the other hand, the isolation or synthesis of catalytic centers similar to those existing in these enzymes might serve as inspiration for the development of either homogeneous or heterogeneous catalyst. Homogeneous or heterogeneous catalysts might be an alternative approach to broaden the substrate specificity of the enzymes and thus enabling the conversion of the massive number of compounds present in crude oil and its fractions. This bio-inspired catalyst development (using biomimetic approaches) has not been reported. Similarly, the development of bioreactors or the exploration of mechanisms, which could lead to the improvement of the mass transfer limitations found with the microbial treatment of oil liquids has not been considered in this area.

9.3.2 BIOTREATMENTS

Heavy oils, bitumen, and their derived products have to be treated to upgrade their quality to commercialization standards and specifications. Catalytic HDP is the current applied technology at crude oil refineries for the removal of heteroatomic contaminants present in petroleum. These processes, requiring high temperatures and high hydrogen pressure are not suitable for the treatment of heavy oils and bitumen. Biocatalysis have been considered as an option, as well as for the conversion of the heavy asphaltenic molecules. However, in order to be competitive with conventional heterogeneous catalytic processes, the alternative biocatalytic option must remove the heteroatoms from a large variety of compounds, in a significant proportion. As a point of reference, HDP operates at almost total conversion, by removing the heteroatoms at levels higher than 98%.

Biocatalytic treatments of crude oils, including BDS, BDN, BDM and bioconversion (BDC) were very prolific R&D topics in the 1990s. More than three decades of effort was described in books published in 2004 (Vazquez-Duhalt, 2004) and 2007 (Ramirez-Corredores and Borole, 2007). Although, there were attempts for technology licensing and implementation, no success has been reported. In this chapter, a brief description of the technology status is presented, together with the remaining challenges that need to be addressed.

9.3.2.1 PROCESS CHALLENGES

The biological treatment of crude oil and intermediate refining streams has been stalled by catalyst and process development failures, despite incentives associated with the mild conditions of the process and negligible energy requirements of the biocatalysts. When considering the use of living cells as substitutions for currently used chemical catalysts, refiners become hesitant, given their negative experiences with microbial induced corrosion (MIC) and biological product contamination. These and the fears about pathogenicity have made refiners reluctant to handle bacterial cultures in chemical reactor plants. Even though live whole-cell biocatalysts have the benefit of being self-regenerative, they have to be employed in an aqueous phase that introduces additional mass transfer challenges. Two phase bioreactors, such as the impeller-based stirred reactors were thought as a potential solution. However, the mechanical stirring of the mixing device in an industrial scale bioreactor increases the energy requirements of the process. This led Oak Ridge National Laboratory to move towards emulsion-operated bioreactors. The developed system was based on the dispersion of the biocatalyst aqueous phase into the hydrocarbonaceous phase (Borole and Kaufman, 1999). Others have opted to the addition of surfactants when using immobilized cell biocatalysts (Dinamarca et al., 2014). Another approach was based on an electrically driven emulsion phase contactor, suggested for efficiently contacting immiscible phases in the absence of surfactants (Scott et al., 1994). Even with two-phase systems, if the catalytic enzymes are within the cell, mass transfer across the cell wall will be rate-limiting. Hence, surface-associated or extracellular enzymes became preferred catalysts.

Although the mass transfer issues may be resolved by using molecular biocatalysts, enzymes or active proteins, other problems arise. While the food and pharmaceutical industries value enzymes for their substrate specificity, petroleum refiners require catalysts that have high activity with a broader range of compounds. Thus, a multi-functional or multi-enzymatic catalyst might be needed, potentially limiting economic feasibility.

Therefore, a biocatalytic process useful in the petroleum industry must:

- Address the mass transport between two liquid phases and between the gas and liquid phases;
- Operate continuously under high throughput;
- Minimize capital investments (CapEx) and operating costs (OpEx);
- Improve product separation and biocatalyst recovery.

9.3.2.2 FINDINGS AND LESSONS LEARNED ON BIOCATALYSIS

The co-existence of a microflora with the crude oil in reservoirs is clear evidence of the catalytic and or metabolic action of the microorganisms in the presence of a HC matrix. The oxidative reactions mediated by enzymes such as dehydrogenases, ureases, catalases, polyphenol oxidases, and peroxidases produce intermediate metabolic products such as: alcohols, aldehydes, ketones, fatty, and carboxylic acids (Fonken and Johnson, 1972; Svarovskaya and Altunina, 1999; Head et al., 2003). Ultimate biodegradation, or mineralization, leads to the formation of CO_2 and other inorganic compounds, the objective for the bioremediation of petroleum contaminated soils (Altunina et al., 1994; Herman et al., 1994; Meredith et al., 2000; Clemente et al., 2004; Clemente and Fedorak, 2005; Kim et al., 2005; Quagraine et al., 2005; Scott et al., 2005; Biryukova et al., 2007; Paslawski, Headley, et al., 2009; Paslawski et al., 2009; Tahhan and Abu-Ateih, 2009; Quesnel et al., 2011; Huang et al., 2012; Pan et al., 2013; Johnson et al., 2013; Toor et al., 2013; Gunawan et al., 2014; McKenzie et al., 2014). As mentioned above, the activity of some microorganisms collected from oil wells has been the basis for the development of enhanced recovery technologies (microbial enhanced oil recovery (MEOR)) specifically designed for heavy oils and bitumen (see for instance (Premuzic and Lin, 1994; Bryant et al., 1998; Lazar et al., 2007)). Similarly, some oil well microorganisms have been studied in the search for appealing catalytic activities. The ultimate goal of BDC in the petroleum industry is the reduction of average MW through breakage of C-C and other covalent bonds, and the removal of heteroatoms (X), which is typically achieved via either C-rejection or H-addition pathways, due to the interest in producing HC products. The removal of any heteroatom leads to a reduction in MW, but for macromolecules with MW exceeding 2,000 amu, the reduction is marginal.

BDS, BDN, and BDM, together with asphaltene conversion have been discussed in upgrading and BDC works (Engineering Resources, Inc., 1992; Premuzic and Lin, 1994; Ferrari et al., 1996; Bosecker, 1998; Nash and Palmer, 1999; Premuzic and Lin, 1999c; Premuzic et al., 1999; Shong, 1999; Yuste et al., 2000; Ginosar et al., 2002; Kilbane and Le Borgne, 2004; Kirkwood et al., 2004; Leon and Kumar, 2005; Borole, 2006; Hao and Lu, 2007; Steinbakk, 2011; Bachmann et al., 2014; Strubinger et al., 2015). BDN and BDM have been investigated to a lesser extent than BDS and BDC. Most of the studies on conversion of N-containing compounds were concerned with degradation and very little is known on the actual removal of nitrogen.

BDS has received most of the R&D attention. Topics such as, biocatalysts tested and developed, reactions, and reactivity, and metabolic pathways, have been reviewed (Le Borgne and Quintero, 2003; Mohebali and Ball, 2008; Tuli and Kumar, 2008; Xu et al., 2009; Debabov, 2010; Liu et al., 2010; Chandra Srivastava, 2012; Sohrabi et al., 2012; Wei and Ning, 2012; Mužic and Sertić-Bionda, 2013; Srinivas Raju et al., 2013; Boniek et al., 2014; Mjalli et al., 2014).

Process improvements have been achieved by the incorporation of other technologies and some of the published examples include: the emulsification of the reacting system (Dinamarca et al., 2014), sonochemistry (Tang et al., 2013; Bhasarkar et al., 2015), biocatalyst immobilization (Xu et al., 2015), and electrokinetic studies (Boshagh et al., 2014). Unfortunately, the R&D activities in BDS started to decline as a consequence of two concurring factors:

- Most of the tested biocatalysts partially oxidized the S-moiety of the S-containing compounds producing sulfoxides rather than S-free compounds; and
- The production of ultra-low sulfur fuels (ULSF) required desulfurization of the feed in a much higher level than that achievable by BDS.

Early BDS studies, mostly on DBT, relied on gas chromatography for measuring conversion based on the disappearance of the peak corresponding to the reactant. In these studies, the area reduction of the S-compound peak was taken as S-removal, when in reality it was only its partial oxidation. The 4S pathway for microbial DBT desulfurization was first reported in 1993, by Kayser et al. (Kayser et al., 1993) and involves three enzymes: two mono-oxygenases to generate a sulfoxide and then a sulfone, and a desulfinase for C-S cleavage (Olson et al., 1993; Monticello, 1998). The BDS of DBT to 2-hydroxybiphenyl (HBP) and sulfite via the 4S pathway (shown in Figure 9.10 (Ramirez-Corredores and Borole, 2007)) consists of four reaction steps:

i. DBT is oxidized by DszC (DBT monooxygenase, DBT-MO) to DBT sulfoxide (dibenzothiophene-5-oxide, DBTO);
ii. DBTO is converted to the corresponding sulfone (dibenzothiophene-5,5-dioxide, DBTO2) by DszC;
iii. DBTO2 forms hydroxyphenyl benzene sulfonate (phenol-phenyl sulfinate, HBPS) on DszA; and
iv. Sulfite is eliminated from HBPS to produce 2HBP (2-hydroxybiphenyl) by HPBS desulfinase (DszB).

DszC and DszA are flavin-dependent and require the co-catalysis of a third enzyme, a flavin reductase, DszD, to perform the reactions described above. The catalytic microorganism was initially identified as *R. rhodochrous* sp. IGTS8, later classified as *R. erythropolis* sp. IGTS8. Additional intermediates (sultine and sultone) were identified (Olson et al., 1993) and a detailed metabolic pathway was proposed (Oldfield et al., 1997).

Substrate specificity and level of desulfurization have limited the application of BDS reactions. Additionally, the effect of the sulfoxides on the viscosity of the treated crude oil has not been indubitably established. However, for upstream operations where the desired level of upgrading is just a reduction in viscosity, these reactions could be potentially interesting if post-biocatalysis demulsification issues can be addressed.

Biodegradation and BDC of asphaltenes have been the subject of several reviews (Shong, 1999; Pineda-Flores and Mesta-Howard, 2001; Le Borgne and Quintero, 2003; Ramirez-Corredores and Borole, 2007; Ramirez, 2008). The bio-catalyzed reactions undergone by the high MW components of crude oils include oxidation of aliphatic and aromatic carbon groups, oxidation of naphthenic acids, and oxidation and desulfurization of aromatic and aliphatic sulfur groups (Kirkwood et al., 2004; Pineda-Flores et al., 2004; Pineda et al., 2007). Some of these reactions have been discussed above, in this chapter. BDC of asphaltenes, as a downhole "in situ" method has been suggested to represent a potential and promising solution for reducing viscosity (Ramirez, 2008). An appreciable conversion of asphaltenes has been reported to occur on exposure to certain biocatalysts, namely, *Neosartorya fischeri* (Uribe-Alvarez et al., 2011), (chloro)peroxidase (Ayala et al., 2007, 2012; Zhi et al., 2007; Juarez-Moreno et al., 2015) and cytochrome C (Torres et al., 1995; Tinoco and Vazquez-Duhalt, 1998; Garcia-Arellano et al., 2002; Borole et al., 2004).

Brookhaven National Laboratory, jointly with academic groups through the ASS Universities Inc., and led by Premuzic, developed a series of bioprocess technologies, which were initially oriented towards enhancing heavy oil recovery. The intellectual property comprises the use of the biocatalysts for the treatment of HC feedstocks; the method of adapting the microorganism to perform the desired function; and the particular bioprocesses for a given application. These bioprocesses were described in five patents:

- Biochemically enhanced oil recovery and oil treatment (Premuzic and Lin, 1994);
- Process for producing modified microorganisms for oil treatment at high temperatures, pressures, and salinity (Premuzic and Lin, 1996);
- Biochemical upgrading of oils (Premuzic and Lin, 1999a);
- Biochemical transformation of coals (Premuzic and Lin, 1999b); and
- Biochemical transformation of solid carbonaceous material (Premuzic and Lin Mow, 2001).

The first three inventions (Premuzic and Lin, 1994, 1996; Premuzic and Lin, 1999a) concerned applications for the processing of heavy crude oils, while the other two (Premuzic and Lin, 1999b; Premuzic and Lin Mow, 2001) are applicable to coal.

A wide range of microorganisms have been tested and were included in the Premuzic patents. A set of 67 strains were selected and tested for the MEOR application, from which the preferred organisms were *Achromobacter* sp. BNL-4-23 (ATCC 55021), *Sulfolobus solfataricus* BNL-TH-29 (ATCC

55022), *Sulfolobus solfataricus* BNL-TH-31 (ATCC 55023), *Sulfolobus acidocaldarius* BNL-TH-1, *Pseudomonas* sp. BNL-4-24 (ATCC 55024), *Leptospirillum ferrooxidans* BNL-5-30 (ATCC 53992), *Leptospirillum ferrooxidans* BNL-5-31 (ATCC 53993), *Acinetobacter calcoaceticus* BNL-4-21 (ATCC 53996) and *Arthrobacter* sp. BNL-4-22 (ATCC 53997). Among these, the best performance was attained with BNL-TH-1, BNL-TH-29, and BNL-TH-31, obtained from the parent strains *Sulfolobus acidocaldarius* (ATCC # 33909) and *Sulfolobus solfataricus* (ATCC # 35091 and 35092), respectively (Premuzic and Lin, 1994). The biocatalyst was subjected to an adaptation method for adjusting its capabilities to the extreme conditions of the oil well. Upgrading as well as sulfur and metals removal from the crude oil took place during the "in situ" MEOR process (Premuzic and Lin, 1994).

The developed technologies were modified for processing different types of feeds: heavy oils, coals or solid carbonaceous material, as well as to provide different processes that operate downhole (MEOR) or at surface level (BDS, BDM, and BDC). The development of most of the microbial catalyst was undertaken by a selective enrichment mechanism, in which the microorganisms adjust to perform at the extreme temperature, pressure, pH, and salt concentrations of an oil reservoir or of the surface process. The adaptation leads to the selection of thermophilic microorganisms or a mixture of thermophilic strains, which are also oil-tolerant and capable of growth under limited nutrient conditions. One of these adaptation methods modified microorganisms for their application in oil treatment at high temperatures, pressures, pH (1.5–7.5) and salinity (Premuzic and Lin, 1996). This process comprised steps, which successively limit the carbon source and increase the temperature (from 70°C to 90°C), pressure (from 2,000 to 2,500 psi) and salinity of the media (from 1.3 to 35%), until microbial growth occurs essentially on crude oil as a sole carbon source at the established conditions. A list of 38 bacterial strains were considered for the development of potential biocatalysts for this upgrading application. From this list, the set of preferred biocatalysts was based on the strains of *Thiobacillus ferrooxidans* BNL-2-49s (ATCC 55530), *Thiobacillus ferrooxidans* BNL-2-48s (*ATCC 55529*), *Thiobacillus ferrooxidans* BNL-2-47s (ATCC 55528), *Thiobacillus ferrooxidans* BNL-2-46s (ATCC 55527), *Thiobacillus ferrooxidans* BNL-2-45s (ATCC 55526), *Thiobacillus ferrooxidans* BNL-2-44s (ATCC 55525), *Leptospirillum ferrooxidans* BNL-5-31s (ATCC 55524), *Leptospirillum ferrooxidans* BNL-5-30s (ATCC 55523). The developed biocatalysts were capable of promoting BDS, BDM, and BDC of heavy crudes, producing an upgraded oil with overall changes in its composition such as, increases in the light (less than C20 chains) content; containing treatment-derived emulsifying agents that dispersed heavy molecules; decreases in the content of S-compounds (e.g., thionaphthalene), in the C20 to C40 alkanes, and in organometallic compounds (Premuzic and Lin, 1996). The upgraded oil showed increased content of saturated HCs as well as decrease in S (20% to 50%), N (15% to 45%), and metals (20% to 45%) content (Premuzic and Lin, 1999a).

The bioupgrading of several heavy oils (i.e., off-shore California (OSC), midway sunset (MWS), Cerro Negro, and A851 Monterey) reduced the MW of aromatic and resins fractions, with a concomitant reduction in the proportion of these fractions in the product. Other composition changes included increases of the oil saturate fraction (20% to 66%) and reduction of heteroatom content: nitrogen (15–45%), sulfur (15–30%), and metals (nickel and vanadium, 16 to 36%). All these findings require biocatalysts with a very broad spectrum for substrate targeted preferences, indicating that most likely, such a biocatalyst may be constituted by a mixture of bacterial strains. However, no details were provided on the microbial community composition, or on the potential reactions that may have taken place (Premuzic et al., 1999).

The process they described for the biochemical transformation of coals is carried out in a slurry phase at temperatures between 40°C to 85°C, pressures in the range from atmospheric to 2,500 psi, pH values of 2–10 and a salinity range from about 1.5–35 wt.%. Although the range of conditions might look broad for a microbial application, considering the patent language, the claimed ranges are rather narrow. A wide variety of biocatalysts were suggested for this process: *Achromobacter* sp. BNL-4-23 (ATCC 55021), *Sulfolobus solfataricus* BNL-TH-29 (ATCC 55022), *Sulfolobus solfataricus* BNL-TH-31 (ATCC 55023), *Sulfolobus acidocaldarius* BNL-TH-1 (ATCC 35091), *Pseudomonas* sp. BNL-4-24 (ATCC 55024), *Leptospirillum ferrooxidans* BNL-5-30 (ATCC 53992), *Leptospirillum ferrooxidans* BNL-5-31 (ATCC 53993), *Acinetobacter calcoaceticus* BNL4-21 (ATCC 53996), *Arthrobacter* sp. BNL-4-22 (ATCC 53997), *Acinetobacter calcoaceticus* BNL-4-21s (ATCC 55489), *Arthrobacter* sp. BNL-4-22s (ATCC 55490), *Achromobacter* sp. BNL-4-23s (ATCC 55491), *Pseudomonas* sp. BNL4-24s (ATCC 55492), *Leptospirillum ferrooxidans* BNL-5-30s (ATCC 55523), *Leptospirillum ferrooxidans* BNL-5-31s (ATCC 55524), *Thiobacillus ferroxidans* BNL-2-44s (ATCC 55525), *Thiobacillus ferrooxidans* BNL-2-45s (ATCC 55526), *Thiobacillus ferrooxidans* BNL-2-46s (ATCC 55527), *Thiobacillus ferrooxidans* BNL-2-47s (ATCC 55528), *Thiobacillus ferrooxidans* BNL-2-48s (ATCC 55529), *Thiobacillus ferrooxidans* BNL-2-49s

(ATCC 55530), Unknown BNL-NZ-3 (ATCC 55488), and a series of Mixed Cultures labeled R.I.-1 through R.I.-14 (ATCC 55501 through ATCC 55514) (Premuzic and Lin, 1999b). The preparation method of the biocatalytic system was reported in a separate patent (Premuzic and Lin Mow, 2001).

In summary, from Premuzic et al.'s findings, and due to the numerous compounds present in the carbonaceous materials and crude oils subjected to upgrading, the requirement of a complex biocatalytic system could be inferred. A detailed examination of these results indicated the *Thiobacillus*, *Leptospirillum*, *Sulfolobus*, *Pseudomonas*, and *Arthrobacter* genera as the best candidates for biocatalysts. *Thiobacillus*, *Leptospirillum*, *Sulfolobus* are well known for their capability in transforming inorganic sulfur while *Thiobacillus* and the *Pseudomonas* are known for nitrogen conversion. The latter two are known to be involved in the breakdown of HCs (Ramirez-Corredores and Borole, 2007). Nevertheless, further studies, particularly on reaction mechanisms or metabolic pathways are needed for supporting and understanding all these findings.

A BDC hybrid process was performed in a cyclic manner by running a first aerobic stage, followed by a second anaerobic stage, and repeating the cycles as many times as to reach the desired conversion level. The performance is monitored for defining the number of alternating cycles needed. During each hemicycle, either a carboxylation or a decarboxylation reaction took place. Accordingly, the process was named repetitive alternating carboxylation-decarboxylation cycle (RACDC) (Maure et al., 2005). The alternating hemicycles are carried out with one or more aerobic microorganisms and with one or more anaerobic microorganisms, at temperatures ranging from 4°C to 150°C and pressures from 1 to 600 atm. A surface application can take place in conventional operating units, e.g., storage tanks, refining processes, oil/water emulsion transportation pipelines and petrochemical plants. The process operation is a time- and redox-dependent methodology based on a systematic bio-inoculation with co-metabolic and symbiotic microorganisms on qualified oil-base substrates. Figure 9.11 depicts this cyclic process (Maure et al., 2005) and a brief description of the process steps follows, the reader is referred to the patent for further details:

1. **Pre-Inoculation:** It is the optimization of an adaptive co-metabolic and symbiotic biocatalytic system to achieve both, hemicycle A (carboxylation/oxidative stage) and hemicycle B (decarboxylation/reductive stage) and the determination of the sequence and timing of events within RACDC process.

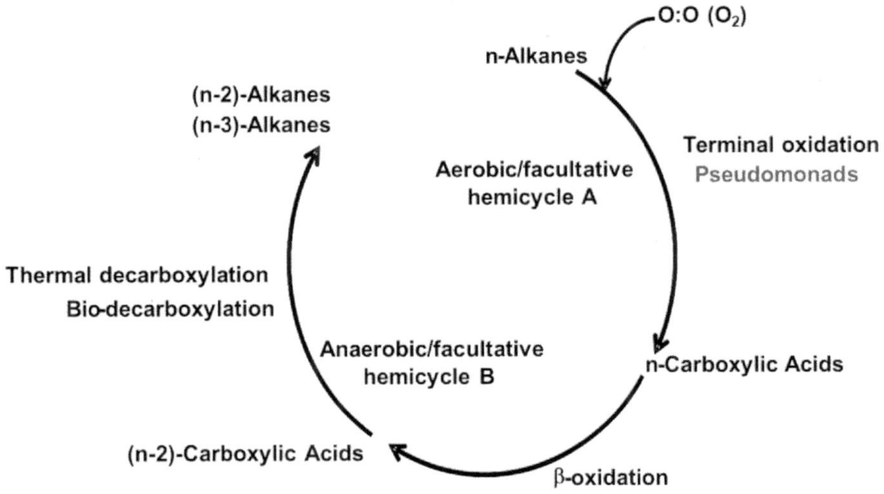

FIGURE 9.11 Repetitive alternating carboxylation-decarboxylation cycle (RACDC process). *Source:* Adapted from: Maure et al. (2005).

2. **Hemicycle A:** a cycle is initiated by inoculating the feedstock mixture with the aerobic/facultative biocatalytic system (pseudomonads/denitrifying blend of microorganisms, nitrate/nitrite molybdenum salts). The water-base microbial concentrates used as inoculum should have a minimal colony-forming unit (CFU) of 10^7 units per ml, in a range of 10 to 10,000 ppm of microbial concentrates on relevant system total fluid volume in discontinuous bioreactors, for instance. Other process parameters need to be controlled depending on reactor type and configuration. During this step a "terminal oxidation" is promoted by bubbling oxygen through the aqueous phase of the reacting mixture. Typical conditions for this step are: contact time of 4 to 72 h, surfactant/co-surfactant additives between 0.1 to 2 of critical micellar concentration of surfactant, inhibitors in the range of 10 to 1,000 ppm, dose of salts between 10 to 100 ppm and biocatalysts 1 to 100 ppm.
3. **Intermediate:** An inhibitory intermediate stage is needed to change from aerobic/facultative environment of hemicycle A to the progressively anaerobic conditions required for the next hemicycle of low redox potential. Depletion of oxygen and reduction of nitrate/nitrite salts are attained by adding oxygen scavengers. Elimination of inhibiting role of sodium bisulfite is obtained with a cobalt catalyst,

added in the range of 10 to 1,000 ppm. Main reaction to promote is the β-oxidation.

4. **Hemicycle B:** For this half-reaction, the carboxylated product mixture is then inoculated or stimulated with the indigenous syntrophic system of *Archaea* and *Syntrophomonas* (low redox potential), in the presence of nickel or cobalt salts to support enzyme synthesis. Nickel coenzymes, known under the trade designations of F420 and F430 could be used. Most of the conditions of Hemicycle A are kept and the contact time is extended to 5 to 60 days.

A process for the remediation of contaminated sites produces lower MW HCs from heavy oils. Conversion and product distribution depend on incubation time. Potential biocatalysts for a BDC upgrading process might be developed based on those microorganisms that were tested for this bioremediation application, which comprise bacterial strains isolated from seawater and belonging to the genera of *Acinetobacter* (FERMBP-7046 and FERMBP-7049), *Pseudomonas* (FERMBP-7047), *Alcaligenes* (FERMBP-7048), *Flavobacterium* (FERMBP-7050, FERMBP-7051, FERMBP-7052) and *Moraxella* (FERMBP-7053). Both individual bacteria and consortia were tested at room temperature (Fujita, 2003).

A couple of thermophilic microorganisms, *Bacillus thermoleovorans* B23 (FERM-P-17696) and *Bacillus thermoleovorans* H41 (FERM-P-17715) were found to be capable of decomposing heavy HCs at 60–65°C, under aerobic conditions. A conversion of 60% was observed for C10–C18 alkanes (Morikawa and Kato, 2001).

Upgrading biocatalysts based on *Thiobacillus aquesullis*, *Thiobacillus thioparus* and/or *Thiobacillus denitrificans* (Kurashov, 2002) and on *Vibrio furnissii* M1 (FERM P-18382) (Miyamoto et al., 2003) have been identified. The *Thiobacillus* strains converted heavy PAHs into branched alkanes and low MW aromatics, boiling within a range of less than 200°C; and simultaneously removed sulfur, nitrogen, and oxygen compounds. The process is carried out at temperatures of 4°C to 60°C and a pH ranging approximately from 6.0 to 9.0, under shaking (Kurashov, 2002).

Biocatalytic demetallization (BDM) has not been studied as broadly and deeply as the other bioprocesses for removing the heteroatomic contaminants from crude oils and oil products (Le Borgne and Quintero, 2003). Metals tend to concentrate towards the BotB, i.e., the metals concentration profile increases with increasing MW of the crude oil fraction. Consequently, the changes in metal-bearing compounds have been considered in connection with asphaltene conversion.

In terms of biocatalysts, metals removal has been evaluated using whole cells as well as molecular biocatalysts. As with other biotreatment studies, model compounds were typically used as substrates though in a few instances, extracts from crude oil had been considered.

The demetallization of nickel and vanadium petroporphyrins using peroxidase biocatalysts, such as heme oxygenase and cytochrome c was reported. These oxidizing enzymes were able to release the metals by degrading the porphyrin molecules but leaving the carbon skeleton intact (Fedorak et al., 1993b).

The work previously discussed in the context of enzymatic conversion of PAH using chloroperoxidase and the handicap of chlorine incorporation in the product, also reported the removal of 53% of the Ni and 27% of the V present in crude oil (Mogollon et al., 1998). In this case, when porphyrins were used as model compounds, the BDM of nickel octaethylporphine and vanadium octaethylporphine was studied using a ternary solvent system containing toluene, isopropanol, and an aqueous buffer of pH of 3.0, which was required to dissolve the petroporphyrin molecules. In addition to the model compounds, an asphaltene fraction from Cold lake heavy oil was also studied. The reported BDM for the model compounds was much higher than that observed for the asphaltene fraction. In fact, it was 93% for nickel octaethylporphine, 53% for vanadium octaethylporphyrin but only 20% for the asphaltene fraction. It is important to emphasize once more that besides its feebleness, the chloroperoxidase enzyme presents the inconvenience of needing the concerted action of hydrogen peroxide and chloride. The potential production of chlorinated organic molecules represents an additional drawback for this enzymatic system. In fact, besides its high peroxidase activity and its versatility, chloroperoxidases are well known as biocatalysts for the halogenation of aromatic molecules, as in the case of PAHs will be an undesirable reaction of the aromatic core of the metal-bearing compounds (Ayala et al., 2000; Vázquez-Duhalt et al., 2001).

Cytochrome c is a soluble hemeprotein, which catalyzes oxidation reactions on a number of hydrophobic compounds such as substituted phenols, organic sulfides, thiophenes, and PAHs. Cytochrome c has some advantages as a biocatalyst when compared with peroxidases for its application. It is stable in organic solvents, and even water-immiscible organic solvents, over a wide range of pH (2 to 11). Furthermore, Ryu et al. indicated economic advantages (Ryu et al., 2002). The catalytically interesting features of cytochrome c are centered in its heme active site (Figure 9.12). The functions of heme (iron-protoporphyrin IX) prosthetic groups in nature include a wide range of reactions such as electron transfer, oxygen transport and storage, catalysis, gas sensing, and gene regulation. The variable that defines the type of heme group is the nature of substituents on the porphyrin macrocycle. In the case of heme c, it is characterized by the presence of two covalent thioether bonds formed between Cys side chains and the heme vinyl groups at positions 2 and 4. The stereochemistry of heme attachment is the same in all known examples of heme c, and the vinyl groups at positions 2 and 4 are attached to the N- and C-terminal Cys of the CXXCH motif, respectively (Bowman and Bren, 2008).

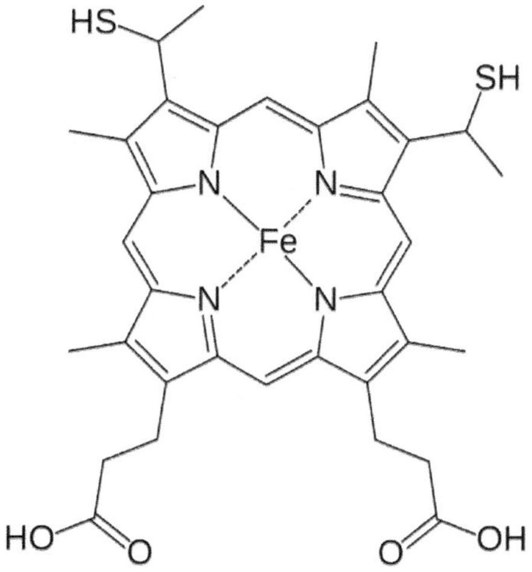

FIGURE 9.12 Heme C structure.

Modified cytochrome c (Torres et al., 1995; Villegas et al., 2000) has been tested for BDM. A series of synthetic porphyrins and a petroporphyrin-rich fraction were demetallized at mild conditions, only a marginal amount of water (5% v/v) was needed for enabling the reaction. A higher level of demetallization was attained with the synthetic porphyrins such as, 95% for V and 74% for Ni. Conversion of asphaltenes may increase the potentialities of the development of an upgrading bioprocess with this biocatalyst (García-Arellano et al., 2004).

Metals removal from fossil fuels employing heme oxygenase and cytochrome c reductase was claimed in a patent (Xu et al., 1998). The cytochrome c reductase biocatalyst was isolated from *Bacillus megaterium*, *Catharanthus roseuse*, *Escherichia coli*, animal cells, plant cells, and yeast cells. The biocatalyst may be used as a substantially cell-free preparation or a whole-cell preparation.

A microbial biocatalyst comprising *Pseudomonas azelaica* YA1 strain (FERM P-15356) was tested for the decomposition of porphyrin metal complexes (e.g., vanadium octaethylporphyrin) (Tomita et al., 1997). Its use in the decomposition of porphyrin-based compounds, in a meat extract medium containing the whole cells spawns was reported, as well (Tomita et al., 1999). However, metal removal has not been proven for these experiments. Again, the reaction progress was followed by the disappearance of the UV-Vis absorption band of the porphyrin, but the real metal content in the organic product was not analyzed.

Metal compounds concentrate in the heaviest part of the crude oil and particularly in the asphaltene fraction, which is well known to flocculate. In this regard, advances in protein engineering have been applied for the development of flocculating agents. Thus, amphipathic proteins are used as flocculating agents in water treatments (Marison and Doerries, 2003). The question is whether these proteins or the phenomenological principle could be applied for the development of agents that work on porphyrin and non-porphyrin metal compounds in organic media.

9.4 CONCLUSIONS AND FUTURE RESEARCH DIRECTIONS

The development of a bioupgrading process requires the concourse of genetic engineering tools to be applied on two ends of the technology. On one side, the biocatalyst has to be modified to improve its affinity to the oil phase. On the other side, the substrate specificity and the catalytic activity need to be broadened and increased, respectively, on the whole-cell level. The engineering efforts on the bacterial cultures should address the broadening of the functionality and specificity, as well as enhancing enzymatic overexpression. For the latter, the considered common approach has been the development of biocatalysts based on microbial consortium or including more than one biocatalyst. Additionally, enzyme costs are prohibitive for this type of application, which favors the use of whole cells. However, for opportunity crude oil upgrading, the concerned inherent molecules exhibit a size too large that causes severe limitations on transport through the cell membrane and would prompt for a cell-free application. This controverted situation has precluded the development of process technologies for the biocatalytic processing of opportunity crudes. The definition and careful design of an R&D strategy is required to overcome these barriers. Currently, the economic benefits of whole cells make this a preferred option. Meanwhile, enzyme production costs should take advantage of the advances in molecular biology technologies and improvements in protein production and bulk manufacturing. A refining application would expedite the growth of the enzymes market and may drive a price decrease, bringing the possibility of enzyme-based catalysis closer to reality in the future. Nevertheless, enzymes also need engineering for tailor-designed biocatalysts adjusted to the needs of the applications in oil refining. So far, biotechnological improvements have taken a very long time. However, the time has come for making the desired impact on the oil industry, by taking full advantage of the achievements attained and the cumulative knowledge. A more proactive involvement of the oil industry is needed, especially for the synergistic addition of the process development skills into the biotechnology competencies, for building a multidisciplinary team to address the development of a bioupgrading process.

The research needs for the development of bioprocessing alternatives have not changed during the last five or six decades and reside on the development of improved biocatalysts. As a summary from the previous paragraph, the biocatalyst has to be:

1. **Highly Active:** Multi-functional with capabilities for BDS, BDN, BDM, and BDC.
2. **Non-Selective:** Active towards multiple molecules (broader substrate specificity).
3. **Organic Solvent-Tolerant:** Adaptable to an oil environment, with minimum requirement for water.

Whether the biocatalyst is a whole cell or an enzyme, it is important to fulfill the hetero-atom removal targets and the suitability for process integration under feasible economics. The rapid evolution of the oil production technologies is in clear contrast with the slow path of refining processes. The implementation of upstream technologies has changed the quality of the produced crude oil and bitumen, and the dynamics of new crude oil incorporation in the market. Refineries are not prepared enough for this new dynamic since refining technologies have not evolved in response to those advances. In consequence, refiners keep struggling with their margins. Biocatalysis as a low-temperature conversion process represents a very attractive and potential solution for bitumen and extra-heavy oil, and in general for opportunity crudes. The development of a bio-upgrading process proven at surface level could later on be evolved and adapted to a subsurface, downhole or "in situ" treatment application. While upstream operations have not found enough economic incentives to pursue this approach, for downstream, the incorporation of a bioprocess in a refinery is seen as a cost increment. The incorporation of a refining application in upstream operations suffers from the lack of mutual knowledge of each other's operations and businesses. This is particularly relevant for the formation of the required R&D teams.

In the future, the opportunities for downhole or in-situ bioprocessing of heavy oils and bitumen may be greater than they were in the past. The focus and targets for in-situ bioprocessing are different because the level of upgrading required upstream is much lower than that demanded for fuels production in the refinery. This latter refining application has captured most of the R&D efforts since the 1970s' and has provided significant contributions in knowledge growth. An upstream bioprocessing application shifts research to focus on reservoir conditions and to applied lessons learnt from the past research. For instance, the bacterial strains present in the well or in tailings pond sediments have been probably metabolizing some of the molecules of interest, but clearly many other compounds remain unconverted for being refractory to these strains. Some of the biocatalytic reactions that were irrelevant to refining could be probably better suited for an upstream application. These reactions include: oxidation of high MW aliphatics, aromatics, and S- and N-compounds, as well as hydrogenation and dehydrogenation of lower-MW compounds. If a biotreatment like this is applied downhole or at well surface, the process as well as the biocatalyst requirements for a refining application would change and most likely refining bioupgrading would be facilitated.

KEYWORDS

- **asphaltenes**
- **bioconversion**
- **biological upgrading**
- **bioupgrading**
- **bitumen**
- **heavy oils**
- **metal-compounds**
- **processing technologies**
- **upgrading**

REFERENCES

Ali, M. F., & Abbas, S., (2006). A review of methods for the demetallization of residual fuel oils. *Fuel Proc. Technol., 87*(7), 573–584.

Altunina, L. K., Svarovskaya, L. I., Rozhenkova, Z. A., & Golovko, A. K., (1994). Oil degradation by reservoir microorganisms. *Proc. Int. Symposium Adv. Prod. Appl. Lube Base Stocks*, 354–361.

Ancheyta, J., & Rana, M. S., (2007). Future technology in heavy oil processing. In: Pedro De, A. P. F., & Adolfo, P. P., (eds.), *Petroleum Engineering - Downstream* (p. 41). Unesco EOLSS.

Angeles, M. J., Leyva, C., Ancheyta, J., & Ramírez, S., (2014). A review of experimental procedures for heavy oil hydrocracking with dispersed catalyst. *Catal. Today, 220–222*, 274–294. doi: 10.1016/j.cattod.2013.08.016.

Ayala, M., Hernandez-Lopez, E. L., Perezgasga, L., & Vazquez-Duhalt, R., (2012). Reduced coke formation and aromaticity due to chloroperoxidase-catalyzed transformation of asphaltenes from Maya crude oil. *Fuel., 92*(1), 245–249. doi: 10.1016/j.fuel.2011.06.067.

Ayala, M., Robledo, N. R., Lopez-Munguia, A., & Vazquez-Duhalt, R., (2000). Substrate specificity and ionization potential in chloroperoxidase- catalyzed oxidation of diesel fuel. *Environ. Sci. Technol., 34*(13), 2804–2809. doi: 10.1021/es991270o.

Ayala, M., Verdin, J., & Vazquez-Duhalt, R., (2007). The prospects for peroxidase-based biorefining of petroleum fuels. *Biocatal. Biotransform., 25*(2), 114–129.

Bachmann, R. T., Johnson, A. C., & Edyvean, R. G. J., (2014). Biotechnology in the petroleum industry: An overview. *Int. Biodeterior. Biodegradation, 86*(Part C), 225–237. doi: 10.1016/j.ibiod.2013.09.011.

Bacon, R., & Tordo, S., (2004). Crude oil prices. *Predicting Price Differentials Based on Quality*. Report No. Note 275, Washington, DC, USA.

Bacon, R., & Tordo, S., (2005). *Crude oil Price Differentials and Differences in Oil Qualities: A Statistical Analysis*. Report No. WB.34906. Technical paper No. 81, Washington, DC, USA.

Barbosa, A. L., Vega, A. F., & Amador, E. R., (2014). Hydrodesulfurization of crude oil: Basis for improving fuel. A review. [hidrodesulfuración de crudos de petróleo: Base para el mejoramiento de combustibles. Una revisión]. *Avan. Cienc. Ing., 5*(3), 37–60.

Bhasarkar, J. B., Dikshit, P. K., & Moholkar, V. S., (2015). Ultrasound assisted biodesulfurization of liquid fuel using free and immobilized cells of *Rhodococcus rhodochrous* MTCC 3552: A mechanistic investigation. *Bioresour. Technol., 187*, 369–378. doi: 10.1016/j.biortech.2015.03.102.

Biasca, F. E., Chang, E., Bailey, R. T., Dickenson, R. L., Johnson, H. E., & Simbeck, D. R., (2003). *Upgrading Heavy Crude Oils and Residues to Transportation Fuels: Technology, Economics and Outlook*. Report No. Phase 7. Multiclient Review Report, Mountain View, California.

Biryukova, O. V., Fedorak, P. M., & Quideau, S. A., (2007). Biodegradation of naphthenic acids by rhizosphere microorganisms. *Chemosphere., 67*(10), 2058–2064.

Boduszynski, M. M., (1987). Composition of heavy petroleums. 1. Molecular weight, hydrogen deficiency, and heteroatom concentration as a function of atmospheric equivalent boiling point up to 1400°f (760c). *Energy Fuels, 1*(1), 2–11.

Boniek, D., Figueiredo, D., Dos, S. A. F. B., & De Resende, S. M. A., (2014). Biodesulfurization: A mini review about the immediate search for the future technology. *Clean Technol. Environ. Policy, 17*(1), 29–37. doi: 10.1007/s10098-014-0812-x.

Borole, A. P., & Kaufman, E. N., (1999). Process considerations in crude oil biodesulfurization. *ACS Spring Meet.* Anaheim, CA. USA.

Borole, A. P., (2006). *Biological Upgrading of Petroleum - Recent Developments and Research Needs*. AIChE Spring Nat. Meet., Orlando, FL.

Borole, A. P., Cheng, C. L., & Davison, B. H., (2004). Substrate desolvation as a governing factor in enzymatic transformations of PAHs in aqueous-acetonitrile mixtures. *Biotechnol. Prog., 20*(4), 1251–1254.

Bosecker, K., (1998). Microbial deterioration of hydrocarbons. In: *6th International Conference on Stability Handling of Liquid Fuels*. Vancouver, BC. Canada.

Boshagh, F., Mokhtarani, B., & Mortaheb, H. R., (2014). Effect of electrokinetics on biodesulfurization of the model oil by *Rhodococcus erythropolis* PTCC1767 and bacillus subtilis DSMZ 3256. *J. Hazard. Mater., 280*, 781–787. doi: 10.1016/j.jhazmat.2014.09.006.

Bowman, S. E. J., & Bren, K. L., (2008). The chemistry and biochemistry of heme c: Functional bases for covalent attachment. *Nat. Prod. Rep., 25*(6), 1118–1130. doi: 10.1039/b717196j.

BP International Plc. *BP Statistical Review of World Energy*. https://www.bp.com/en/global/corporate/energy-economics/statistical-review-of-world-energy.html (accessed on 25 October 2021).

Bryant, R. S., Bailey, S. A., Stepp, A. K., Evans, D. B., Parli, J. A., & Kolhatkar, A. R., (1998). Biotechnology for heavy oil recovery. In: *7th UNITAR Intern. Conf. 'Heavy Crude Tar Sands,'* Beijing, China.

Cai, X., & Tian, S., (2011). Review and comprehensive analysis of composition and origin of high acidity crude oils. *China Pet. Proc. Petrochem. Technol., 1*, 6–15.

Canadian Association of Petroleum Producers (CAPP). *Crude Oil Forecast, Markets and Pipeline Expansions* (p. 52). Available from: https://www.capp.ca/resources/crude-oil-forecast/ (accessed on 25 October 2021).

Chandra, S. V., (2012). An evaluation of desulfurization technologies for sulfur removal from liquid fuels. *RSC Adv., 2*(3), 759–783. doi: 10.1039/c1ra00309g.

Chem System, (2003). *Refinery of the Future as Shaped by Environmental Regulations*. Report No. PERP 02/03S11, White Plains, NY.

Clark, B., Graves, W. G., Lopez-de-Cardenas, J. E., Gurfinkel, M. E., & Peats, A. W., (2007). Heavy oil. In: Raymond, L. R., (ed.), *Facing the Hard Truths About Energy* (p. 56). Topic paper #22. Washington, DC. USA: National Petroleum Council, Heavy Oil Subgroup of the Technology Task Group.

Clemente, J. S., & Fedorak, P. M., (2005). A review of the occurrence, analyses, toxicity, and biodegradation of naphthenic acids. *Chemosphere, 60*(5), 585–600.

Clemente, J. S., Mackinnon, M. D., & Fedorak, P. M., (2004). Aerobic biodegradation of two commercial naphthenic acids preparations. *Environ. Sci. Technol., 38*(4), 1009–1016.

Coleman, J. K., Brown, M. J., Moses, V., & Burton, C. C., (1992). *Enhanced oil recovery*. Patent No. WO9215771.

Compere, W. A., (1974). *Process for Producing Product from Fossil Fuel*. Patent No. US3826308.

Coyle, C. L., Ferrughelli, D. T., Logan, M. S P., Siskin, M., & Zylstra, G., (2000). *Biological Activation of Aromatics for Chemical Processing and/or Upgrading of Aromatic Compounds, Petroleum, Coal, Resid, Bitumen and other Petrochemical Streams*. Patent No. US6156946.

Crude Quality Inc. *Crude Monitor*. http://www.crudemonitor.ca/home.php (accessed on 25 October 2021).

De Bruijn, T., (2002). *Critical Review of Existing and Emerging Approaches to Moderate Upgrading of Extra Heavy Crudes*. Report No. 99-35 (CF), Devon, Alberta, Canada.

Debabov, V. G., (2010). Microbial desulfurization of motor fuel. *Appl. Biochem. Microbiol., 46*(8), 733–738. doi: 10.1134/S0003683810080016.

Delbianco, A., & Montanari, R., (2007). Emerging technologies for the conversion of residues. In: Treccani, P., (ed.), *Encyclopaedia of Hydrocarbons* (pp. 137–170). Roma: Istituto Della Enciclopedia Italiana, Fondata Da Giovanni Treccani S.P.A., Marchesi Grafiche Editoriali S.P.A.

Deshannavar, U. B., Rafeen, M. S., Ramasamy, M., & Subbarao, D., (2010). Crude oil fouling: A review. *J. Appl. Sci., 10*, 3167–3174.

Dinamarca, M. A., Rojas, A., Baeza, P., Espinoza, G., Ibacache-Quiroga, C., & Ojeda, J., (2014). Optimizing the biodesulfurization of gas oil by adding surfactants to immobilized cell systems. *Fuel, 116*, 237–241. doi: 10.1016/j.fuel.2013.07.108.

Dusseault, M. B., (2007). Hydrocarbons from non-conventional and alternative fossil resources. In: Treccani, P., (ed.), *Encyclopaedia of Hydrocarbons* (pp. 21–55). Roma: Istituto Della Enciclopedia Italiana, Fondata Da Giovanni Treccani S.P.A., Marchesi Grafiche Editoriali S.P.A.

Engineering Resources, Inc., (1992). *Biological Upgrading of Coal Liquids*. Quarterly report. Report No. DOE/PC/91059 T4, Fayetteville, AR. USA.

Enomoto, T., Sekimoto, M., Nasuno, K., & Yoshimura, T., (1998). *Low-Sulfur Gas Oil*. Patent No. JPH10017874.

Fahim, M. A., Alsahhaf, T. A., & Elkilani, A., (2010). *Fundamentals of Petroleum Refining*. Oxford, Great Britain: Elsevier.

Fedorak, P. M., Semple, K. M., Vazquez-Duhalt, R., & Westlake, D. W. S., (1993a). Chloroperoxidase-mediated modifications of petroporphyrins and asphaltenes. *Enzyme Microb. Technol., 15*, 429–437.

Fedorak, P. M., Semple, K. M., Vazquez-Duhalt, R., & Westlake, D. W. S., (1993b). Chloroperoxidase-mediated modifications of petroporphyrins and asphaltenes. *Enzyme Microb. Technol., 15*(5), 429–437. doi: 10.1016/0141-0229(93)90131-K.

Ferrari, M. D., Neirotti, E., Albornoz, C., Mostazo, M. R., & Cozzo, M., (1996). Biotreatment of hydrocarbons from petroleum tank bottom sludges in soil slurries. *Biotechnol. Lett., 18*(11), 1241–1246.

Filatov, D. A., El'chaninova, E. A., Ovsyannikova, V. S., & Altunina, L. K., (2016). Biochemical oxidation of polyaromatic compounds by hydrocarbon-oxidizing bacteria in a liquid medium. *Pet. Chem., 56*(7), 657–662. doi: 10.1134/S0965544116070069.

Fischer, F., & Fetzner, S., (2000). Site-directed mutagenesis of potential catalytic residues in 1H-3-hydroxy-4-oxoquinoline 2,4-dioxygenase, and hypothesis on the catalytic mechanism of 2,4-dioxygenolytic ring cleavage. *FEMS Microbiol. Lett., 190*, 21–27.

Flint, L., (2004). *Bitumen and Very Heavy Crude Upgrading Technology: A Review of Long Term R&D Opportunities*. Report No. Available at http://www.ptac.org/links/dl/osdfnlreport.pdf (accessed on 25 October 2021).

Fonken, G. S., & Johnson, R. A., (1972). *Chemical Oxidation with Microorganisms*. New York, USA: Marcel Dekker.

Frost and Sullivan, (2004). *Oil Refining Technologies--New Developments and Growth Opportunities*. Report No. D274, San Antonio, TX.

Fujita, Tokio, (2003). *Bacteria Mixture Having Heavy Oil Degrading Ability and Method of Treating Oil Components*. Patent No. US6649400.

Garcia-Arellano, H., Buenrostro-Gonzalez, E., & Vazquez-Duhalt, R., (2004). Biocatalytic transformation of petroporphyrins by chemical modified cytochrome C. *Biotechnol. Bioeng., 85*(7), 790–798.

Garcia-Arellano, H., Valderrama, B., Saab-Rincon, G., & Vazquez-Duhalt, R., (2002). High temperature biocatalysis by chemically modified cytochrome c. *Bioconjugate Chem., 13,* 1336–1344.

Gary, J. H., & Handwerk, G. E., (2001). *Petroleum Refining: Technology and Economics.* New York: CRC Press, Marcel Dekker.

Gibbons, J. H., (1980). Materials program. *An Assessment of Oil Shale Technologies.* Washington, DC, USA: U.S. Government, Office of Technology Assessment.

Ginosar, D. M., Bala, G. A., Anderson, R. P., Fox, S. L., & Stanescue, M. A., (2002). *Bioprocessing of High-Sulfur Crudes Via Application of Critical Fluid Biocatalysis.* Report No. INEEL/EXT-02-00711, Idaho, US.

Gray, M. R., Tykwinski, R. R., Stryker, J. M., & Tan, X., (2011). Supramolecular assembly model for aggregation of petroleum asphaltenes. *Energy Fuels, 25*(7), 3125–3134. doi: 10.1021/ef200654p.

Gray, R. M., (1994). *Upgrading Petroleum Residues and Heavy Oils.* New York: Marcel Dekker / CRC Press.

Guerreiro, O., Alves, S. P., Costa, M., Cabo, Â., Duarte, M. F., Jerónimo, E., & Bessa, R. J. B., (2016). Effects of extracts obtained from *Cistus ladanifer* L. On in vitro rumen biohydrogenation. *Anim. Feed Sci. Technol., 219,* 304–312. doi: 10.1016/j.anifeedsci.2016.06.024.

Gunawan, Y., Nemati, M., & Dalai, A., (2014). Biodegradation of a surrogate naphthenic acid under denitrifying conditions. *Water Res., 51,* 11–24. doi: 10.1016/j.watres.2013.12.016.

Guo, H., Li, Y., Yiran, Z., Wang, F., Wang, Y., Yu, Z., Haicheng, S., et al., (2015). Progress of microbial enhanced oil recovery in China. *SPE Asia Pacific Enhanced Oil Recovery Conference, EORC.* Kuala Lumpur, Malaysia.

Guo-Wei, X., Mitchell, K. W., & Monticello, D. J., (1998). *Process for Demetalizing a Fossil Fuel.* Patent No. US5726056.

Hammel, K. E., (1996). Extracellular free radical biochemistry of ligninolytic fungi. *New J. Chem., 20*(2), 195–198.

Haney, III., William, H., & Monticello, D. J., (1992). *Microbial Process for Reduction of Petroleum Viscosity.* Patent No. WO9211343.

Hao, R., & Lu, A., (2007). Conversion and degradation of crude oil by bacillus sp3. *Chin. J. Geochem., 26*(2), 201–206.

Hartmann, J. C. M., (2003). Crude valuation for crude selection. *PTQ, 8*(1), 123–127.

Head, I. M., Jones, D. M., & Larter, S. R., (2003). Biological activity in the deep subsurface and the origin of heavy oil. *Nature, 426,* 344–352. doi: 10.1038/nature02134.

Heck, R. H., & Degnan, T. F., (1989). *Fundamentals of Resid Upgrading.* Edited by American Institute of Chemical Engineers. New York: AIChE.

Herman, D. C., Fedorak, P. M., MacKinnon, M. D., & Costerton, J. W., (1994). Biodegradation of naphthenic acids by microbial populations indigenous to oil sands tailings. *Can. J. Microbiol., 40*(6), 467–477.

Hofrichter, M., & Fakoussa, R., (2001). Microbial degradation and modification of coal. In: Hofrichter, M., & Steinbuchel, A., (eds.), *Biopolymers: Lignin, Humic Substances and Coal* (Vol. 1, pp. 393–427). Wiley.

Hofrichter, M., & Fritsche, W., (1996). Depolymerization of low rank coal by extracellular fungal enzyme systems.1. Screening for low rank-coal-depolymerizing activities. *Appl. Microbiol. Biotechnol., 46*(3), 220–225.

Hsu, C. S., & Robinson, P. R., (2006). *Practical Advances in Petroleum Processing.* New York: Springer.

Huang, J., Nemati, M., Hill, G., & Headley, J., (2012). Batch and continuous biodegradation of three model naphthenic acids in a circulating packed-bed bioreactor. *J. Hazard. Mater., 201, 202*, 132–140. doi: 10.1016/j.jhazmat.2011.11.052.

Hydrocarbon Processing, (2008). *Refining Process Handbook.* Houston, TX. USA: Gulf Publishing Co.

Hydrocarbon Publishing Co., (2006). *Opportunity Crudes: Technical Challenges and Economic Benefits.* Report No. Multi-Client Strategic Report, Southeastern, PA.

Ingram, C., (1996). *Heavy Oil Upgrading.* Report No. DOE/MT/95014-2, Atlanta, GA, USA.

Jayanegara, A., Kreuzer, M., & Leiber, F., (2012). Ruminal disappearance of polyunsaturated fatty acids and appearance of biohydrogenation products when incubating linseed oil with alpine forage plant species in vitro. *Livest. Sci., 147*(1–3), 104–112. doi: 10.1016/j.livsci.2012.04.009.

Jean-Pierre, W., (1995). Petroleum refining V.1: Crude oil. Petroleum products. *Process Flowsheets, V.2: Separation Processes.* Paris, France: Editions Technip.

Johnson, R. J., Smith, B. E., Rowland, S. J., & Whitby, C., (2013). Biodegradation of alkyl branched aromatic alkanoic naphthenic acids by *Pseudomonas putida* kt2440. *Int. Biodeterior. Biodegradation., 81*, 3–8. doi: 10.1016/j.ibiod.2011.11.008.

Jones, D. S. J., (2015). Introduction to crude oil and petroleum processing. In: Treese, S. A., Pujadó, P. R., & Jones, D. S. J., (eds.), *Handbook of Petroleum Processing* (pp. 3–51). Springer International Publishing.

Juarez-Moreno, K., Díaz De, L. J. N., Zepeda, T. A., Vazquez-Duhalt, R., & Fuentes, S., (2015). Oxidative transformation of dibenzothiophene by chloroperoxidase enzyme immobilized on (1d)-γ-Al2O3 nanorods. *J. Mol. Catal. B: Enzym., 115*, 90–95. doi: 10.1016/j.molcatb.2015.02.004.

Kayser, K. J., Bielagajones, B. A., Jackowski, K., Odusan, O., & Kilbane, J. J., (1993). Utilization of organosulfur compounds by axenic and mixed cultures of *Rhodococcus-rhodochrous* igts8. *J. Gen. Microbiol., 139*, 3123–3129.

Kilbane, J. J. II., & Le Borgne, S., (2004). Petroleum biorefining: The selective removal of sulfur, nitrogen, and metals. In: Vazquez-Duhalt, R., & Quintero-Ramirez, R., (eds.), *Petroleum Biotechnology: Developments and Perspectives, Studies in Surface Science and Catalysis* (Vol. 151, pp. 29–65). Elsevier, Amsterdam.

Kim, S., Stanford, L. A., Rodgers, R. P., Marshall, A. G., Walters, C. C., Qian, K., Wenger, L. M., & Mankiewicz, P., (2005). Microbial alteration of the acidic and neutral polar NSO compounds revealed by Fourier transform ion cyclotron resonance mass spectrometry. *Org. Geochem., 36*(8), 1117–1134.

Kirkwood, K. M., Foght, J. M., & Gray, M. R., (2004). Prospects for biological upgrading of heavy oils and asphaltenes. In: Vazquez-Duhalt, R., & Quintero-Ramirez, R., (eds.), *Petroleum Biotechnology: Developments and Perspectives* (Vol. 151, pp. 113–143). Studies in surface science and catalysis, Elsevier, Amsterdam.

Kurashov, V. M., (2002). *Microbiological Method for Enriching Petrol and Petroleum Products by Isoparaffin and Aromatic Hydrocarbons with Simultaneous Removal of Injurious Additives and Means for Carrying Out Said Method.* Patent No. WO0246445.

Lazar, I., Petrisor, I. G., & Yen, T. F., (2007). Microbial enhanced oil recovery (MEOR). *Pet. Sci. Technol., 25*(11), 1353–1366. doi: 10.1080/10916460701287714.

Le Borgne, S., & Quintero, R., (2003). Biotechnological processes for the refining of petroleum. *Fuel Proc. Technol., 81*(2), 155–169.

Le Page, J. F., Chatila, S. G., & Davidson, M., (1990). Petrole (Refining and Conversion of Heavy Products from Oil). Paris: Editions Technip.

Leon, V., & Kumar, M., (2005). Biological upgrading of heavy crude oil. *Biotechnol. Bioprocess Eng., 10*(6), 471–481.
Liu, L., Lu, H., Qian, J., & Xing, J., (2010). Progress in the technology for desulfurization of crude oil. *China Pet. Proc. Petrochem. Technol.*, (4), 1–6.
Loganathan, S., (2011). Biohydro-fined diesel (BHD) and biodiesel (BOD) production process and property review. In: *Innovations in Fuel Economy and Sustainable Road Transport* (pp. 97–107).) Institution of Mechanical Engineers, Woodhead Publishing Limited.
Magomedov, R. N., Popova, A. Z., Maryutina, T. A., Kadiev, K. M., & Khadzhiev, S. N., (2015). Current status and prospects of demetallization of heavy petroleum feedstock (review). *Pet. Chem., 55*(6), 423–443. doi: 10.1134/S0965544115060092.
Marcotte, Brian, W. G., Sheehy, A. J., Hill, C. K., & Carroll, M. T., (2018). *Method of Microbially Assisted Water Alternating Gas Injection as a Means of Enhanced Oil Recovery by Supplementing the Immiscible Water Injection Cycle with Nutrients to Improve Oil Release in Oil-Containing Rock Formations*. Patent No. US10018023.
Marison, I. W., & Doerries, S. C., (2003). *Flocculating Agents*. Patent No. US2003008343 (Abandoned).
Martinez, A. R., (1984). Report of working group on definitions. Vol.1, pp. xvii–xviii. In: R. F. Meyer, J. C. Wynn, & J. C. Olson (eds.), The Future of Heavy Crude and Tar Sands, Second International Conference, McGraw-Hill, New York, NY, USA.
Maure, A., Warren, A., & Brown, F., (2005). *Microbial-Induced Controllable Cracking of Normal and Branched Alkanes in Oils*. Patent No. US6905870.
McKenzie, N., Yue, S., Liu, X., Ramsay, B. A., & Ramsay, J. A., (2014). Biodegradation of naphthenic acids in oils sands process waters in an immobilized soil/sediment bioreactor. *Chemosphere, 109*, 164–172. doi: 10.1016/j.chemosphere. 2014.02.001.
McRae, S., (2017). Crude oil price differentials and pipeline infrastructure. *National Bureau of Economic Research Working Paper Series*. No. 24170. doi: 10.3386/w24170.
Meredith, W., Kelland, S. J., & Jones, D. M., (2000). Influence of biodegradation on crude oil acidity and carboxylic acid composition. *Org. Geochem., 31*(11), 1059–1073.
Miyamoto, K., Hirata, K., Boku, M., & Miyasaka, H., (2003). *Microorganism and Method for Producing Alternative Oil for Petroleum Using the Same*. Patent No. JP2003000229.
Mjalli, F. S., Ahmed, O. U., Al-Wahaibi, T., Al-Wahaibi, Y., & AlNashef, I. M., (2014). Deep oxidative desulfurization of liquid fuels. *Rev. Chem. Eng., 30*(4), 337–378. doi: 10.1515/revce-2014-0001.
Mogollon, L., Rodriguez, R., Larrota, W., Ortiz, C., & Torres, R., (1998). Biocatalytic removal of nickel and vanadium from petroporphyrins and asphaltenes. *Appl. Biochem. Biotechnol., 70–72*, 765–777. doi: 10.1007/BF02920187.
Mohamed, M. E. S., Al-Yacoub, Z. H., & Vedakumar, J. V., (2015). Biocatalytic desulfurization of thiophenic compounds and crude oil by newly isolated bacteria. *Front. Microbiol., 6*. doi: 10.3389/fmicb.2015.00112.
Mohan, P. K., Nakhla, G., & Yanful, E. K., (2006). Biokinetics of biodegradation of surfactants under aerobic, anoxic and anaerobic conditions. *Water Res., 40*(3), 533–540. doi: https://doi.org/10.1016/j.watres.2005.11.030.
Mohanty, S., Kunzru, D., & Saraf, D. N., (1990). Hydrocracking: A review. *Fuel, 69*(12), 1467–1473. doi: http://dx.doi.org/10.1016/0016-2361(90)90192-S.
Mohebali, G., & Ball, A. S., (2008). Biocatalytic desulfurization (BDS) of Petro diesel fuels. *Microbiology (London, U. K.), 154*(8), 2169–2183. doi: 10.1099/mic.0.2008/017608-0.
Montaghi, M., Ulrich, B., & Subramanian, A., (2011). Slurry-phase hydocracking possible solution to refining margins. *Hydrocarbon Proc.*, (2), 37–43.

Monticello, D. J., (1998). Riding the fossil fuel biodesulfurization wave. *Chemtech., 28*(7), 38–45.

Monticello, D. J., (2000). Biodesulfurization and the upgrading of petroleum distillates. *Curr. Opin. Biotechnol., 11*(6), 540–546.

Morel, F., Kressmann, S., & Kasztelan, S., (1997). Processes and catalysts for hydrocracking of heavy oil and residues. In: Delmon, B., Grange, P., & Froment, G. F., (eds.), *Hydrotreatment and Hydrocracking of oil Fractions* (pp. 1–16). Elsevier Science B. V., New York.

Morikawa, M., & Kato, T., (2001). *New Microorganism Decomposing Alkanes and Method for Decomposing Alkanes*. Patent No. JP2001224360.

Mullins, O. C., (2008). Review of the molecular structure and aggregation of asphaltenes and petroleomics. *SPE J., 13*(1), 48–57.

Murty, M. V. S., Aleem, M. I. H., Kermode, R. I., & Bhattacharyya, D., (1994). Biotreatment of coals and coal-related compounds by hydrogen-utilizing microorganisms. *J. Chem. Technol. Biotechnol., 60*(4), 359–367. doi: 10.1002/jctb.280600405.

Mužic, M., & Sertić-Bionda, K., (2013). Alternative processes for removing organic sulfur compounds from petroleum fractions. *Chem. Biochem. Eng. Quarterly., 27*(1), 101–108.

Nash, L. J., & Palmer, S. R., (1999). *Biocatalytic Upgrading of Petroleum*. ACS Spring Meet. Anaheim, CA. USA.

National Energy Technology Laboratory, (2006). *A Literature Review on Cold Cracking of Petroleum Crude Oil.* Report No. Unnumbered.

Nordblad, M., Gomes, M. D., Meissner, M. P., Ramesh, H., & Woodley, J. M., (2018). Scoping biocatalyst performance using reaction trajectory analysis. *Org. Process Res. Dev., 22*(9), 1101–1114. doi: 10.1021/acs.oprd.8b00119.

Oelderik, J. M., Sie, S. T., & Bode, D., (1989). Progress in the catalysis of the upgrading of petroleum residue: A review of 25 years of R&D on shell's residue hydroconversion technology. *Appl. Catal., 47*(1), 1–24. doi: http://dx.doi.org/10.1016/S0166-9834(00)83258-3.

Office of Technology Assessment Materials Program, (1996). *An Assessment of Oil Shale Technologies: Chapter 5. Technology*. Report No. 80-600101. PB80-210115, Washington, DC, USA.

Oldfield, C., Pogrebinsky, O., Simmonds, J., Olson, E. S., & Kulpa, C. F., (1997). Elucidation of the metabolic pathway for dibenzothiophene desulphurization by *Rhodococcus sp.* Strain IGTS8(ATCC 53968). *Microbiology (London, U. K.), 143*, 2961–2973.

Olson, E. S., Stanley, D. C., & Gallagher, J. R., (1993). Characterization of intermediates in the microbial desulfurization of dibenzothiophene. *Energy Fuels, 7*, 159–164.

Orr, A. S., & Van, S. R. W., (1960). Residual fuel oils. In: Guthrie, V. B., (ed.), *Petroleum Products Handbook* (pp. 8.1–8.31). McGraw-Hill, New York.

Ortiz-Leon, M., Velasco, L., & Vazquez-Duhalt, R., (1995). Biocatalytic oxidation of polycyclic aromatic-hydrocarbons by hemoglobin and hydrogen-peroxide. *Biochem. Biophys. Res. Commun., 215*(3), 968–973.

Pan, Y., Liao, Y., Shi, Q., & Hsu, C. S., (2013). Acidic and neutral polar NSO compounds in heavily biodegraded oils characterized by negative-ion ESI FT-ICR MS. *Energy Fuels, 27*(6), 2960–2973. doi: 10.1021/ef400191h.

Paslawski, J. C., Headley, J. V., Hill, G. A., & Nemati, M., (2009). Biodegradation kinetics of trans-4-methyl-1-cyclohexane carboxylic acid. *Biodegradation., 20*(1), 125–133. doi: 10.1007/s10532-008-9206-2.

Paslawski, J., Nemati, M., Hill, G., & Headley, J., (2009). Biodegradation kinetics of trans-4-methyl-1-cyclohexane carboxylic acid in continuously stirred tank and immobilized cell bioreactors. *J. Chem. Technol. Biotechnol., 84*(7), 992–1000. doi: 10.1002/jctb.2122.

Patel, J., Borgohain, S., Kumar, M., Rangarajan, V., Somasundaran, P., & Sen, R., (2015). Recent developments in microbial enhanced oil recovery. *Renew. Sust. Energ. Rev., 52*, 1539–1558. doi: 10.1016/j.rser.2015.07.135.

Petersen, M. B., & Jensen, S. K., (2014). Biohydrogenation of fatty acids is dependent on plant species and feeding regimen of dairy cows. *J. Agric. Food Chem., 62*(16), 3570–3576. doi: 10.1021/jf405552m.

Pickard, M. A., Roman, R., Tinoco, R., & Vazquez-Duhalt, R., (1999). Polycyclic aromatic hydrocarbon metabolism by white rot fungi and oxidation by *Coriolopsis gallica* UAMH 8260 laccase. *Appl. Environ. Microbiol., 65*(9), 3805–3809.

Pineda, L. A., Trejo, F., & Ancheyta, J., (2007). Correlation between properties of asphaltenes and precipitation conditions. *Pet. Sci. Technol., 25*(1 and 2), 105–119.

Pineda-Flores, G., & Mesta-Howard, A. M., (2001). Petroleum asphaltenes: Generated problematic and possible biodegradation mechanisms. *Rev. Latinoam. Microbiol., 43*(3), 143–150.

Pineda-Flores, G., Boll-Arguello, G., Lira-Galeana, C., & Mesta-Howard, A. M., (2004). A microbial consortium isolated from a crude oil sample that uses asphaltenes as a carbon and energy source. *Biodegradation, 15*(3), 145–151.

Premuzic, E. T., & Lin, M. S., (1994). *Biochemically Enhanced Oil Recovery and Oil Treatment*. Patent No. US5297625.

Premuzic, E. T., & Lin, M. S., (1996). *Process for Producing Modified Microorganisms for Oil Treatment at High Temperatures, Pressures and Salinity*. Patent No. US5492828.

Premuzic, E. T., & Lin, M. S., (1999a). *Biochemical Upgrading of Oils*. Patent No. US5858766.

Premuzic, E. T., & Lin, M. S., (1999b). *Biochemical Transformation of Coals*. Patent No. US5885825.

Premuzic, E. T., & Lin, M. S., (1999c). Induced biochemical conversions of heavy crude oils. *J. Pet. Sci. Eng., 22*(1–3), 171–180.

Premuzic, E. T., & Lin, M. S., (2001). *Biochemical Transformation of Solid Carbonaceous Material*. Patent No. US6294351.

Premuzic, E. T., Bohenek, M. S., Lin, M., & Zhou, W. M., (1999). Bioconversion reactions in asphaltenes and heavy crude oils. *Energy Fuels, 13*(2), 297–304.

Quagraine, E. K., Headley, J. V., & Peterson, H. G., (2005). Is biodegradation of bitumen a source of recalcitrant naphthenic acid mixtures in oil sands tailing pond waters?. *J. Environ. Sci. Health, Part A: Toxic/Hazard. Subst. Environ. Eng., 40*(3), 671–684.

Quesnel, D. M., Bhaskar, I. M., Gieg, L. M., & Chua, G., (2011). Naphthenic acid biodegradation by the unicellular alga *Dunaliella* tertiolecta. *Chemosphere, 84*(4), 504–511. doi: 10.1016/j.chemosphere.2011.03.012.

Rabinow, R. A., (2001). *Quality Bank Policy Applying on Crude Petroleum from Points Offshore Texas to Points Onshore Texas*. Report No. F.E.R.C 102, Houston, TX. USA.

Rahimi, P. M., & Gentzis, T., (2006). The chemistry of bitumen and heavy oil processing. In: Hsu, C. S., & Robinson, P. R., (eds.), *Practical Advances in Petroleum Processing* (pp. 597–634). Springer, New York.

Ramirez, M., (2008). Biorefining of crude oil. In: Sandi, S., Tina, M., & Boonchai, B., (eds.), *Bioinspired Chemistry for Energy: A Workshop* (p. 3). National Academies Press.

Ramirez-Corredores, M. M., & Borole, A. P., (2007). *Biocatalysis in Oils Refining*. Amsterdam Elsevier.

Ramirez-Corredores, M. M., (2017). *The Science and Technology of Unconventional Oils: Finding Refining Opportunities* (1st edn.). NY, USA: Elsevier Science.

Rana, M. S., Sámano, V., Ancheyta, J., & Diaz, J. A. I., (2007). A review of recent advances on process technologies for upgrading of heavy oils and residua. *Fuel., 86*(9 SPEC. ISS.), 1216–1231. doi: 10.1016/j.fuel.2006.08.004.

Reis, R. S., Pacheco, G. J., Pereira, A. G., & Freire, D. M. G., (2013). Biosurfactants: Production and applications. In: Chamy, R., & Rosenkranz, F., (eds.), *Biodegradation - Life of Science* (pp. 31–61). InTech Open.

Robinson, P. R., (2006). Petroleum processing overview. In: Hsu, C. S., & Robinson, P. R., (eds.), *Practical Advances in Petroleum Processing* (pp. 1–78). Springer, New York.

Rojas-Avelizapa, N. G., Cervantes-Gonzalez, E., Cruz-Camarillo, R., & Rojas-Avelizapa, L. I., (2002). Degradation of aromatic and asphaltenic fractions by *Serratia* liquefasciens and bacillus sp. *Bull. Environ. Contam. Toxicol., 69*(6), 835–842.

Rontani, J. F., Bosserjoulak, F., Rambeloarisoa, E., Bertrand, J. C., Giusti, G., & Faure, R., (1985). Analytical study of asthart crude-oil asphaltenes biodegradation. *Chemosphere, 14*(9), 1413–1422.

Ryu, K., Heo, J., & Yoo, I. K., (2002). Removal of dibenzothiophene and its oxidized product in anhydrous water-immiscible organic solvents by immobilized cytochrome C. *Biotechnol. Lett., 24*(2), 143–146. doi: 10.1023/A:1013806830105.

Safdel, M., Anbaz, M. A., Daryasafar, A., & Jamialahmadi, M., (2017). Microbial enhanced oil recovery, a critical review on worldwide implemented field trials in different countries. *Renew. Sust. Energ. Rev., 74*, 159–172. doi: 10.1016/j.rser.2017.02.045.

Saha, P., & Sarkar, S., (2018). Microbial degradation of coal into a value added product. *Int. J. Coal Prep. Util.,* 1–19. doi: 10.1080/19392699.2018.1448800.

Sahu, R., Song, B. J., Im, J. S., Jeon, Y. P., & Lee, C. W., (2015). A review of recent advances in catalytic hydrocracking of heavy residues. *J. Ind. Eng. Chem., 27*, 12–24. doi: http://dx.doi.org/10.1016/j.jiec.2015.01.011.

Sakai, K., Nakazawa, A., & Kondo, K., (1985). Microbial hydrogenation of nitroolefins. *Agricultural Biol. Chem., 49*(8), 2331–2335. doi: 10.1271/bbb1961.49.2331.

Salacier, J. L., (2009). Integrated process to transport heavy crude oils as an o/w emulsion: Recent advances. In: *8th World Congress of Chemical Engineering: Incorporating the 59th Canadian Chemical Engineering Conference and the 24th Interamerican Congress of Chemical Engineering*. Montreal, QC. Canada.

Santos, V. E., Almudena, A., Martín, A. B., Emilio, G., & Felix, G. O., (2007). Desulfurization of dibenzothiophene using the 4S enzymatic route: Influence of operational conditions on initial reaction rates. *Biocatal. Biotransform., 25*(2–4), 286–294. doi: 10.1080/10242420701422773.

Sawarkar, A. N., Pandit, A. B., Samant, S. D., & Joshi, J. B., (2007). Petroleum residue upgrading via delayed coking: A review. *Can. J. Chem. Eng., 85*(1), 1–24.

Sayles, S., & Routt, D. M., (2011). Unconventional crude oil selection and compatibility. *NPRA Annual Meet.* San Antonio, TX.

Schoemaker, H. E., Meijer, E. M., Leisola, M. S. A., Haemmerli, S. D., Waldner, R., Sanglard, D., & Schmidt, H. W. H., (1989). Oxidation and reduction in lignin biodegradation. *ACS Symposium Series, 399*, 454–471.

Scott, A. C., MacKinnon, M. D., & Fedorak, P. M., (2005). Naphthenic acids in *Athabasca* oil sands tailings waters are less biodegradable than commercial naphthenic acids. *Environ. Sci. Technol., 39*(21), 8388–8394. doi: 10.1021/es051003k.

Scott, T. C., DePaoli, D. W., & Sisson, W. G., (1994). Further development of the electrically driven emulsion-phase contactor. *Ind. Eng. Chem. Res., 33*(5), 1237–1244. doi: 10.1021/ie00029a020.

Self, F., Ekholm, E., & Bowers, K., (2000). *Refining Overview-Petroleum, Processes and Products*. NY, USA: AIChE, Fuels and Petrochemical Divisions.

Sheehy, A. J., Govreau, B. R., Hill, C. K., Carroll, M. T., & Marcotte, B. W. G., (2014). *Method of Modification in Oil Reservoirs Using Resident Microorganisms and Nutrient Supplementation Through Biological Wettability Alteration*. Patent No. US20140367087.

Sheehy, A. J., Govreau, B. R., Hill, C. K., Carroll, M. T., & Marcotte, B. W. G., (2018a). *Biological Augmentation of Low Salinity Water Flooding to Improve Oil Release Using Nutrient Supplementation of Injected Low Salinity Water*. Patent No. US10077393.

Sheehy, A. J., Govreau, B. R., Hill, C. K., Carroll, M. T., & Marcotte, B. W. G., (2018b). *Method of Gravity Assisted Microbiologically Enhanced Oil Recovery by Improving the Distribution of Nutrients Introduced into Oil Producing Rock Formations*. Patent No. US9868895.

Sheehy, A. J., Marcotte, B. W. G., Carroll, M. T., & Hill, C. K., (2018). *Method of Monitoring the Flow of Natural or Injected Water During Oil Field Recovery Processes Using an Organic Tracer*. Patent No. US10030514.

Sheehy, A., (1992a). *Oil Recovery Using Microorganisms*. Patent No. WO9201780.

Sheehy, A., (1992b). *Recovery of Oil from Oil Reservoirs*. Patent No. US5083610.

Shong, R. G., (1999). *Bioprocessing of Crude Oils*. ACS Spring Meet. Anaheim, CA. USA.

Slavcheva, E., Shone, R., & Turnbull, A., (1999). Review of naphthenic acid corrosion in oil refining. *Br. Corrosion J., 34*, 125–131.

Sohrabi, M., Kamyab, H., Janalizadeh, N., & Huyop, F. Z., (2012). Bacterial desulfurization of organic sulfur compounds exist in fossil fuels. *J. Pure Appl. Microbiol., 6*(2), 717–729.

Solari, & Rodolfo, B., (2000). Asphaltene hydroconversion. In: Fu, Y. T., & George, V. C., (eds.), *Asphaltenes Asphalts 2: Developments in Petroleum Science* (pp. 149–171). Elsevier.

Speight, J. G., (1999). *The Chemistry and Technology of Petroleum* (4th edn.). Basel, Switzerland: Marcel Dekker AG.

Speight, J. G., (2004). Petroleum asphaltenes - part 1: Asphaltenes, resins and the structure of petroleum. *Oil Gas Sci. Technol., 59*(5), 467–477. doi: 10.2516/ogst:2004032.

Speight, J. G., (2010). *The Refinery of the Future*. Amsterdam: Elsevier.

Speight, J. G., Wernick, D. L., Gould, K. A., Overfield, R. E., Rao, B. M. L., & Savage, D. W., (1985). Molecular weight and association of asphaltenes: A critical review. *Oil Gas Sci. Technol., 40*(1), 51–61.

Speight, James G., (2001). *Handbook of Petroleum Analysis* (2nd edn.). NY, USA: Wiley.

Srinivas, R. K., Sridevi, V., Chandana, L. M. V. V., & Thomas, K. K., (2013). Biodesulfurization of fossil fuels- an overview. *Int. J. Pharm. Biol. Sci., 4*(3), B719–B727.

Steinbakk, S., (2011). *Bioconversion of Heavy Oil*. MSc, Biotechnology, Norwegian University of Science and Technology.

Steudle, A. K., Nestl, B. M., Hauer, B., & Stubenrauch, C., (2015). Activity of squalene-hopene cyclases in bicontinuous microemulsions. *Colloids Surf. B: Biointerfaces., 135*, 735–741. doi: 10.1016/j.colsurfb.2015.08.034.

Strausz, O. P., Mojelsky, T. W., & Lown, E. M., (1992). The molecular structure of asphaltene: An unfolding story. *Fuel, 71*, 1355–1363.

Strubinger, A., Ehrmann, U., León, V., DeSisto, A., & González, M., (2015). Changes in Venezuelan Orinoco belt crude after different biotechnological approaches. *J. Pet. Sci. Eng., 127*, 421–432. doi: 10.1016/j.petrol.2015.01.039.

Svarovskaya, L. I., & Altunina, L. K., (1999). Biodegradation of petroleum hydrocarbons by formation microflora. *Pet. Chem., 39*(2), 128–131.

Tahhan, R. A., & Abu-Ateih, R. Y., (2009). Biodegradation of petroleum industry oily-sludge using Jordanian oil refinery contaminated soil. *Int. Biodeterior. Biodegradation., 63*(8), 1054–1060.

Tahmassebi, H., (1986). Crude oil and product value differentials: Historical perspective and outlook. *Energy, 11*(4), 343–359. doi: https://doi.org/10.1016/0360-5442(86)90121-0.

Tang, Q., Lin, S., Cheng, Y., Liu, S. J., & Xiong, J. R., (2013). Enhanced biodesulfurization of bunker oil by ultrasound pre-treatment with native microbial seeds. *Biochem. Eng. J., 77*, 58–65. doi: 10.1016/j.bej.2013.05.004.

Thakur, D. S., & Thomas, M. G., (1985). Catalyst deactivation in heavy petroleum and synthetic crude processing: A review. *Appl. Catal., 15*(2), 197–225. doi: 10.1016/S0166-9834 (00)81837-0.

Tinoco, R., & Vazquez-Duhalt, R., (1998). Chemical modification of cytochrome C improves their catalytic properties in oxidation of polycyclic aromatic hydrocarbons. *Enzyme Microb. Technol., 22*(1), 8–12.

Tomita, F., Yokota, A., Saito, K., & Abe, A., (1999). *Method for Decomposing Porphyrin-Based Compound.* Patent No. JP11187890.

Tomita, F., Yokota, A., Saito, K., Yamamoto, C., & Abe, A., (1997). *Decomposition of Porphyrin.* Patent No. JP9176168.

Toor, N. S., Han, X., Franz, E., Mackinnon, M. D., Martin, J. W., & Liber, K., (2013). Selective biodegradation of naphthenic acids and a probable link between mixture profiles and aquatic toxicity. *Environ. Toxicol. Chem., 32*(10), 2207–2216. doi: 10.1002/etc.2295.

Torres, E., & Vazquez-Duhalt, R., (2000). Chemical modification of hemoglobin improves biocatalytic oxidation of pahs. *Biochem. Biophys. Res. Commun., 275*(2), 713–714.

Torres, E., Baeza, A., & Vazquez-Duhalt, R., (2002). Chemical modification of heme group improves hemoglobin affinity for hydrophobic substrates in organic media. *J. Mol. Catal. B: Enzym., 19*, 437–441.

Torres, E., Sandoval, J. V., Rosell, F. I., Mauk, A. G., & Vazquez-Duhalt, R., (1995). Site-directed mutagenesis improves the biocatalytic activity of iso-1-cytochrome-c in polycyclic-hydrocarbon oxidation. *Enzyme Microb. Technol., 17*(11), 1014–1020.

Torres, E., Tinoco, R., & Vazquez-Duhalt, R., (1996). Solvent hydrophobicity predicts biocatalytic behavior of lignin peroxidase and cytochrome c in aqueous solution of water-miscible organic solvents. *J. Biotechnol., 49*(1–3), 59–67.

Torres, E., Tinoco, R., & Vazquez-Duhalt, R., (1997). Biocatalytic oxidation of polycyclic aromatic hydrocarbons in media containing organic solvents. *Water Sci. Technol., 36*(10), 37–44.

Trambouze, P., (2000). *Petroleum Refining: Materials and Equipment.* Paris, France: Institut Francais du Petrole Publications / Editions Technip.

Treccani, P., (2008). Emerging technologies for the conversion of residues. In: Poli, R., (ed.), *Encyclopaedia of Hydrocarbons* (pp. 137–160). Roma, Italy: Istituto Della Enciclopedia Italiana.

Tuli, D., & Kumar, A., (2008). Non-conventional technologies for fuel desulphurization. In: *19th World Pet. Congress.* Madrid, Spain.

Uribe-Alvarez, C., Ayala, M., Perezgasga, L., Naranjo, L., Urbina, H., & Vazquez-Duhalt, R., (2011). First evidence of mineralization of petroleum asphaltenes by a strain of neosartorya fischeri. *Microbial Biotechnol., 4*(5), 663–672. doi: 10.1111/j.1751-7915.2011.00269.x.

Van, H. J. D., Fedorak, P. M., Foght, J. M., Gray, M. R., & Dettman, H. D., (2004). Use of a novel fluorinated organosulfur compound to isolate bacteria capable of carbon-sulfur bond cleavage. *Appl. Environ. Microbiol., 70*(3), 1487–1493.

Vazquez-Duhalt, R., (2004). *Petroleum Biotechnology: Developments and Perspectives.* Elsevier.

Vázquez-Duhalt, R., Ayala, M., & Márquez-Rocha, F. J., (2001). Biocatalytic chlorination of aromatic hydrocarbons by chloroperoxidase of caldariomyces fumago. *Phytochem., 58*(6), 929–933. doi: 10.1016/S0031-9422(01)00326-0.

Vazquez-Duhalt, R., Fedorak, P. M., & Westlake, D. W. S., (1992). Role of enzyme hydrophobicity in biocatalysis in organic-solvents. *Enzyme Microb. Technol., 14*(10), 837–841.

Vazquez-Duhalt, R., Semple, K. M., Westlake, D. W. S., & Fedorak, P. M., (1993). Effect of water-miscible organic-solvents on the catalytic activity of cytochrome-C. *Enzyme Microb. Technol., 15*(11), 936–943.

Vazquez-Duhalt, R., Westlake, D. W. S., & Fedorak, P. M., (1994). Lignin peroxidase oxidation of aromatic-compounds in systems containing organic-solvents. *Appl. Environ. Microbiol., 60*(2), 459–466.

Villegas, J. A., Mauk, A. G., & Vazquez-Duhalt, R., (2000). A cytochrome C variant resistant to heme degradation by hydrogen peroxide. *Chem. Biol., 7*, 237–244.

Walia, D. S., & Srivastava, K. C., (1997). *Biological Production of Humic Acid and Clean Fuels from Coal.* Patent No. US5670345.

Walia, D. S., & Srivastava, K. C., (1998). *Biological Production of Humic Acid and Clean Fuels from Coal.* Patent No. US5854032.

Wei, Z., & Ning, X., (2012). Development and application of the bio-desulfurization technology. *Adv. Mater. Res., 518–523*, 178–182. doi: 10.4028/www.scientific.net/AMR.518-523.178.

Weissman, J. G., (1997). Review of processes for downhole catalytic upgrading of heavy crude oil. *Fuel Proc. Technol., 50*(2, 3), 199–213.

Wondrack, L., Szanto, M., & Wood, W. A., (1989). Depolymerization of water-soluble coal polymer from subbituminous coal and lignite by lignin peroxidase. *Appl. Biochem. Biotechnol., 20–21*(1), 765–780.

Xu, H., Lee, H. Y., Hwang, B., Nam, J. H., Kang, H. Y., & Ahn, J., (2008). Kinetics of microbial hydrogenation of free linoleic acid to conjugated linoleic acids. *J. Appl. Microbiol., 105*(6), 2239–2247. doi: 10.1111/j.1365-2672.2008.03937.x.

Xu, P., Feng, J., Yu, B., Li, F., & Ma, C., (2009). Recent developments in biodesulfurization of fossil fuels. *Adv. Biochem. Engin./Biotechnol., 113*, 255–274.

Xu, X. H., Qiu, X. L., Mu, T. Z., Zhou, J. M., Bai, B., Song, Z. Y., Wu, D., et al., (2015). Biodesulfurization of immobilized sulfur oxidation bacteria thioalkalivibrio versutus d301 with Fe_3O_4 magnetic nanoparticles. *Guocheng Gongcheng Xuebao/The Chin. J. Proc. Eng., 15*(2), 301–306.

Yuste, L., Corbella, M. E., Turiegano, M. J., Karlson, U., Puyet, A., & Rojo, F., (2000). Characterization of bacterial strains able to grow on high molecular mass residues from crude oil processing. *FEMS Microbiol. Ecol., 32*(1), 69–75.

Zhang, S., Liu, D., Deng, W., & Que, G., (2007). A review of slurry-phase hydrocracking heavy oil technology. *Energy Fuels, 21*(6), 3057–3062. doi: 10.1021/ef700253f.

Zhi, L., Jiang, Y., Wang, Y., Hu, M., Li, S., & Ma, Y., (2007). Effects of additives on the thermostability of chloroperoxidase. *Biotechnol. Prog., 23*(3), 729–733.

CHAPTER 10

Hydrogen and Electricity Production from Oil and Gas Wastes

SCOTT J. SATINOVER[1] and ABHIJEET P. BOROLE[1,2]

[1]*Bredesen Center for Interdisciplinary Research and Education, The University of Tennessee, Knoxville–37996, Tennessee, USA*

[2]*Chemical and Biomolecular Engineering, The University of Tennessee, Knoxville–37996, Tennessee, USA*

10.1 INTRODUCTION

Oil and gas wastes and wastewater are primarily generated from two sources: the wells producing hydrocarbons (HCs), and the refineries that further process the raw resources. Cementing and drilling wastes make up a much smaller percentage of the waste, and in some instances, may be included in recovered fluids depending on the completion procedures used on the well. The production of oil and gas from wells results in wastewater called produced water, while the downstream processes generate so-called refinery wastewater. Today, indefinite disposal is often more commonplace than repurposing, though oil and gas companies have recently expressed interest in finding new means of repurposing their wastewater.

Of the wastewater the oil and gas industry produces, the overwhelming majority is produced water generated directly from the oil and gas wells. The exact quantity of produced water is not regularly recorded; however, a few recent estimates have been conducted. According to the groundwater protection council (GWPC), the US produced approximately 21.2 billion barrels of water in 2012 for both offshore and onshore wells (GWPC, 2015). This equates to roughly 58 million barrels of produced water per day. In the same year, the United States produced 11.1 million barrels of crude oil per day, and approximately 65.8 billion cubic feet of natural gas per day (EIA, 2017), or 11.4 million barrels of oil equivalent by energy content. Thus, produced water makes up most of the recovered fluid from oil and gas wells.

Produced water generated from oil and gas wells, in general, has little value. While a small fraction of the wastewater is reused for other oil and gas operations, such as in hydraulic fracturing, most of the fluid is disposed of into underground injection wells. The EPA specifies wells that take fluids produced directly by oil and gas wells as Class II injection wells. The EPA estimates that, "over 2 billion gallons of produced water are injected in the US every day," using Class II wells (EPA, 2017a). Chesapeake Energy claims that 80% of all Class II wells are used for enhanced oil recovery (EOR), restoring reservoir pressure as the wells produce oil by pumping the waste fluid back underground (McCurdy). However, the remaining 20% of wells are used to permanently dispose of the fluid, and these wells are often referred to as saltwater disposal wells (SWDs). Produced water is either trucked or piped to SWDs. These wells can be scarce or plentiful depending on the geographic region's accessibility, and the maturity of the plays nearby.

By comparison, refinery wastewater is more process dependent than produced water. Many of the individual refining processes produce their own wastewater source, and each source produces a different type and percentage of pollutants. The EPA published a report in 2004 that evaluates the origins of these wastewater sources in detail. To summarize, distillation accounts for the majority of total wastewater, followed by fluid catalytic cracking (FCC) and catalytic reforming, which together account for 78% of the total wastewater produced by a typical refinery (Leavitt et al., 2004). Refineries consume a regular stream of water for each gallon of refined product. On average, refineries use 1–2.5 gallons of water per gallon of refined product (EPA, 2017b). Not all of the water used is recovered. For instance, some of the water used in distillation columns is vented as steam. One presentation at the Water/Energy Sustainability Forum in 2009 suggested that of the water used, only 41% is discharged, and 5% is contained in sludge (Ostenberg, 2009). This introduces an entirely different problem for refineries than for production operations. Even if refineries could reuse all of the wastewater they produce, they still cannot recoup the water lost. As a result, the motives are different. Refineries have some control over the amount of water they use and would like to rely on less additional water to generate less waste. Produced water, however, is an inevitability that varies not only with the formation from which HCs are being produced, but also depends on the fracturing operations employed.

Refinery wastewater is also controlled by different regulatory requirements. Refinery wastewater cannot be injected into Class II injection wells, as it is not a product associated with oil and gas production. Rather, refineries must use Class I injection wells, which is suitable for several additional industrial wastewaters, including food and metal industrial wastewaters (EPA, 2015). Despite an otherwise subtle distinction between wastes, this regulatory difference has been a significant point of contention, as operating a Class I well is traditionally more expensive than Class II wells (Paque, 1997).

Produced water and refinery wastewater aren't entirely different in scope. Refinery wastewater is still a product that has little commercial value and it is still a waste that is regulated by the EPA. Treating it for efficient recycling, regardless of current disposal practices, seems wise if the practice can be made viable. Both wastewaters have a variety of treatment options however, these options have limitations. Energy requirements for treating these wastes are not trivial and the composition they have hinders economically viable options to repurpose this water. With the surge of oil and gas production in the US via horizontal drilling and hydraulic fracturing, production of these wastes has increased, and are likely to continue to increase for the foreseeable future. New technologies must be developed to not only improve ease of reuse and accelerate treatment times, but also reduce treatment costs to make recycled waters attractive products, rather than low-value waste they are currently designated. The next section outlines the composition of these wastes to illustrate the challenges they present.

10.1.1 COMPOSITION OF PRODUCED WATER AND REFINERY WASTEWATER

Produced water varies wildly in composition from source to source, but most wells produce water with some salinity, total dissolved solids (TDS), and total suspended solids (TSS). Neff et al. (2011) determined that salt concentrations found in produced water could vary from a few thousand parts per million to near the point of saturation. Ionic species, while generally dominated by sodium and chloride, can also be quite diverse. Notable species include divalent ions like magnesium, calcium, strontium, bromides, and sulfates (Roach et al., 1993; Neff et al., 2011; Emam et al., 2014). The impacts of these species are frequently relevant to treatment potential, and also contribute to the hazardous character of the waste.

Perhaps one of the most important inorganic species adversely affecting produced water quality is hydrogen sulfide (H_2S). H_2S is often found in produced water, though its concentrations can vary by source. Depending on the concentration of the overall sulfur content detected in their wells, which includes H_2S, producers may use the nomenclature "sweet" or "sour," where sweet implies low or no detectable sulfur concentration, and sour designates the opposite. Sour wells are not desirable since they can result in added cost primarily due to the processing required to remove the excess sulfur. For fracturing and surface reuse, sour wells generate produced water with unique problems. H_2S in produced water can cause rapid catastrophic failure in high-pressure pumping equipment via, "sulfide stress cracking" (Ogden, 2005). To prevent this, H_2S is treated and monitored regularly at suspected sources. However, H_2S removal is slightly more difficult than other inorganics in produced water. H_2S is unique from other inorganic hazards in that its presence is linked directly to the activity of sulfate-reducing bacteria (SRB) (Cord-Ruwisch et al., 1987). Because of this biological origin, H_2S can develop once produced water is brought to the surface if sufficient organics and sulfates are present. If the water is allowed extended contact with equipment, biological activity can also induce biocorrosion (Enning and Garrelfs, 2014) without adequate treatment. Produced water pretreatment therefore often involves the use of biocides such as glutaraldehyde, chlorine dioxide, etc., but pretreatment may not always be sufficient or performed adequately. For example, H_2S may form at the bottom of settling ponds where the produced water is intermediately stored, despite there being no detection of H_2S after treatment near the surface. In these ponds, the oxygen at the bottom is consumed faster than it can diffuse from the surface. This produces an anaerobic environment ripe for SRBs to begin metabolizing organics, inevitably producing H_2S.

As suggested, produced water often contains a significant amount of organic content, including emulsified oil and greases as well as dissolved organic content (Neff et al., 2011). A particular group of compounds of interest includes organic acids, such as acetic acid, which can be regarded as a nuisance compound. Acetic acid, among other organic acids, can cause corrosion in transportation pipelines (Joosten et al., 2002). These organic compounds can be added by hydraulic fracturing operations, may be present in the formation, may be the result of chemical reactions occurring downhole, or can be caused directly by biological activity at the surface. Acetic acid is often added as a pH buffer for fracturing fluids to ensure fracturing fluid integrity. It can also be generated by HC degradation by various fermenters and strict anaerobes, which can also function as an intermediate to methane production via microbial methanogenesis under the right thermodynamic conditions (Dolfing et al., 2007; Callbeck et al., 2013). Both routes play a role in total acetic acid found in produced water.

The chemical additions vary considerably from well to well, from as little as a few to hundreds of individual compounds (Stringfellow et al., 2017). Other organic content can also include ingredients found in gelling agents and friction reducers (guar gum, derivatized guar gums, cellulose, polyacrylamide, etc.), biocides (glutaraldehyde, tributyl tetradecyl phosphonium chloride, chlorine dioxide, etc.), and surfactants (glycerin, ethoxylated alcohols, ethoxylated fatty acids, poly(oxy-1,2-ethanediyl), alpha-isodecyl-omega-hydroxy, etc.). Some of these compounds are more degradable than others.

The presence of a large range of organics and TDS makes treatment of produced water complicated. The challenges, however, do not end there. Produced water quality from hydraulically fractured wells changes over time. One study conducted on three hydraulically fractured Marcellus shale horizontal wells determined that initial chloride concentration in early time recovered water, often referred to as flowback water, started at less than 10,000 mg/L, but water produced after HCs were recovered reached near 100,000 mg/L over 340 days (Cluff et al., 2014). That same study also concluded that dissolved organic carbon dropped to less than 50 mg/L from initial concentrations between 550 and 300 mg/L. Still, these changes in composition are often rapid compared to a well's overall lifespan. Another study estimated that only 4–8% of the recovered fluid was actually injected, while the remaining percentage of fluid was found to originate from the reservoirs studied (Kondash et al., 2017).

Much like chemical composition, biological composition of produced water can be diverse and may change over time. The same study conducted on the Marcellus shales found that injected fluids had higher Shannon diversity than late-stage produced waters, where late-stage fluids were almost entirely dominated by the genus *Halanaerobium* (Cluff et al., 2014). The temporal dynamics of microbial diversity of oil and gas wells have also been studied elsewhere (Daly et al., 2016). Remarkably, the findings of Daly et al. (2016) suggest that microbes introduced at the surface can flourish in reservoir conditions despite biocide use. This could imply that a variety of newly introduced microbes can survive in produced water at surface temperatures and pressures as well, as atmospheric conditions are significantly less hostile to microbial growth.

Today, salinity, organic content, and microbial activities represent the biggest challenges in treating produced water, which is exacerbated with the addition of fracturing fluid recovery in unconventional wells. Eliminating all the contaminants may be ideal, however abiotic pretreatment for desalination may be impractical. Generating "clean brines" devoid of microbial activity and organics that retain high salinity may be a less complex and viable option. This recycled water may not have the reusability of desalinated waters, but their reuse at the surface in additional oil and gas operations may prove beneficial to other drilling and completion operations. Many companies already produce products that can be used in high salinity fluids, such as friction reducers used in hydraulic fracturing operations (Halliburton, 2014). To formulate fracturing suitable brines at the surface, biocide use for pretreatment remains common, in addition to various physicochemical processes like filtration, cyclonic separation, and use of oil-water separators. These processes have little to no energy recovery, and less emphasis is made on the removal of soluble and emulsified organics.

Much of the information regarding produced water applies to refinery wastewater, as refinery wastewater poses many of the same challenges discussed for produced water. Wastewater derived from refining processes also may contain H_2S, dissolved organics, oil, and grease, and high salinity (Leavitt et al., 2004). Like produced water, the concentrations of these compounds need to be reduced before the water can be reused and may need physiochemical separation processes as a pretreatment. Some more mature wastewater technologies have also been proposed for surface reclamation, including the use of activated sludge for HC removal, and this has been more commonly adopted by refineries in recent years. Still, these old techniques have deficiencies. Activated sludge requires oxygen and provides no energy recovery post-treatment. To improve the value of these waste products and to ensure they are more sustainably treated, value must be recovered from them. This can be accomplished by using an emerging technology discussed in the next section.

10.2 OVERVIEW OF BIOELECTROCHEMICAL SYSTEMS (BES)

HCs are energy-rich molecules, therefore, conceptually they have high potential to produce energy. Because spills and waste products have a significant amount of HC content (as well as other organics), many studies have investigated the mechanisms of HC degradation. That topic is covered elsewhere in this book, but of particular interest for this chapter is the utility of anaerobic biodegradation mechanisms. Some of these mechanisms can be referred to from earlier chapters. Here, we focus on energy recovery from the waste streams. As such, oxidative degradation is not useful, since aerobic oxidation results in an energy gain for the microorganisms and few opportunities to extract energy or other valuable products. However, anaerobic conversion can result in the production of intermediates such as hydrogen, organic acids, etc., which can be used for generation of bioenergy. Below, we discuss how these anaerobic pathways can be integrated with electrochemically driven processes to produce energy products using bioelectrochemical systems (BES).

A BES degrades organics and produces electrons, which are used for the production of electricity, hydrogen, or other products (Figure 10.1). A BES contains an anode and a cathode. Traditionally, the anode harbors a biologically compatible electrode, where anode-respiring bacteria, or exoelectrogens, oxidize organic material (such as HCs or intermediates from their degradation) and transfer electrons to the electrode, using it as a terminal electron acceptor. As a result, the biological composition of the anode is primarily colonized by obligate or facultative anaerobes. The cathode then takes the electrons, combines them with positive ionic species, performing a balanced reduction reaction, completing the redox reaction. To prevent the crossover of the microbes and compounds, the anode and cathode are sometimes separated by a barrier, such as an ion-exchange membrane. Like all electrochemical cells, this electrochemical process can be described by the Nernst equation:

$$E_{cell} = E_{cell}^{\ominus} - \frac{RT}{zF} \ln(Q_r)$$

where, E_{cell} is the cell potential, $E_{cell}^{\ominus}$ is the cell potential at standard conditions and reactant and product concentrations of 1M, R is the universal gas constant, T is temperature, z is the amount of charge transferred in the oxidation-reduction reaction, F is the Faraday constant, and Q_r is the reaction quotient of the oxidation-reduction reaction being investigated. Modeling of the biological processes can result in more integrated equations capable of further explaining the overall bioelectrochemical processes occurring in the system (Marcus et al., 2007; Hamelers et al., 2011; Rimboud et al., 2015).

Exoelectrogenic bacteria use several mechanisms to transport electrons to the electrode, and investigation within this field is ongoing. Logan (2009) published a comprehensive review of these bacteria, which discussed these mechanisms. To summarize, three primary mechanisms exist: (1) Electron transporting molecules called mediators that shuttle electrons between the microbes and the anode, (2) cytochromes or other electron transfer proteins, which contact the anode or other organisms directly, and (3) appendages known as nanowires, which form an electrically conductive conduit for electron transfer (Logan, 2009). Nanowires are particularly novel because of their ability to transfer electrons over long distances via direct connection (Reguera et al., 2005; Gorby et al., 2006). Genera like *Geobacter* are the most frequently studied exoelectrogens that produce nanowires; however, other microbes may also use these structures, including *Shewanella, Synechocystis*, and *Aeromonas* (Kumar et al., 2015; Sure et al., 2015).

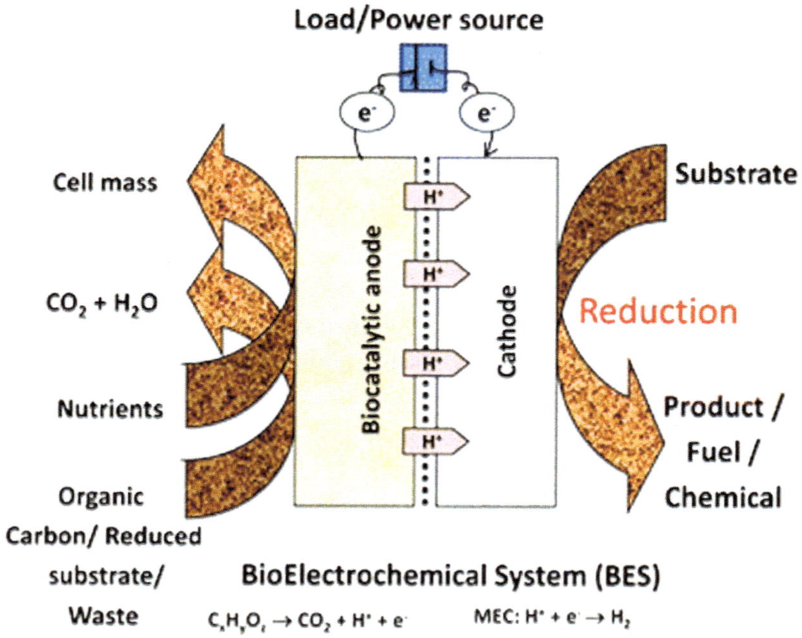

Type of BES	Cathode substrate	Product
MFC	Oxygen	Electricity
MEC	Protons	Biohydrogen
BES	Acetate	Ethanol
BES	Oxygen	Hydrogen peroxide

FIGURE 10.1 A schematic of a bioelectrochemical system.
Source: Reprinted with permission from Borole et al., 2011b. © 2011 Royal Society of Chemistry.

Three types of BES reactors are discussed in this chapter: microbial fuel cells (MFCs), microbial electrolysis cells (MECs), and microbial desalination cells (MDCs). For MFCs, the potential difference between the anode and the cathode is positive, and thus the reaction is spontaneous. This results in the production of a current, which can be used to power electrical devices. In an MEC, the desired reaction has a negative redox potential, and thus a voltage is applied to a cell to overcome the electrical potential needed. An MEC's final product may involve the use of a biological cathode as well. A variety of end products have been generated in BESs, including hydrogen, bases, methane, and hydrogen peroxide, often resulting in unique communities (Logan et al., 2008; Rozendal et al., 2009, 2010b; Pant et al., 2012; Dykstra and Pavlostathis, 2017). Finally, an MDC uses the energy produced to drive desalination processes. MDCs also have a variety of different configurations (Saeed et al., 2015; Carmalin Sophia et al., 2016; Jingyu et al., 2017), but the end goal always involves the simultaneous removal of salt ions and organics.

Many studies and reviews have discussed the current status of BESs designs, their concepts, the microbes of interest, and the electrochemical parameters needed to investigate cell performance (Logan et al., 2006; Rozendal et al., 2008; Logan et al., 2008, 2015; Logan, 2009; Borole et al., 2011b; Sleutels et al., 2012). To characterize cells, several performance parameters have been established that have become regularly used in BES studies. The performance parameters of importance in BESs typically include current density (CD), power density (in the case of MFCs), hydrogen productivity (in the case of MECs), and Coulombic efficiency. Removal percentage of COD is also traditionally calculated for all BESs; however, MDCs will also typically include removal percentages of ions like sodium, calcium, magnesium, and chlorides. Less commonly, calculated metrics exclusive to MECs include cathode conversion efficiency, hydrogen recovery, overall energy efficiency, and electrical efficiency. The definitions of these parameters are given below.

CD is typically calculated by taking the current observed (I_{obs}) by the cell and dividing it by the projected surface area of the anode (A):

$$CD = \frac{I_{obs}}{A}$$

In addition to the CD based on surface area, a volumetric CD can also be determined using the anode volume instead (A/m^3). Power density (PD), a useful metric for MFCs, uses the CD and the measured voltage of the cell to calculate the power based on Ohm's law.

$$P = IV \rightarrow PD = \frac{VI_{obs}}{A}$$

Hydrogen productivity applies specifically to MECs and takes the amount of measured hydrogen produced from the cathode per unit time and divides it by the anode volume (V_{ano}) (L/L-day).

$$H_{prod} = \frac{V_H}{V_{ano}t}$$

Hydrogen productivity can alternately be based on total reactor volume, including the volume of the anode and cathode. Coulombic efficiency (CE) is calculated from I_{obs}, the moles of substrate n_s consumed and the number of electrons available per mole of substrate (based on complete conversion to CO_2) n_e, using the following equation:

$$CE_{anode} = \frac{I_{obs}t}{n_s n_e F}$$

where; F is the Faraday constant and t is the time duration of the experiment. The substrate concentration is typically measured as chemical oxygen demand (COD). However, not all compounds are removed during a conversion process, therefore, a COD measurement before and after the treatment is necessary to determine the substrate consumed. Percent removal, be it for specific compounds, ions, or COD, is typically calculated by the following:

$$\%removal = \frac{conc.in - conc.out}{conc.in} \times 100$$

where; conc. in and conc. out represent the concentration of the content of interest before and after being processed by the reactor, respectively.

Cathodic conversion efficiency (CCE) is calculated as the ratio of the amount of hydrogen produced and the maximum amount of hydrogen that could be produced based on the average current generated over the experiment, I_{obs}. This is calculated by the following equation:

$$CCE = \left(\frac{V_{H2}}{\frac{I_{obs}tRT}{2FP}} \right)$$

Hydrogen recovery (Y_{H2}) is calculated as moles of hydrogen produced per mole of hydrogen expected from the conversion of substrate utilized in the anode:

$$Y_{H2} = \frac{PV_{H2} / RT}{2\Delta nCOD}$$

where; P is the atmospheric pressure; V_{H2} is the volume of hydrogen produced; R is the gas constant; T is the temperature in Kelvin and ΔnCOD is the moles of COD consumed during the experiment. Overall energy efficiency (η_{E+S}) calculations are done using the ratio of the combustion energy of hydrogen (W_{H2}), over the net energy produced to the system, which is the difference between the substrate energy input (W_S), and the electrical energy input (W_E):

$$\eta_{E+S} = \frac{-W_{H2}}{W_E - W_s}$$

The electrical efficiency (η_E) is the ratio of total energy recovered as hydrogen (W_{H2}) to the electrical energy input (W_E):

$$\eta_E = \frac{-W_{H2}}{W_E}$$

As for the biological composition, most studies use 16S rRNA gene sequencing for classification of microbial communities, and more recently omics techniques, to characterize anode microbial consortia. This practice is typically done to understand the molecular processes occurring in the anode and the cathode.

Many studies on BESs use simple and ideal substrates, such as acetate and glucose (Pant et al., 2010; Kadier et al., 2014), however recent studies have investigated more complex waste streams (Ullery and Logan, 2014; Borole, 2016; Marone et al., 2017). BESs were developed originally for the treatment of waste streams with simultaneous energy production, however, recently they have been investigated for much broader areas of application including energy-efficient nitrogen and phosphorous recovery, metal recovery, chemical production, desalination, and dechlorination (Rabaey et al., 2010a; Borole, 2012; Cusick and Logan, 2012; Ichihashi and Hirooka, 2012; Kuntke et al., 2012; Pant et al., 2012; Kelly and He, 2014). Nevertheless, the BES's ability to utilize complex waste organics with simultaneous energy production gives them great promise for use in waste treatment, including applications in the oil and gas industry.

10.2.1 BIOLOGICAL COMPOSITION OF BES

Most of the interest in the oil and gas industry to date for using BESs have been for bioremediation, not as a means of creating energy, even though energy is an inherent product of the process. Oil and gas spills generate significant attention in the public and academic press. This reaction is perfectly rational; spills are inherently dangerous and can have lasting environmental, social, and political impacts. The Deepwater Horizon Oil Spill immediately comes to mind. The aftermath of the incident resulted in a total cost of $61.6 billion to BP (Bomey, 2016) following the aftermath of multiple civil and criminal charges. But even if catastrophic spills on the order of Deepwater Horizon are never planned, precautions, emergency action plans, and cleanup strategies are essential.

Relative to the amount of HCs that are successfully processed and transported for consumption without error, accidents and spills of all types are rare. However, solutions to treat waste streams effectively are still needed. Even without these disasters, oil, and gas wastes regularly generate significant quantities of HCs. As underground injection becomes a less attractive option, new alternative treatment strategies are required. Without a treatment protocol, these environmental implications are left unchecked.

For the most part, HC degradation is much faster in aerobic conditions than in anaerobic ones (Salminen et al., 2004). However, there are some instances where anaerobic conditions are required, but are not very favorable to HC degradation, or are simply energetically expensive. Within those constraints, BESs offer a unique advantage for bioremediation and water treatment applications.

The biological composition of a BES is either a consortium or a specifically selected microbe, often an exoelectrogen. Ideally, all microbes used in a BES would be exoelectrogenic, however, for complex waste streams; this is rarely the case, especially when a consortium is used as an initial inoculum for the reactor. Using a case from outside the petroleum industry, Lewis et al. (2015) investigated the hydrogen production from a switchgrass derived pyrolysis bio-oil aqueous phase and found that the anode was dominated by a variety of non-exoelectrogenic microbes. Further work characterized additional diversity of microbes present in the anode including various kinds of fermenters, which can be limiting to the conversion process compared to the flux through exoelectrogens (Lewis et al., 2018). For BES applications in the oil and gas industry, similar limitations can exist in the communities responsible for the degradation of complex HCs and other organics.

Specific microbial compositions of BESs that use oil and gas products will be discussed in the following sections.

10.2.2 HYDROCARBON (HC)-DRIVEN BESS AND RESULTANT ANODE BIOLOGY

Microbial communities that are responsible for HC conversion in nature and at spill sites are diverse. These often include γ-Proteobacteria and α-Proteobacteria, and these communities consume both gaseous and liquid HCs (Hazen et al., 2010; Kessler et al., 2011; Kostka et al., 2011). Communities in BESs can include these microbes but must include exoelectrogens. Together, HC degraders and exoelectrogens need to be synergistic with each other. This section focuses on the microbes that have been reported in BESs. Occasionally process conditions and reactors are included here to provide context, however, Section 3 in this chapter provides a comprehensive overview of the BES reactors and operating conditions relevant to conversion of the waste streams into energy products.

10.2.2.1 TREATMENT OF PETROLEUM AND HYDROCARBON (HC) MIXTURES

Perhaps the first study to consider using HCs as a feedstock for BESs was conducted by Davis and Yarbrough (1962). In that study, they used *Nocardia* and *E. coli* as the microbes, but HC degradation in the cell was not observed. Although the study was unsuccessful at converting the HCs, it did contribute to the idea of using electron mediators found in future BESs. Since then, the body of work within this field has grown considerably. The first successful MFC to run on a mixture of HCs as a feedstock was published by Morris and Jin (2007). More studies have since been published using new reactor configurations and inoculum sources, and their performance metrics will be discussed later. Morris et al. (2009) followed their first study with an investigation of the degradation of diesel in an MFC. Here, they found that closed circuit (CC) conditions (i.e., the reactor being run as a fuel cell) outperformed equivalent open circuit (OC) conditions (no circuit applied to cell) over the span of 21 days. Analysis of the microbial community via 16S rRNA clone library sequencing identified the predominant species, which included denitrifiers such as *Citrobacter* sp., *Pseudomonas* sp., and *Stenotrophomonas* sp. as well as the exoelectrogen *Shewanella* sp.

Intuitively, we expect the ecology of a BES to change from the initial inoculum source, even if the source is from another BES. In particular, we would expect exoelectrogenic populations to grow, as their terminal electron acceptor is essentially limited only by the available surface area of the anode, linked to the organic energy source provided. One study attempted to investigate this by replicating a spill location on soil, taking the soil samples from the location and using those in an MFC, described as a soil MFC, or SMFC (Li et al., 2014). The authors claimed that some level of enrichment occurred, due to observed increases in the Shannon-Wiener Index and community richness. Additionally, observed bands on DGGE (denaturing gradient gel electrophoresis) fingerprinting gels indicated an increase in taxonomic diversity after applying close circuit conditions compared to OC conditions over an operating period of 180 days. Notable prominent microbes belonged to the genus *Escherichia*, somewhat complimenting the work conducted by Davis and Yarbrough (1962). These findings are illustrated in Figure 10.2.

Demonstrating the potential for BESs is valuable, but comparisons between identically constructed cells using different anode materials sources can also yield important insights. Another soil-based BES study by Lu et al. (2014a) attempted to make a comparison between reactors and soil content by changing the anode material. Three materials were used, carbon cloth, biochar, and a surfactant modified carbon cloth electrode, before comparing the microbial community in each reactor to that observed in the soil. The authors determined that Proteobacteria dominated all samples from soil and MFCs, and the dominant genera on the carbon cloth, biochar, and surfactant modified carbon cloth anodes were *B. petrii, Comamonas* spp., and *B. petrii.* respectively, supplemented with a variety of exolectrogenic and HC-degrading microbes. *B. petrii* is a potentially pathogenic facultative anaerobe, and is the only species of the *Bordetella* genus that is a facultative anaerobe (von Wintzingerode et al., 2001). *Escherichia* was not detected in the study. Lu et al. (2014b) conducted a follow-up study using a larger reactor with a 50 L volume for the same application, but the anode biological composition was not investigated.

MDC experiments that characterize the microbial community composition using oil and gas products are rare, and HC mixtures are even less likely. Sabina et al. (2014) is one of the few studies that utilized 16S rRNA gene amplicon sequencing to characterize an MDC. The authors demonstrated that engine oil could be degraded effectively in an MDC, where the best MDC used a strain of *Bacillus*, specifically, *B. subtilis* moh3. Further work is needed in MDCs to understand the effect of substrates and design on microbial species.

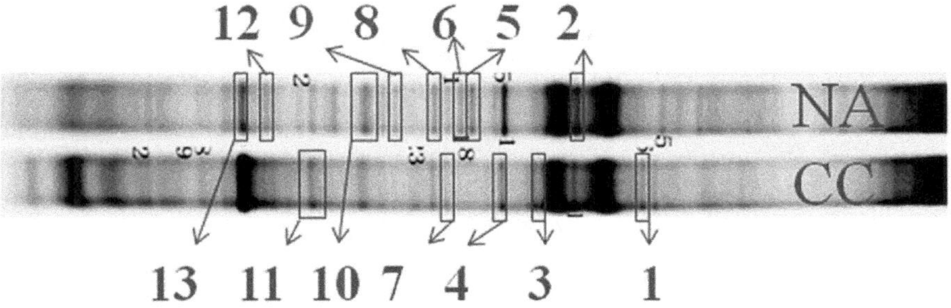

Table 1 Overview of the sequencing results of the bands obtained from the DGGE profiles

Band	Accession	Name	Organism	Similarity
1	CP007391	*Escherichia coli*	*Proteobacteria, Gammaproteobacteria, Enterobacteriales, Enterobacteriaceae*	100%
2	EF668606	*Geobacteraceae* sp.	*Proteobacteria, Deltaproteobacteria, Desulfuromonadales*	98%
3	FJ440032	UF-1	Uncultured bacterium (*Firmicutes*, environmental sample)	98%
4	FN548084	UF-2	Uncultured bacterium (*Firmicutes*, environmental sample)	97%
5	JN221495	*Escherichia* sp.	*Proteobacteria, Gammaproteobacteria, Enterobacteriales, Enterobacteriaceae*	100%
6	CP001383	*Shigella*	*Proteobacteria, Gammaproteobacteria, Enterobacteriales, Enterobacteriaceae*	99%
7	JF344166	UG-1	*Proteobacteria, Gammaproteobacteria*	96%
8	KF767890	*Escherichia* sp.	*Proteobacteria, Gammaproteobacteria, Enterobacteriales, Enterobacteriaceae*	99%
9	GU477928	UR-1	Uncultured bacterium (RDX degrading microorganisms)	94%
10	KF851241	*Escherichia* sp.	*Proteobacteria, Gammaproteobacteria, Enterobacteriales, Enterobacteriaceae*	100%
11	HQ857728	*Salinimicrobium*	*Bacteroidetes, Flavobacteriia, Flavobacteriales, Flavobacteriaceae*	92%
12	KF851241	*Escherichia* sp.	*Proteobacteria, Gammaproteobacteria, Enterobacteriales, Enterobacteriaceae*	100%
13	DQ768632	*Alcanivorax*	*Proteobacteria, Gammaproteobacteria, Oceanospirillales, Alcanivoracaceae*	100%

FIGURE 10.2 Overview of the difference in bands between control closed-circuit (CC) conditions and open-circuit (OC) conditions and the species associated with each band in the DGGE profile.

Source: Reprinted with permission from Li et al., 2014. © 2014 Royal Society of Chemistry.

10.2.2.2 BIOCONVERSION (BDC) OF INDIVIDUAL HYDROCARBON (HC) SUBSTRATES

Many of the BES studies conducted on petroleum products utilize specific compounds found in HC sources, such as toluene, benzene, and naphthalene. Adelaja et al. (2014) tested the degradation of phenanthrene in one experiment using an MFC using seven different inocula. They found that reactors running on an inoculum derived from an anaerobic digester sludge supplemented with an additional strain of *P. aeruginosa* had the best performance compared to the other reactors. The authors reported that these systems were more efficient and had higher degradation rates than OC cells or other anaerobic systems.

Toluene has been used in BESs. Zhang et al. (2010) investigated the degradation of toluene by *G. metallireducens* and a mixed consortium. For both isolated and mixed cultures, most of the toluene was degraded. Additionally, the authors concluded that the degradation of toluene, benzene, and naphthalene was faster under CC conditions using a graphite rod than in OC conditions. *G. metallireducens* oxidized toluene faster when using an electrode than when using Fe(III) as the terminal electron acceptor. This suggests that a boundless terminal electron acceptor can increase HC degradation capacity beyond typical anaerobic metabolism that relies on free ions, while also producing energy in the process. Several years later, Wu et al. (2014) successfully used toluene in an MFC but relied on pyocyanin as an electron mediator in the anode media to improve electricity production. Thus, mediators can still help increase electron production while using electrodes.

Benzene has been studied often because it was once considered recalcitrant under anaerobic conditions and has proven to be a potential substrate for MFCs (Rakoczy et al., 2013; Wu et al., 2013; Wei et al., 2014; Chang et al., 2017). Rakoczy et al. (2013) conducted experiments with MFCs to degrade benzene and oxidize sulfides and found that cells inoculated with benzene- and sulfide-contaminated groundwater had anodes dominated by δ-Proteobacteria, while cathodes were dominated by α-Proteobacteria. Not all HCs need to be liquid to be degraded in an MFC. Chen and Smith (2018) built a methane driven MFC and found that in addition to methanotrophs, *Geobacter* spp. appeared to have a high level of activity. The authors concluded that *Geobacter* spp. used intermediates produced by the methanotrophs to generate current, further illustrating a novel syntrophic relationship between the two organisms.

MFCs are not the only BESs that have been used to degrade HCs. Yan and Reible (2015) demonstrated that microbially driven electrolysis could be used to remediate sediment contaminated with HCs, where they used phenanthrene as the primary substrate. While they did not analyze the microbial community, power requirements, or hydrogen productivity of the system, they did investigate the genes expressed by the community. They found that that the phenanthrene concentrations decreased by 50% in reactors with applied voltage, but not in OC controls, and HC degradation genes were expressed only in the experimental reactors.

10.2.3 REFINERY WASTEWATER BES

Refinery wastewater treatment via BESs has also received considerable attention. Like HCs, a variety of reactor designs and inocula have been used in the study of this topic. The former is discussed in detail in Section 10.3. The work done to characterize the biological communities found in these reactors has been limited. Most of the focus has been on substrate degradation and electrochemical performance metrics. The few studies which discuss microbiology are included below.

Refinery wastewater degradation has been investigated using combinations of known microbes. Guo et al. (2014) used a two-chamber MFC and tested eight reactors with a variety of strain combinations isolated from a previously matured MFC, including *Paenibacillus* sp., *Deinococcus* sp., and *Microbacterium* sp., for a total of seven reactor communities. Of the communities established, the authors claimed that the best performing reactor used a combination of *Paenibacillus* sp. and *Deinococcus* sp., further suggesting that none of the reactors operated using isolated strains investigated in this study outperformed the reactor using multiple microbes. Community synergies may therefore be essential for other complex mixtures of HCs and oil and gas wastewaters.

The effect of anode material selection has also been investigated for refinery wastewater. Guo et al. (2016) studied the biology of the anodes in an MFC using three different materials; a non-packed anode made out of a carbon rod used as a performance control, a packed granule graphite anode, a packed granule-activated carbon. The authors found that the packed granule-activated carbon removed less of the COD compared to the granule graphite MFC, but oil removal was roughly equal for both packed cells. Based on 16S rRNA gene sequence analysis, the dominant species in the packed granule-activated carbon anode belonged to the following genera: *Acinetobacter* sp., *Bacillus* sp., and *Microbacterium* sp. Conversely, the granule-graphite cell had six dominant genera: *Arthrobacter* sp., *Microbacterium* sp., nicotinic acid metabolizing *Bacillus* sp., *Paenibacillus* sp., *Deinococcus* sp., and *Bacillus* sp. These microbes are similar to those found in their earlier study (Guo et al., 2014) further indicating the importance of inoculum source. Not all studies used a consortium for the starting inoculum. One study opted to treat refinery wastewater using a culture of *P. putida* as the sole microbe for the anode catalyst. The degradation of organics was found to be limited to 30% removal efficiency of COD (Majumder et al., 2014), indicating significant residual organics were not converted. This is not necessarily surprising based on the complexity of refinery wastewater. Higher organic removal percentages have been reported using consortia from domestic wastewater as inoculum (Mohanakrishna et al., 2018).

10.2.4 PRODUCED WATER BES

BESs for the conversion of the organics in produced water focus primarily on organic degradation exclusively or simultaneous desalination of produced water. Several studies have combined MFCs with capacitive deionization to construct bioelectrochemically driven desalination units, either as an MFC powering a separate capacitive deionization unit for desalination, being processed in a single MDC, or through a similar device known as a microbial osmotic fuel cell (MOFC) (Ismail and Ibrahim, 2015; Stoll et al., 2015; Monzon et al., 2017).

Studies documenting the biological composition of the anode in produced water-driven BESs are less common. Using a previously described MFC design (Monzon et al., 2014), Monzon et al. (2017) documented the microbial make-up of the MFCs used in their study and found that *Halanaerobium praevalens* and *M. hydrocarbonoclasticus* were the dominant species found on the MFC anode, even though *H. praevalens* was not found in significant quantities in the produced water. The low amount of *Halanaerobium* found in this water sample is not unexpected. While *Halanaerobium* has been found to be dominant in produced water samples, it is not always dominant in samples acquired from the same wells at different times of production, as early well production has shown considerably less *Halanaerobium* than later stage production (Cluff et al., 2014; Daly et al., 2016). The authors suggest that the *Halanaerobium* genus, in which many species are fermentative and halotolerant, may also have exoelectrogenic metabolic pathways. The exact mechanisms by which this microbe survived the extreme conditions were not determined, but the use of quorum sensing signaling molecules such as quinolones has been attributed to improvement in exoelectrogenic potential in hypersaline cells (Monzon et al., 2016). Additionally, *H. hydrogeniformans* was first shown to have exoelectrogenic activity by Paul et al. (2014). Thus, the conclusion by Monzon et al. (2017) is not entirely unrealistic, and *Halanaerobium* may be a valuable microbe genus for saline BESs like those operating on produced water.

A study by Roustazadeh Sheikhyousefi et al. (2017) compared the performance of a MFC by altering the membrane and the type of cathode catalyst. They found that the working cells were dominated by members belonging to the genera *Aeribacillus*, *Caldimonas*, *Methylobacter*, and *Methylotenera* among others, but only found small percentages of *Marionbacter*, which contrasts with the findings in Monzon et al. (Monzon et al., 2017). This study did introduce some potential for oxygen diffusion from the cathode to the anode, as evidenced by the growth of aerobic species. If this is true, a small amount of oxygen may have contributed to increased HC degradation while also producing substrates for exoelectrogens more frequently than in more anaerobic conditions, altering the community makeup and performance. This conjecture has been postulated for HC sources elsewhere (Morris and Jin, 2007; Rakoczy et al., 2013; Adelaja et al., 2014), as well as non-HC substrates (Oh et al., 2009; Xie et al., 2013).

More recently, flowback water from unconventional reservoirs was used in a sulfur-sulfate mediated MFC for removal of organics and iron by using a combined community or sulfide oxidizing and sulfate-reducing bacteria (Zhang et al., 2018). The study focused primarily on microbial community and chemical species reduction but did not necessarily document performance, making comparisons between other MFCs from an energy generating perspective challenges. The authors found that the phyla *Bacteroidetes*, *Firmicutes*, *Proteobacteria*, *Actinobacteria*, and *Chloroflexi* dominated the anodes. Interestingly, the authors suggested the large quantity of unclassified *Desulfuromonadales* detected contributed to electricity production and sulfate oxidation in some fashion. While Fe(III) reduction can be valuable for precipitating iron and sulfate from produced waters, the requirement for sulfate-sulfur mediation may inhibit the viability of this technology since most produced waters contain sulfate. Mediator-less BESs may promote different microbial communities as well.

Only one study used a microbial electrolysis cell to treat produced water. Ghasemi et al. (2015) constructed a so-called SMXC that could function as both an MFC and an MEC. The authors suggested that the dominant species in the biofilm was *L. macroides* but did not discuss other contributors.

10.2.5 STATE OF ANODE BIOCATALYSTS FOR TREATING OIL AND GAS WASTES

Most studies investigating oil and gas industry wastes have used microbial inocula from the environment, either from pristine or contaminated sites, and used them directly as a catalyst in the anode without significant enrichment. However, this practice ignores a critical aspect of BES performance documented by several researchers (Borole et al., 2009b, 2011a; Miceli III et al., 2012): the enrichment procedure, as well as inoculum, appears to play a critical role in the substrate degradability and BES performance.

One study investigated the effect of a pre-enriched microbial consortium on degradation of waste streams from the petroleum industry. Venkidusamy et al. (2016) pre-enriched a microbial culture on diesel range HCs and studied the treatment process at a concentration of about 800 mg/L. They report that the pre-enriched consortium was much better than the freshly inoculated anode resulting in complete removal of the diesel contaminants within 30 days at a power density of 90.9 mW/m^2, while the freshly inoculated anode produced only 15 mW/m^2 and removed only 83.4% of the contaminants. The process was also tested at a higher diesel concentration of 8,000 mg/L, resulting in 93.5% removal using the MFC with pre-enriched culture, but only 79.5% removal in the MFC receiving the fresh inoculum. The control experiment operated at OC conditions showed a 32% removal of the diesel HCs using the pre-enriched MFC. The authors determined that the new anode had higher complexity than the mature electrode and was dominated by Proteobacteria. Various species of *Pseudomonas* were mainly prevalent in the new biofilm; however, the mature film did not share these characteristics, having some dominant unique organisms like those belonging to the class *Clostridia*. Strictly identifying the anode taxonomy only painted some of the picture here, so the authors investigated further. Genomics analysis determined that the pre-enriched culture was enriched with microbes containing a biodegradative gene (*alkB*), which was thought to be responsible for the greater catabolic action against diesel range HCs.

MFCs and MECs may generate different anode communities using similar inocula and substrates, as MECs may have significant operational criteria that affect anode biology. In an MFC configuration, the cathode is fed air to supply oxygen as the electron acceptor. In contrast, an MEC is anaerobic at both electrodes. Oxygen has been shown to diffuse across the membrane used in MFCs (Kim et al., 2007), which likely affects the anode community. As suggested earlier, this possibility for MFC HC degradation utilizing some oxygen has also been investigated by a few researchers (Morris and Jin, 2007; Rakoczy et al., 2013; Adelaja et al., 2014). Investigations of communities resulting from oxygen intrusion in oil and gas-related BESs are sparse. A comparison of anaerobic and microaerophilic conditions for BESs treating non-petroleum waste streams was briefly reviewed by Borole et al. (2011b). However, BES studies with microbial characterization and degradation rate comparisons within the same study are uncommon. Further comparison between the communities established by an MEC and MFC and the degradation speed of substrates (complex or simple) for each BES would provide insights into the microbial processes occurring in these systems. Similar work can be done for BESs treating oil and gas industry wastes when sufficient volume of MEC studies using strictly anaerobic conditions become available.

10.2.6 LOOKING BEYOND MICROBIOLOGY

The previous discussion provides a few insights into the evolution of anode biology in BESs. It can be tempting to compare studies based solely on the biology and the substrates used, however, it is known that operating conditions have a significant impact on the biological composition of the anode, and can be just as important for the BES performance as the microbes present. Furthermore, while it is important to enrich the right microbes, certain BES conditions can also enrich the wrong microbes. Reactor inocula often contain methanogens and SRBs, which actively compete against exoelectrogens. These microbes decrease BES productivity by robbing electrons from contributing to current production. Specifically, methanogens are notorious for reducing Coulombic efficiency and have received a lot of attention in BES experiments. Many strategies have been investigated to inhibit methanogen growth, including adjusting MFC external resistances, brief oxygen introduction to anode media, actively harvesting hydrogen from MEC cathodes, and temporary anode starvation (Chae et al., 2010; Kaur et al., 2014; Lu et al., 2016). Methanogen inhibition protocols may be important for deploying BES for treatment of HC spills and wastewaters as well.

From the studies discussed, it becomes clear that inoculum source, substrate, enrichment procedure, and anode material selection all contribute to mature anode taxa and cell performance. Focusing strictly on anode taxa, therefore, leads to a myopic view of the complexities. The most important goal of any BES involves not only HC degradation potential, but also maximizing HC degradation and efficient energy production. The diversity of the inocula used in BES shows that perhaps many consortia may be suitable for application in BESs. However, to push this technology towards commercial feasibility, one needs to also study other parameters that affect BES performance metrics.

Optimal BES design is governed by three-parameter categories: process/operating parameters, system design parameters, and biological parameters (Figure 10.3). Several review papers on the assessment of the parameters and mechanisms of the biocatalysts for BESs are available (Borole et al., 2011b; Logan, 2009; Lovley, 2012). Advanced designs of BESs have implemented these concepts for optimization of electro-active biofilms leading to current densities in the range of 6–50 A/m^2 (Fan et al., 2007; Jeremiasse et al., 2010; Ichihashi et al., 2014; Lewis et al., 2018).

10.3 BIOREACTORS FOR HYDROCARBON (HC) CONVERSION

Treatment of HC-contaminated water or soil with production of bioenergy in the form of bio-electricity or hydrogen requires optimization of not only the biology of the anode, but also the reactor design and process conditions. Improvements can be made concerning the reactor design and process parameters. These include anode and cathode volume and surface area, type of separator between anode and cathode, charge transfer characteristics, electrode conductivity, electrode porosity, and distance between electrodes. The process parameters of importance include ionic strength or conductivity of the anolyte and catholyte, solution pH, mode of operation: batch vs. continuous substrate addition, flow rate/retention time, substrate loading rate, and substrate concentration. As discussed in Section 2, the biocatalyst is an important component of the system. However, for the biocatalyst to perform optimally, the process and reactor parameters must be maintained at optimal values to maximize energy extraction from the waste stream.

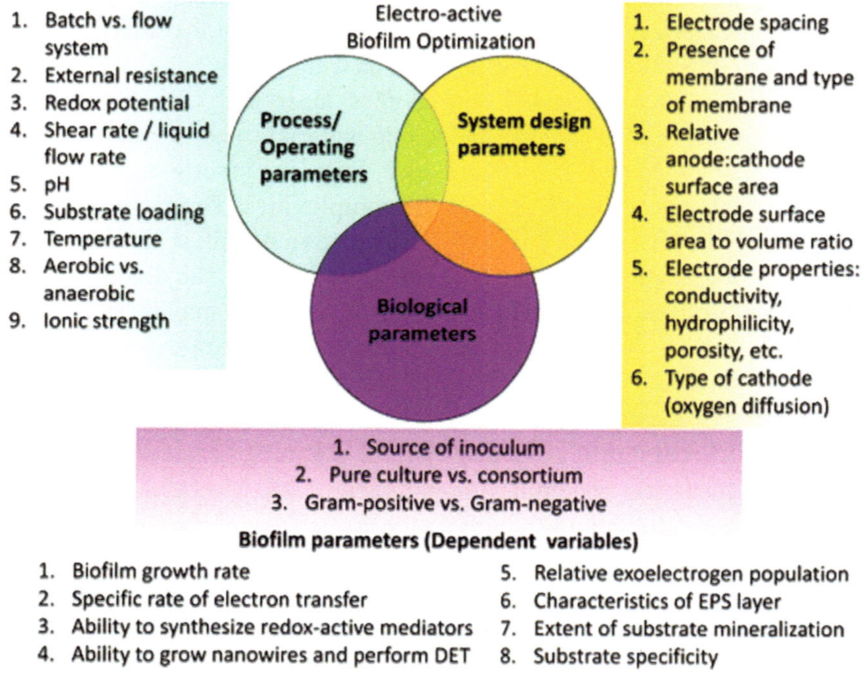

FIGURE 10.3 Parameters for optimization of BESs performance via development of suitable biocatalyst for substrates relevant to individual waste streams.

Source: Reprinted with permission from Borole et al., 2011b. © 2011 Royal Society of Chemistry.

As suggested earlier, the application of BESs for treating oil and gas industry wastes is relatively new. Since the state-of-the-art of BES technology is more advanced for non-petroleum waste streams, insights from parallel studies using other waste streams can be useful. Several reviews have been published, which focus on specific parameters relevant for reactor design and process operation, reactor materials, or type of substrates used previously in BESs (Rabaey et al., 2010a; Kadier et al., 2014, 2016; Sharma et al., 2014; Pandey et al., 2016; Popat and Torres, 2016). These references are mined for knowledge to discuss potential improvements in the performance of BESs for applications in the oil and gas industry. The sections below focus on the bioreactor design parameters for the treatment of oil and gas industry wastes for different cell configurations, with particular attention to those important for optimization of bioelectrochemical reactor performance. In addition to BESs mentioned in the prior section, other forms of biological reactors are also included to compare with the BES rectors, providing perspective on the relevance of BESs for treatment of oil and gas sector waste streams.

10.3.1 MICROBIAL FUEL CELLS (MFCS)

The primary goal of MFC design is to maximize electricity production from organic waste. However, bioconversion of these molecules does not always proceed at high rates, requiring the integration of physicochemical methods to improve removal rates. This is particularly relevant to oil and gas industry wastes due to the low solubility of HCs, which limits the transfer of these molecules from the organic phase to the liquid phase, followed by intracellular transport to the regions within the microorganisms where the respective metabolic processes are active. Reactor and process design can help alleviate these problems by increasing HC solubility.

Several MFC configurations have been investigated as discussed by Logan et al. (2006). An overview of different MFC reactor designs and process conditions investigated for treatment of the oil and gas industry wastes is included below. Figure 10.4 shows various types of reactors used for the treatment of HC-contaminated wastes. These include two-chamber reactors, single-chamber reactors, U-tube reactors, sediment columns, and multi-layered soil reactors (Adelaja et al., 2014; 2017; Guo et al., 2016; Li et al., 2014; Mohan and Chandrasekhar, 2011; Morris and Jin, 2007, 2012; Wang et al., 2012).

10.3.1.1 INITIAL APPROACH TO INTEGRATED ENERGY PRODUCTION AND CONTAMINANT REMOVAL

The earliest investigations conducted on HC-degrading BESs were carried out in simple two-chamber or single-chamber reactors. As suggested in section 2, Morris and Jin (2007) are responsible for building the first MFC to successfully degrade HCs, and the MFC design they used is worth discussion. They used a single-chamber MFC and investigated the effect of electrode distance. They reported a higher current with decreasing distance, which was also reported earlier using non-petroleum substrates (Cheng et al., 2006). The uniqueness of their system was the use of a relatively large distance between electrodes ranging to 9 m, probably reflective of an intent to investigate full-scale field application. They reported a 55% decrease in cell voltage for a 9 m electrode distance compared to the cell with an electrode spacing of 0.8 m. Another study came from Mohan and Chandrasekhar (2011) that investigated a bioelectrochemical approach to treat petroleum sludge. Their cell was a relatively simple two-chamber cell with a working volume of 500 mL, which used a fill-and-draw mode of operation to move the liquid in the anode (Figure 10.4A). The MFC was compared with anaerobic treatment without the use of electrodes, showing a total petroleum hydrocarbon (TPH) removal of 41.1% in the MFC vs. a 20.7% removal in the control reactor without electrodes. A range of HCs, including polyaromatic HCs, aliphatics, heteroatom-containing compounds, and asphaltenes was characterized. Up to 75.5% removal of aromatic HCs was reported in the MFC and higher ring aromatics showed better removal in the MFC compared to the control. A third study was reported by Wang et al. (2012), demonstrating the stimulated degradation of organics provided by MFCs. They used a U-tube design to study the influence of the bioelectrochemical connectivity between anode and cathode over the disconnected control (Figure 10.4B). While the power density was low (0.85 mW/m^2), the degradation of petroleum HCs was higher (15.2%) compared to the control (6.9%). They reported up to 79% degradation of total *n*-alkanes, compared to 66% in the control. Thus, BESs can be used to not only outperform natural degradation conditions but can also remove high percentages of certain HC species while recovering energy, paving their way as a promising technology for applications in oil and gas bioremediation. Later studies continued to test other HC wastes and other configurations.

10.3.1.2 SOIL AND SEDIMENT MFCS

Morris and Jin (2012) have reported the development of a sediment column MFC (SCMFC) to investigate the degradation of petroleum HCs. The anode and cathode sections were filled by petroleum-contaminated sediment and a water column, respectively (Figure 10.4D). The electrodes used were a

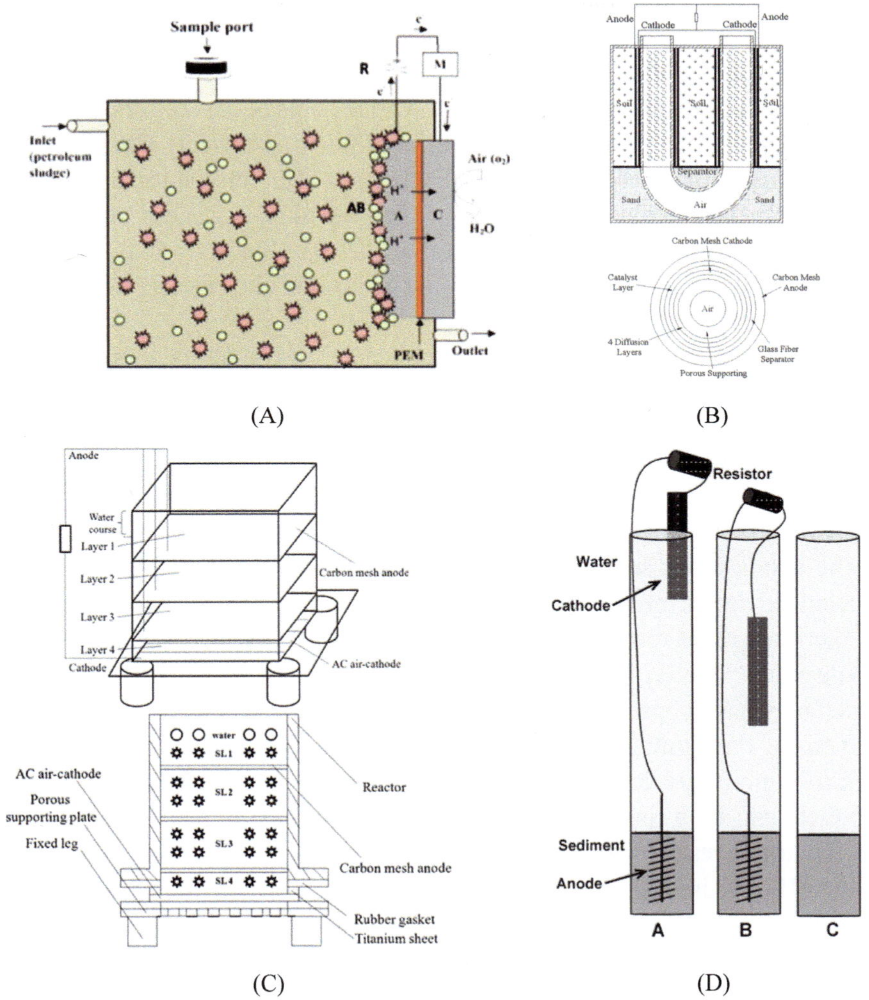

FIGURE 10.4 Microbial fuel cell bioreactors used for treatment of petroleum wastes (A) simple MFC design to test removal of TPH in early designs of BES. (B) U-tube MFC. (C) Soil MFC with multi-stack anodes. (D) Sediment column MFC with wicking cathode option.

Source: Reprinted with permission from A) Mohan and Chandrasekhar, 2011 © 2011 Elsevier; B) Wang et al., 2012 © 2011 Elsevier; C) Li et al., 2014 © 2014 Royal Society of Chemistry; D) Morris and Jin, 2012 © 2012 Elsevier.

brush anode and a platinum-deposited carbon cloth cathode. The researchers reported testing different modes of aeration by keeping the cathode totally submerged or in a wicking position (Reactor I in Figure 10.4D), which maintained the dissolved oxygen concentration 1–2 mg/L above the submerged operation. Using an initial TPH concentration of approximately 15,958 mg/kg, a power density of 2162 mW/m^3 was reported for Reactor I in Figure 10.4D, with 24% removal of TPH, representing performance that was 12-fold higher than the electrodeless operation within 66 days. By combining MFC technology with an *in situ* deployment, this study effectively demonstrated the benefits of the bioelectrochemical approach over conventional bioremediation. As these technologies continued to develop, a variety of different performance metrics were observed. Table 10.1 lists the performance parameters obtained using various reactor configurations and process conditions, which will be further discussed below.

The effect of electrode spacing was studied during the degradation of a wide range of HCs by Li et al. (2014) by placing three anodes at 1 cm, 3 cm, and 5 cm from the cathode in a single reactor (Figure 10.4(C)). The layer closest to the cathode had the highest degradation with 18% removal of TPH, with the other layers contributing less to total TPH removal. The authors reported the removal of aromatic HCs as well as alkanes (C_8–C_{40}). The rate of removal of the alkanes was higher than that of the remaining HCs, but both were accelerated due to the bioelectrochemical process. However, performance was weaker than other reactors, as the maximum power density observed for the whole reactor was 37 mW/m^2. The resistance used was quite high at 1,000 Ω, which may have contributed to the lower power densities observed.

One of the limitations of systems that use bulk soil or sediment is the lack of continuous flow through the reactor. Mass transfer of substrate and products can be a major hurdle in these systems, which can contribute to the electrical resistance of the reactor. By Ohm's law, current production in an MFC is inversely proportional to the external resistance applied to the reactor at a given operating voltage, but the dynamics that contribute to increasing voltage and current production, as well as what contributes to effective mass and charge transfer, must be more accurately understood with other measurement techniques and modeling. Electrochemical impedance spectroscopy has been used to do this in the past, leading to the identification of internal resistance as a primary characteristic that negatively affects current production (He and Mansfeld, 2009; Sekar and Ramasamy, 2013), suggesting that charge and mass transfer is limited in poorly operating devices. Improving charge transfer largely depends on the primary formation of the continuous phase. Treatment of contaminated soil requires handling solids in a reactor where the solids form the bulk of the continuous phase, whereas an MFC used to treat wastewater has primarily a continuous liquid phase. Charge transfer can be improved in a liquid phase simply by increasing anode and cathode ionic strength (Liu et al., 2005; Aaron et al., 2010), but

TABLE 10.1 Comparison of Reactor Configurations and Design Parameters on BESs Performance

Configuration	Anode Volume (L)	Type of Cathode	Substrate	Max % Removal of TPH/COD	Salinity	Dominant Microbes in Anode	Max Current Density (A/m2)	Max Power Density (W/m2)	Reference
Two-chamber and single chamber MFC	0.065 for single/0.45 for dual	Carbon cloth catalyzed with 10% Pt	Refinery wastewater		<200 mM ionic strength in cathode	–	–	0.12	Morris and Jin (2007)
Two-chamber MFC	0.5	Graphite plate	Oil sludge	41%	–	–	0.15	0.053	Mohan and Chandrasekhar (2011)
U-tube MFC	2.736 (total volume including cathode)	0.1 mg/cm^2 Pt-deposited over carbon	Hydrocarbons	15.2%	>14 mS/cm	–	–	0.00085	Wang et al. (2012)
Two-chamber MFC w/ controlled voltage	0.45	Carbon cloth with 0.5 mg/m^2 Pt	Toluene	78%	–	*P. putida F1*	0.094	0.047	Friman et al. (2012)
Single tubular chamber MFC	$\geq$ 0.1	Carbon cloth catalyzed with 10% Pt	Hydrocarbon contaminated sediment	24%	4.54 mS/cm	–	11.4 A/m^3	2.162 W/m^3	Morris and Jin (2012)
Multi-anode soil MFC	0.324	Stainless steel mesh with activated carbon	Contaminated dry soil	18%	1.994 mS/cm	*γ-Proteobacteria*	0.102	0.037	Li et al. (2014)
Single chamber MFC (using mature culture)	0.32	Carbon cloth with PTFE layer, catalyzed with 0.5 g of 10% loading Pt	Hydrocarbons in diesel	100%	–	*Geobacillus, Stenotrophomonas, Gordonia, Flavobacterium* and *Actinobacterium*	0.18	0.0908	Venkidusamy et al. (2016)
Tubular two-chamber MFC	0.5	Carbon felt w/0.35 mg/cm^2 Pt deposited carbon catalyst	Phenanthrene and benzene	>90% for TPH	–	–	–	0.00675	Adelaja et al. (2017)

TABLE 10.1 *(Continued)*

Configuration	Anode Volume (L)	Type of Cathode	Substrate	Max % Removal of TPH/COD	Salinity	Dominant Microbes in Anode	Max Current Density (A/m2)	Max Power Density (W/m2)	Reference
Single chamber MFC	0.5	Non catalyzed Graphite plate	Refinery wastewater sludge	Approx. 35%	–	–	0.149	0.053	Chandrasekhar and Venkata Mohan (2012)
Single Chamber MEC	5 mL	Stainless steel mesh	Refinery wastewater	79%	1.3–6.4 mS/cm	–	2.1	–	Ren et al. (2013)
Two-chamber MFC	0.4	Graphite plate	Refinery wastewater	87%	–	*Paenibacillus* sp., *Deinococcus* sp.	–	0.103 W/m^3	Guo et al. (2014)
Single chamber MFC	0.5	Carbon cloth polished with 40% PTFE	Refinery wastewater	30%	–	*P. putida*	0.15	0.05	Majumder et al. (2014)
Single chamber MFC	0.013	Carbon cloth with 0.5 mg/cm^2 Pt deposited carbon catalyst	Refinery wastewater	84%	>6.6 mS/cm	–	–	0.28	Zhang et al. (2014)
Two-chamber MFC	0.4	Graphite flake	Refinery wastewater	64%	–	*Arthrobacter* sp., *Microbacterium* sp., *Paenibacillus* sp., *Nicotinic acid bacillus* sp. *Deinococcus* sp., and *Bacillus* sp.	–	330.4 mW/cm^3	Guo et al. (2016)
Single chamber MFC	0.2	Carbon cloth with 0.5 mg/cm^2 Pt deposited carbon catalyst	Refinery wastewater	84%	–	–	0.347	0.225	Srikanth et al. (2016)
Single chamber MFC	0.3	0.5 mg/cm^2 Pt deposited carbon cloth	Refinery wastewater	47%	18.9 g/L TDS	–	0.264	0.132	Mohanakrishna et al. (2018)

TABLE 10.1 *(Continued)*

Configuration	Anode Volume (L)	Type of Cathode	Substrate	Max % Removal of TPH/COD	Salinity	Dominant Microbes in Anode	Max Current Density (A/m2)	Max Power Density (W/m2)	Reference
Spiral shaped MEC	0.09	0.5 mg/cm^2 Pt deposited carbon on stainless steel mesh	Produced water	81%	≥40000 ppm	*L. macroides*	310 mA/m^3	–	Ghasemi Naraghi et al. (2015)
Single chamber MFC	0.250	copper foil with 0.5 g/cm Pt deposited carbon catalyst	Produced water	68%	85.7 g/L TDS	*H. praevalens, M. hydrocarbonoclasticus*	–	0.047	Monzon et al. (2017); Monzon et al. (2014)
Single chamber MFC	0.125	Carbon cloth	Produced water	96%	132 mS/cm	*Caldimonas, Aeribacilus*	–	> 0.0035	Roustazadeh Sheikhyousefi et al. (2017)
Two-chamber MFC	0.016	0.5 mg/cm^2 Pt deposited carbon	Glucose and lactate	–	–	*L. raffinolactis, A. hydrophila*	38.4	1.073	Ichihashi et al. (2014)
Two-chamber MFC	0.013	0.5 mg/cm^2 Pt deposited carbon	Corn stover fermentation effluent	> 5 g/L-day	–	Firmicutes and Proteobacteria	10.7	< 0.8	Pannell et al. (2016)

this may not necessarily be true for solid-phase MECs. To work around this, in a study looking at treatment of petroleum-contaminated soil, Li et al. (2015) investigated alternatives to reduce the internal resistance of the reactor via improving charge transfer in solid-phase BESs. Since mass and charge transfer in soils may take place through a solid phase, the researchers investigated increasing the porosity by adding sand to the soil. An increase in porosity from 44.5% to 51.3% reduced the ohmic resistance by 46% and enhanced degradation by 268%.

10.3.1.3 FLOW-THROUGH MFCS

Liquid waste streams such as refinery wastewater, contaminated groundwater, or produced water can be treated in batch systems similar to the soil and sediment MFCs or in continuous, flow-through MFCs. In MFCs that utilize flow-through designs, liquid media is continuously fed and removed from the anode. Flow-through MFCs can be fed by batch addition of substrate or through continuous addition of substrate, which can affect performance. Srikanth et al. (2016) treated petroleum refinery wastewater in flow-through MFCs and compared continuously fed operation with batch operation. Removal of three contaminants including oil, phenol, and sulfide was studied. A total contaminant removal of 84.4% was reported with oil removal of 95%. The continuously fed operation yielded three times higher COD removal, reaching a maximum power density of 225 mW/m^2. Power density for oil and gas-related BESs was also studied as a function of hydraulic retention time by Adelaja et al. (2017), who investigated continuously fed flow-through MFCs for *in situ* and *ex situ* deployment. Here, benzene and phenanthrene were the primary substrates used. Additionally, *in situ* and *ex situ* reactors were different due to the placement of anode and cathode. *In situ* operation represents treatment of the contaminants in the subsurface, while the *ex-situ* operation represents a pump and treat scenario. In the *in situ* reactors, the anode was a carbon cloth placed on the outside of a tubular cathode through which the catholyte was pumped. In the *ex-situ* case, the cathode was outside a tubular anode and both solutions were pumped through the electrode chambers. They demonstrated effective removal of both contaminants at a maximum power density of 6.75 mW/m^2 for both reactor configurations. They further investigated a shock loading operation and compared it with the continuous operation showing that the latter was favorable with a hydraulic retention time of 30 hours resulting in 77% removal of the contaminants.

Continuous flow systems and continuously fed reactors have been reported to give good, and often better, performance in many different studies with BESs compared to batch feeding in anode flow-through systems, including those for treatment of non-petroleum substrates (Borole et al., 2009a, 2013; Sharma et al., 2014; Pannell et al., 2016). The differences between continuous and batch operational performance can be attributed to hydrodynamics as well as kinetic factors. Continuous operation can deliver the substrate at a constant rate to the biocatalyst, enabling a steady state to be reached, whereas batch or fed-batch operations cause substrate delivery concentration to change over time. Achieving steady-state relies on microbial growth or development of an anode electroactive biofilm, which can take anywhere from days to several months (Ramasamy et al., 2008; Borole et al., 2010). The establishment of thick biofilms is certainly a function of the rate of supply of the carbon, energy source, and other process parameters (Borole et al., 2011b), however, studies investigating their growth in MFCs have found biofilm growth can continue for months (Borole et al., 2010). This poorly optimized growth process might be attributed to the novelty of BES research. The field is, in general, only about two decades old, and not enough focus has been paid to increase and optimize biofilm growth rate, so there is a significant potential to reduce this growth period via improving process conditions and understanding growth parameters.

Based on the current understanding of flow-through systems and the studies by Srikanth et al. (2016) and Adelaja et al., (2017), it is possible that these systems can be further improved by optimizing organic loading rate, anode hydraulic retention time, and feeding by continuous addition. The use of high concentrations of substrate like those typically found in batch addition systems can lead to the growth of non-exoelectrogens, which can consume the substrate without producing current, leading to lower performance (Sleutels et al., 2016). Continuously fed reactors, by contrast, feed a much smaller concentration of organics than batch addition at a given time to the MEC, and should therefore encourage less non-exoelectrogen growth as long as steady-state has been reached, and the community has adapted to the waste stream being provided. A flow-through system further helps with the performance due to improved mass transfer and delivery of substrate to the biocatalyst (Borole et al., 2009b, 2011a; Sleutels et al., 2011). Batch-operated systems rarely promote convective mass transfer, which severely limits the ability to transfer charge from the anode to the cathode. Without adequate charge transfer, either by promoting convective flow or by using more conductive materials, these systems will not perform well. Thus, continuous flow-through operations can offer better performance and stability for waste stream treatment than batch-fed or batch-operated BESs.

10.3.1.4 EFFECT OF ELECTRODE MATERIALS AND OTHER PARAMETERS

Electrode materials can also impact contaminant removal in BESs. As discussed previously, Guo et al. (2016) investigated the effects of granular graphite and granular activated carbon on MFC cell biology using petroleum refinery wastewater as the carbon source, but the study also investigated the resulting performance differences. The authors found that the conductivity of the granular activated carbon was less than the granular graphite, and further suggested that this may have explained the differences in performance when treating the substrate. A high level of electricity generation was reported with the granular graphite reaching a power density of 330 mW/cm^3 (Table 10.1). The granular activated carbon, on the other hand, was better at removing volatile phenols and benzene, potentially due to its higher adsorption capability. Thus, each material has pros and cons in terms of treatment beyond the microbial composition, so process and reactor design have to be optimized that largely depend on the targets for treatment.

The use of a separator between the anode and cathode vs. placing the anode and cathode at a specified distance was compared by Zhang et al. (2014). These scenarios essentially result in different internal resistance of the cell. Using refinery wastewater as a substrate, it was demonstrated that the use of a separator resulted in the highest power density of 280 mW/m^3. This was due to the minimal distance between the electrodes afforded by the use of a separator. The authors concluded that the minimal distance decreased the internal resistance, a metric discussed earlier in this chapter that has been known to affect BES performance. The researchers also reported a comparison of the performance using domestic wastewater as the substrate, which reached a power density of about 350 mW/m^2. However, the Coulombic efficiency for the domestic wastewater substrate was several-fold higher than the refinery wastewater (49% vs. 11%). One cause for this difference could be the poor biodegradability of the refinery wastewater compared to the domestic wastewater. A significant fraction of the COD from the refinery wastewater was removed (84%) despite low performance, indicating that much of the carbon source was diverted away from electricity production. Tying in with the findings of the previous study by Guo et al. (2016), it may be possible that more of the organics in the substrate had adsorbed to the anode rather than being degraded, however, this was not determined by Zhang et al. (2014). Adsorption of substrate may be a benefit or a disadvantage depending on if and when the adsorbed compounds degrade, and the impedance to BES operation chemical adsorption may produce. Choosing materials that optimize compound adsorption while also minimizing electrode separation could assure more viable systems in the future.

10.3.1.5 LARGE SCALE BES STUDIES

One of the largest BES studies conducted used 50 L storage containers for two-column type BESs designed to treat diesel-contaminated soil by vertically embedding the devices in the contaminated soil (Lu et al., 2014b), as shown in Figure 10.5(A). One BES used a graphite granule anode while another used an anode made of biochar. The authors used a large contaminated soil bed to test the radius of influence of the BESs on HC degradation, which extended up to 34 cm from the cathode. The graphite granule outperformed the biochar electrochemically, producing a maximum CD of 70.4 mA/m^2 and a maximum power density of 8.8 mW/m^2. Compared to the other studies summarized in Table 10.1, this CD is quite low. Scale-up of BESs is currently a topic of interest, and lower normalized performance metrics introduce additional barriers to more widespread adoption of this technology. However, HC degradation potential appeared promising in this system. The authors reported that their design enhanced biodegradation in the tested radius around the BES up to 241% than was observed in their OC control system within the same range around the anode. The radius of influence is an important parameter considering that this technology has potential application for bioremediation in the subsurface. The greater the radius over which a single well can act, the more economical the operation will be. Therefore, designing the process to maximize the radius of influence is key to the success of this technology in the field. Wang et al. (2015) provided a review that provides additional perspective on scale-up limitations for *ex situ* and *in situ* applications, some of which can be applied to remediating contaminated soils and HC wastes.

10.3.2 MICROBIAL ELECTROLYSIS CELLS (MECS)

As described earlier, MECs are devices similar to MFCs but with an applied potential between the anode and cathode. MECs can offer some advantages over MFCs. One includes the ability to control the reactions via the manipulation of the anode potential or the whole-cell potential. Additionally, the degradation mechanism may be primarily anaerobic, as both the anode and cathode are anaerobic, unlike MFCs. However, MECs have not been investigated as extensively for treatment of oil and gas industry wastes. One of the earliest studies used a small-scale MECs to treat de-oiled refinery wastewater Ren et al. (2013). Treatability was demonstrated with a maximum CD of up to 2.1 A/m^2 and 70% COD removal, compared to an OC experiment that removed only ~ 36–58% of the initial COD. The HC component was not specifically characterized, but the results are indicative of the potential removal of these contaminants. While the current densities reported were high, the small scale of reactors used in the study makes the performance parameters less consequential.

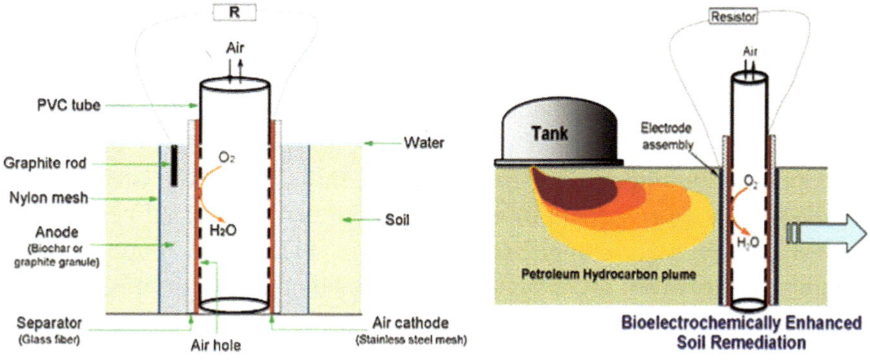

(A) (B)

FIGURE 10.5 Large volume reactors investigating conversion of petroleum hydrocarbons (A) 50 L bedding with embedded BES surrounded by soil. The diameter of the cathode was 15 cm, however, that of the anode surrounding the cathode was about 100 cm. (B) A schematic of field-scale implementation of BES technology for remediation of petroleum contaminates subsurface. *Source:* Reprinted with permission from A) Lu et al., 2014bB) (Lu et al., 2014a. © 2014 Elsevier.

High performance is not always observed for oil and gas wastewater operated MECs even at small scales. The study mentioned earlier by Ghasemi et al. (2015) also tracked the electrochemical performance of their SMXC device in both MFC and MEC mode. Using a 90-mL reactor, the authors reported a maximum power density of 0.65 mW/m^3 in MFC mode. In MEC mode, they achieved a maximum CD of 0.31 A/m^3, a COD removal of 81%, and a hydrogen production rate of 400 mL/m^3-day. Only a hydrogen recovery of 0.18% was observed despite high organic removal. Compared to the study by Ren et al. (2013), this performance is less significant. Perhaps the biggest limitation observed by Ghasemi et al. (2015) was their device's internal resistance. The authors reported an internal resistance of 110 kΩ, which is very high and certainly contributed to the low performance.

Daghio et al. (2018) investigated the conversion of benzene, toluene, ethylbenzene, and xylene (BTEX) in a BES operated at different cell voltages (0.8, 1.0, and 1.2 V). While the authors suggested that their devices may have had some MEC-like functionality, they did not track hydrogen production and decided to focus primarily on organic removal. A maximum CD of 0.48 A/m^2 was reported at an applied potential of 0.8 V. While removal of all components was observed, only degradation of toluene, *m*-xylene, and *p*-xylene was linked to the applied voltage. Interestingly, the highest removal was observed at the lowest potential studied (0.8 V). In this study, the cathode and anode were placed in the same container, thus, any hydrogen that would have been produced was not separated from the anode. The authors suggest that this may have affected not only current production but also community behavior by promoting a process called hydrogen recycling, where anode respiring bacteria use the hydrogen produced at the cathode for their metabolism. Hydrogen recycling has been reported as early as 2009 by Lee et al. (2009); and Call et al. (2009), and therefore the framework proposed by Daghio et al. (2018) may also be likely in single-chamber oil and gas wastewater MECs.

Demonstrating high organic removal is only one goal of MEC development. To design highly efficient MECs with high hydrogen recoveries, reactors will need to effectively separate hydrogen generated at the cathodes from the anodes. Anode potential is another important parameter that can affect the terminal respiratory proteins present in the microorganisms interfacing the contaminants and the electrode. However, higher anode potentials may not always be sufficient to alter the members of the community growing on the anode or evolve them towards the desired functionality of higher current production (Torres et al., 2009). Still, anode potential is not trivial, as the length of time a fixed anode potential is applied to MECs can have lasting effects on performance. A recent study reported the development of electroactive consortia with higher performance at more negative anode potentials via a biofilm replacement approach (Lewis and Borole, 2017). The anode potential is therefore an important parameter in the development of treatment options for energy-efficient contaminant removal.

10.3.3 MICROBIAL DESALINATION CELLS (MDCS)

MDCs represent a relatively new configuration of BESs applied to oil and gas wastewaters. Few of the studies that use them on oil and gas wastewaters investigated microbial community structure, opting to track performance, and chemical degradation and migration instead. They were first used on oil and gas wastes by Stoll et al. (2015) in an MDC that combined capacitive deionization, referred to as a microbial capacitive deionization cell (MCDC). The authors were able to treat shale gas produced water, removing 36 mg TDS per g activated carbon, and 6.4 mg of dissolved organic carbon per hour. The use of an MDC to treat petroleum refinery wastewater at a range of salinities was reported by Sevda et al. (2017). The highest COD removal reported was at a salt concentration of 20 g/L with a maximum COD removal of 70.5% using acidified water as the catholyte. Another study investigating MDCs used waste engine oil as a carbon source to drive desalination of salty water (Sabina et al., 2014). The purpose of this study was not to treat wastewater but to use the waste as a resource to treat saltwater. While the power production was low (3.1 mW/m^2), up to 68% desalination was reported with 70.1% waste engine oil degradation. It was also reported that oil degradation was better than the MFC control used in the experiment.

MDCs have been extensively investigated for desalination of salt-contaminated waters with a large variation of the designs for capturing salt ions (Sevda et al., 2015; Jingyu et al., 2017). One main limitation with MDCs is that they transfer salt ions from a high salt-containing solution or seawater to the nutrient media used in other chambers of the MDC. This is a potentially problematic consequence, where one waste brine is desalinated, while another salty brine is produced. Monzon et al. (2017) worked around this by pairing a highly saline tolerant MFC with a capacitive deionization cell, relying on two different devices to prevent the accumulation of additional wastes. To improve MDC performance without relying on multiple devices, the use of membranes and other processes supporting ion transfer have been investigated. The integration of a forward osmosis membrane with a desalination cell has been tested to treat refinery wastewater (Ismail and Ibrahim, 2015). Membrane fouling was found to be an issue, which can significantly impede device longevity. To work around this, use of multiple membrane-separated compartments has been proposed, as well as the use of ion exchange resins or adsorption media for salt removal (ElMekawy et al., 2014; Saeed et al., 2015). These additions, especially the use of membranes, can drive up the cost of these devices.

Additionally, desalination could be practical if energy for it was supplied in part or entirely by the energy produced by this special class of BES. However, as salinity increases, desalination through processes like capacitive deionization becomes unfavorable compared to others like reverse osmosis (Zhao et al., 2013). The energy requirements only increase as salinity goes up. Even some of the most commonly deployed desalination techniques, such as reverse osmosis, lose considerable efficiency and water recovery percentages at the salinities found in produced waters (Thiel et al., 2015). To ensure energy-efficient desalination, an ideal MFC-driven desalination unit would choose desalination processes based on observed salinity, if desalination is reasonable at all. Further work is needed to keep the costs of these devices low, prevent fouling, and increase energy efficiency.

10.3.4 NON-ELECTROCHEMICAL REACTORS

10.3.4.1 ACTIVATED SLUDGE BIOREACTORS

High COD wastes are typically treated using activated sludge reactors, which contain high concentrations of microbes. Wastes from oil and gas production operations can also be treated in activated sludge bioreactors, which has seen some industry deployment in refineries but has had slower adoption for produced water. For both wastewaters, enrichment of HC-degrading bacteria or addition of such bacteria becomes a critical consideration. Also, the addition of salt-tolerant microbes or halo-tolerant microbes is necessary when treating produced water and related wastes from the oil and gas industry. Despite these constraints, success using activated sludge reactors has been documented. One study treated heavy oil wastewater using an activated sludge process, and resulted in 64% removal of the COD and achieved an effluent COD of 75 mg/L (Tong et al., 2013). This study indicated good removal of phenols, alkenes, aldehydes, and organic acids.

Activated sludge processes can also be useful in the treatment of produced water and flowback water generated during hydraulic fracturing. As mentioned earlier, guar gum and its residues can be a significant component of hydraulic fracturing fluids, and its residues may return to the surface. Degradation of this compound from oilfield waste streams has been reported in activated sludge mixed liquor systems at TDS concentration up to 45,000 mg/L, with better degradation percentages at lower TDS (Lester et al., 2013).

However, for higher TDS levels (close to or greater than 100,000 mg/L) activated sludge bioreactor efficacy can be limited. Instead, use of microbial mats has been reported for treatment under such conditions (Akyon et al., 2015). Organic removal rates of 1.45 mg COD/g wet biomass-day were reported using microbial mats at high TDS conditions. Another modification of the activated sludge bioreactor is the sequencing batch reactor. While this is a discontinuous process, it combines other required steps like degradation, sedimentation, and clarification into a single reactor. An update on the use of sequencing batch reactors was given by Jafarinejad (2017) recently. Conversion of petroleum HCs has also been investigated under nitrate-reducing conditions in the presence of salt. Use of an upflow sludge-blanket/fixed-film bioreactor was trialed for removal of TPH at the rate of 60.5 g/m^3-h at a NaCl concentration of 30 g/L (Moussavi and Ghorbanian, 2015). These non-BES reactors may seem to compete with the goals of BESs, often producing methane and additional organics. However, some of these activated sludge pretreatment processes can potentially be integrated with a BES to generate energy. Partial integration of anaerobic digestion with a BES was recently shown to improve energy harvesting (Beegle and Borole, 2017), and a similar approach could be applied for activated sludge.

10.3.4.2 MEMBRANE AND OTHER TYPES OF BIOREACTORS

Activated sludge reactors have high rates of conversion due to their high cell density. Another way of increasing cell density in bioreactors is via the use of membranes bioreactors. A membrane bioreactor was reported for treatment of saline oily waste from marine transportation. Several genera of bacteria were isolated including *Alcanivorax*, *Erythrobacter*, *Marinobacter*, *Microbacterium*, *Muricauda*, *Rhodococcus*, and *Rheinheimera* (Cappello et al., 2016). One important consideration in developing these systems was microbial production of biosurfactants, which can increase the rate of HC degradation. Column bioreactors have been used by Rogers et al. (2017) to study the degradation of compounds used in hydraulic fracturing, such as 2-propanol, ethylene glycol, propargyl alcohol, 2-butoxyethanol, and 2-ethylhexanol. The presence of glutaraldehyde at a concentration as low as 5 mg/L was found to inhibit the biodegradation of these compounds. However, some of the compounds as well as the glutaraldehyde degraded under oxic conditions. The use of algae was studied for treating produced water by Lutzu and Dunford (2019). Using 2-L glass photobioreactors, thirteen species of algae were examined including some cyanobacteria and green algae. Over 65% TDS, 100% nitrates, and over 95% boron reduction in produced water was reported by the algae used.

10.4 CONCLUSIONS AND FUTURE RESEARCH DIRECTIONS

This chapter discussed advancements in bioelectrochemical technology for simultaneous treatment and energy recovery from oil and gas industry wastes. Three types of BES reactors were discussed: MFCs, MECs, and MDCs. Each of these reactors has a slightly different product, although each of them uses wastewater contaminants as the source of energy. Various design parameters including electrode spacing, radius of influence, type of electrode material, flow vs batch operation have been investigated. However, most of the work done is preliminary and significant opportunities exist for further development and optimization. The application of BESs for treatment of oil and gas industry wastes can lead to new sources of energy, environmental, and economic benefits. Understanding anode microbiology, the effect of reactor design, and parameters contributing to long-term stability in performance are areas ripe for further research.

The vast majority of the studies conducted use MFCs, which rely on an air-fed cathode for oxygen supply. This approach may be impractical for some real world applications where oxygen availability is severely limited, such as in subsurface contaminated sites. For surface applications with adequate oxygen supply, MFCs still hold value. Understanding oxygen availability at the anode, and diffusion from the cathode, is an important area of research with potential fruitful directions for HC contaminant removal. Even if oxygen availability is ignored, bioelectrochemically-enhanced treatments are still a potential path to treat oil and gas wastewaters. Even in the absence of oxygen, using an electrode as the terminal electron acceptor in BESs has shown an improvement in the degradation of petroleum products in CC conditions compared to the same settings in OC conditions, demonstrating the primary benefit of the BES approach.

The studies conducted over the last two decades indicate the potential of this technology for simultaneous contaminant removal and energy production. The understanding of the electro-microbiology in the conversion of HCs in the anode is still in its infancy, and more advancements are necessary (Daghio et al., 2017). Still, several insights have been obtained based on the work conducted to date. Nearly all of the studies that reported phylogenetic classification reported the presence of Proteobacteria as the dominant or significant group of the BES anode consortium. Outside of this commonality, community structure diverges significantly. The apparent diversity indicates that many different community members can perform similar functions under the same conditions. It is possible that the nuances that result from complexity may not be well understood, which makes optimal systems more difficult to engineer. The use of advanced techniques such as metagenomics and transcriptomics can help identify key biological mechanisms important for the conversion of HCs at the anode (Ishii et al., 2013). The interrelationship of aerobic degradation with anaerobic pathways under microaerophilic conditions and the resulting consequences on device performance should continue to be investigated as a potential way to modify degradation rates with simultaneous energy recovery.

The comparatively low performance observed by the studies discussed here may be attributed to recalcitrance of the HCs themselves, organic adsorption to the anode material without equivalent degradation, inhibition caused by one or more components in the waste source, or design and process parameters High current densities have been shown for pyrolysis byproducts (which often contain phenolics) and corn stover fermentation products (Lewis et al., 2015; Lewis and Borole, 2016; Pannell et al., 2016, Satinover et al., 2020), and are higher for simple substrates like acetate and glucose for both MFCs and MECs (Jeremiasse et al., 2010; Ichihashi et al., 2014). However, there may be limitations to performance given the salt concentrations of some of the oil and gas waste streams. Salinity is a very important metric that can affect cell performance, which has been investigated for several device configurations and substrates (Rousseau et al., 2014; Adelaja et al., 2015; Khudzari et al., 2016; Borole and Lewis, 2017). Given that HC spills can occur in salty environments (e.g., for offshore applications), and oil and gas wastes are nearly always very saline, this limitation remains pertinent. Recently, halophilic, thermophilic consortia have been shown to produce high current densities in MECs, producing 6.8 ± 2.1 A/m^2 (Shehab et al., 2017), so further improvements to the tolerance of salinity in BES systems is likely. Designing halotolerant BESs with the same performance metrics found in non-saline devices that also use complex feedstocks will significantly improve BES commercial viability.

In terms of reactor design, a range of new configurations have been developed for MFC and MEC optimization by the BES community (Sleutels et al., 2012; Liu and Cheng, 2014; Lu et al., 2015; Bajracharya et al., 2016; Jingyu et al., 2017). Use of tubular or flat plate configurations, which optimize electrode distance while allowing flow-through operation, are potential directions to pursue for reactor advancement. There is also potential to optimize electrode materials as well, either by building better anodes that promote robust and dense electrically active biofilms, or energy-efficient and cost-effective cathode catalysts (Guo et al., 2015).

Finally, process conditions are critical to long term performance. Once the biocatalyst enrichment is done and reactor design has been selected, operating parameters need to be identified for sustained performance. This is a relatively less studied field, even for wastes streams outside of the oil and gas industry. Therefore, additional work needs to be done to understand the effect of process conditions to identify important operating parameters for sustained performance of these systems. As it stands today, the power densities reported in the literature are low, but the body of knowledge derived from BESs can be applied to oil and gas wastes and has the potential to result in significant improvements. Overall, the use of BESs for treating oil and gas industry waste can be considered as an opportunity as well as a challenge, since fossil fuels will continue to be used for the foreseeable future. As such, developing zero-energy solutions for treating these wastes is a worthwhile goal that can be potentially achieved using the BES approach, while minimizing impacts on the environment.

ACKNOWLEDGMENTS

SJS was partially supported by the Institute for Secure and Sustainable Environments and the Bredesen Center for Interdisciplinary Research and Education at the University of Tennessee, Knoxville.

KEYWORDS

- **bioelectrochemical systems**
- **bioreactors**
- **bioremediation**
- **hydraulic fracturing**
- **oily waste**
- **refinery wastewater**

REFERENCES

Aaron, D. S., et al., (2010). Assessment of the impact of flow rate and ionic strength on performance of microbial fuel cells using electrochemical impedance spectroscopy. *Energies, 3*(4), 592.

Adelaja, O., et al., (2014). Enhanced biodegradation of phenanthrene using different inoculum types in a microbial fuel cell. *Eng. Life Sci., 14*(2), 218–228.

Adelaja, O., et al., (2015). The effect of salinity, redox mediators and temperature on anaerobic biodegradation of petroleum hydrocarbons in microbial fuel cells. *J. Hazard. Mater., 283*, 211–217.

Adelaja, O., et al., (2017). Treatment of phenanthrene and benzene using microbial fuel cells operated continuously for possible in situ and ex situ applications. *Int. Biodeterior. Biodegradation, 116*, 91–103.

Akyon, B., et al., (2015). Microbial mats as a biological treatment approach for saline wastewaters: The case of produced water from hydraulic fracturing. *Environ. Sci. Technol., 49*(10), 6172–6180.

Bajracharya, S., et al., (2016). An overview on emerging bioelectrochemical systems (BESs): Technology for sustainable electricity, waste remediation, resource recovery, chemical production and beyond. *Renew. Energy, 98*, 153–170.

Beegle, J., & Borole, A. P., (2017). An integrated microbial electrolysis-anaerobic digestion process combined with pretreatment of wastewater solids to improve hydrogen production. *Environ. Sci. Water Res. Technol., 3*, 1073–1085.

Bomey, N. (2016). *BP's Deepwater Horizon Costs Total $62B*. https://www.usatoday.com/story/money/2016/07/14/bp-deepwater-horizon-costs/87087056/ (accessed on 25 October 2021).

Borole, A. P., (2012). Bioelectrochemical systems, energy production and electrosynthesis. *J. Microb. Biochem. Technol., 4*(7), XV–XVI.

Borole, A. P., (2016). Chapter 17-microbial electrochemical cells and biorefinery energy efficiency. A2-Eckert, Carrie A. In: Trinh, C. T., (ed.), *Biotechnology for Biofuel Production and Optimization* (pp. 449–472). Elsevier: Amsterdam.

Borole, A. P., et al., (2009a). Improving power production from acetate-fed microbial fuel cells via enrichment of exoelectrogenic organisms in continuous flow systems. *Biochem. Eng. J., 48*, 71–80.

Borole, A. P., et al., (2009b). Integrating engineering design improvements with exoelectrogen enrichment process to increase power output from microbial fuel cells *J. Power Sources, 191*(2), 520–527.

Borole, A. P., et al., (2010). Understanding long term changes in microbial fuel cells using electrochemical impedance spectroscopy. *Environ. Sci. Technol., 44*(7), 2740–2745.

Borole, A. P., et al., (2011a). Enhancement in energy conversion efficiency and current density of 3-dimensional MFC anodes using pre-enriched consortium and continuous supply of electron donors. *Bioresour. Technol., 102*(8), 5098–5104.

Borole, A. P., et al., (2011b). Electroactive biofilms: Current status and future research needs. *Energy Environ. Sci., 4*(12).

Borole, A. P., et al., (2013). Conversion of residual organics in corn stover-derived biorefinery stream to bioenergy via microbial fuel cells. *Environ Sci Technol., 47*, 642–648.

Borole, A. P., Lewis, A. J., (2017). Proton transfer in microbial electrolysis cells. *Sustain. Energy Fuels, 1*(4), 725–736.

Call, D. F., et al., (2009). Hydrogen production by geobacter species and a mixed consortium in a microbial electrolysis cell. *Appl. Environ. Microbiol., 75*(24), 7579–7587.
Callbeck, C. M., et al., (2013). Acetate production from oil under sulfate-reducing conditions in bioreactors injected with sulfate and nitrate. *Appl. Environ. Microbiol., 79*(16), 5059–5068.
Cappello, S., et al., (2016). Oil-degrading bacteria from a membrane bioreactor (BF-MBR) system for treatment of saline oily waste: Isolation, identification and characterization of the biotechnological potential. *Int. Biodeterior. Biodegradation, 110*, 235–244.
Carmalin, S. A., et al., (2016). Microbial desalination cell technology: Contribution to sustainable waste water treatment process, current status and future applications. *J. Environ. Chem. Eng., 4*(3), 3468–3478.
Chae, K. J., et al., (2010). Methanogenesis control by employing various environmental stress conditions in two-chambered microbial fuel cells. *Bioresour. Technol., 101*, 5350–5357.
Chang, S. H., et al., (2017). Electricity production and benzene removal from groundwater using low-cost mini tubular microbial fuel cells in a monitoring well. *J. Environ. Manage., 193*, 551–557.
Chen, S., & Smith, A. L., (2018). Methane-driven microbial fuel cells recover energy and mitigate dissolved methane emissions from anaerobic effluents. *Environ. Sci. Water Res. Technol., 4*(1), 67–79.
Cheng, S., et al., (2006). Increased power generation in a continuous flow MFC with advective flow through the porous anode and reduced electrode spacing. *Environ. Sci. Technol., 40*(7), 2426–2432.
Cluff, M. A., et al., (2014). Temporal changes in microbial ecology and geochemistry in produced water from hydraulically fractured Marcellus shale gas wells. *Environ. Sci. Technol., 48*(11), 6508–6517.
Cord-Ruwisch, R., et al., (1987). Sulfate-reducing bacteria and their activities in oil production. *J. Pet. Technol., 39*, 97–106.
Cusick, R. D., & Logan, B. E., (2012). Phosphate recovery as struvite within a single chamber microbial electrolysis cell. *Bioresour. Technol., 107*, 110–115.
Daghio, M., et al., (2017). Electrobioremediation of oil spills. *Water Res., 114*, 351–370.
Daghio, M., et al., (2018). Bioelectrochemical BTEX removal at different voltages: Assessment of the degradation and characterization of the microbial communities. *J. Hazard. Mater., 341*, 120–127.
Daly, R. A., et al., (2016). Microbial metabolisms in a 2.5-km-deep ecosystem created by hydraulic fracturing in shales. *Nat. Microbiol., 1*, 16146.
Davis, J. B., & Yarbrough, H. F., (1962). Preliminary experiments on a microbial fuel cell. *Science, 137*, 615–616.
Dolfing, J., et al., (2007). Thermodynamic constraints on methanogenic crude oil biodegradation. *IMSE J., 2*, 442–452.
Dykstra, C. M., & Pavlostathis, S. G., (2017). Methanogenic biocathode microbial community development and the role of bacteria. *Environ. Sci. Technol., 51*(9), 5306–5316.
EIA International. (2021). https://www.eia.gov/beta/international/ (accessed on 25 October 2021).
ElMekawy, A., et al., (2014). The near-future integration of microbial desalination cells with reverse osmosis technology. *Energy Environ. Sci., 7*(12), 3921–3933.
Emam, E. A., et al., (2014). Evaluating the characteristics of offshore oilfield produced water. *Pet. Coal, 56*, 363–372.
Enning, D., & Garrelfs, J., (2014). Corrosion of iron by sulfate-reducing bacteria: New views of an old problem. *Appl. Environ. Microbiol., 80*(4), 1226–1236.

EPA Water and Energy Efficiency by Sectors. https://www3.epa.gov/region9/waterinfrastructure/oilrefineries.html (accessed on 25 October 2021).

EPA (2017a). *Class II Oil and Gas Related Injection Wells.* US EPA.

EPA (2021). *Class I Industrial and Municipal Waste Disposal Wells.* https://www.epa.gov/uic/class-i-industrial-and-municipal-waste-disposal-wells (accessed on 25 October 2021).

Fan, Y., et al., (2007). Enhanced coulombic efficiency and power density of air-cathode microbial fuel cells with an improved cell configuration. *J. Power. Sources, 171*, 348–354.

Ghasemi, N. Z., et al., (2015). Produced water treatment with simultaneous bioenergy production using novel bioelectrochemical systems. *Electrochim. Acta, 180*, 535–544.

Gorby, Y. A., et al., (2006). Electrically conductive bacterial nanowires produced by *Shewanella oneidensis* strain MR-1 and other microorganisms. *Proc. Natl. Acad. Sci. USA, 103*(30), 11358–11363.

Guo, K., et al., (2015). Engineering electrodes for microbial electrocatalysis. *Curr. Opin. Biotechnol., 33*, 149–156.

Guo, X., et al., (2014). The influence of microbial synergistic and antagonistic effects on the performance of refinery wastewater microbial fuel cells. *J. Power Sources, 251*, 229–236.

Guo, X., et al., (2016). Influence of packing material characteristics on the performance of microbial fuel cells using petroleum refinery wastewater as fuel. *Renew. Energy, 87*, 437–444.

GWPC, (2015). *U.S. Produced Water Volumes and Management Practices in 2012.* Ground Water Protection Council.

Halliburton, (2014). *PERMVIS™ VFR-10 Friction Reducer*. Halliburton.

Hamelers, H. V. M., et al., (2011). Butler-Volmer-Monod model for describing bio-anode polarization curves. *Bioresour. Technol., 102*(1), 381–387.

Hazen, T. C., et al., (2010). Deep-sea oil plume enriches indigenous oil-degrading bacteria. *Science, 330*(6001), 204–208.

He, Z., & Mansfeld, F., (2009). Exploring the use of electrochemical impedance spectroscopy (EIS) in microbial fuel cell studies. *Energy Environ. Sci., 2*(2), 215–219.

Ichihashi, O., & Hirooka, K., (2012). Removal and recovery of phosphorus as struvite from swine wastewater using microbial fuel cell. *Bioresour. Technol., 114*, 303–307.

Ichihashi, O., et al., (2014). High-performance bioanode development for fermentable substrates via controlled electroactive biofilm growth. *ChemElectroChem, 1*(11), 1940–1947.

Ishii, S. I., et al., (2013). A novel metatranscriptomic approach to identify gene expression dynamics during extracellular electron transfer. *Nat. Commun., 4*, 1601.

Ismail, Z. Z., & Ibrahim, M. A., (2015). Desalination of oilfield produced water associated with treatment of domestic wastewater and bioelectricity generation in microbial osmotic fuel cell. *J. Membr. Sci., 490*, 247–255.

Jafarinejad, S., (2017). Recent developments in the application of sequencing batch reactor (SBR) technology for the petroleum industry wastewater treatment. *Chem. Int., 3*(3).

Jeremiasse, A. W., et al., (2010). Ni foam cathode enables high volumetric H-2 production in a microbial electrolysis cell. *Int. J. Hydrog. Energy, 35*(23), 12716–12723.

Jingyu, H., et al., (2017). Microbial desalination cells technology: A review of the factors affecting the process, performance and efficiency. *Desalination Water Treat., 87*, 140–159.

Joosten, M. W., et al., (2002). In: *Organic Acid Corrosion in Oil and Gas Production.* Corrosion 2002, 2002/01/01; NACE International.

Kadier, A., et al., (2014). A review of the substrates used in microbial electrolysis cells (MECs) for producing sustainable and clean hydrogen gas. *Renew. Energy, 71*, 466–472.

Kadier, A., et al., (2016). Recent advances and emerging challenges in microbial electrolysis cells (MECs) for microbial production of hydrogen and value-added chemicals. *Renew. Sust. Energ. Rev., 61*, 501–525.

Kaur, A., et al., (2014). Inhibition of methane production in microbial fuel cells: Operating strategies which select electrogens over methanogens. *Bioresour. Technol., 173*, 75–81.

Kelly, P. T., & He, Z., (2014). Nutrients removal and recovery in bioelectrochemical systems: A review. *Bioresour. Technol., 153*, 351–360.

Kessler, J. D., et al., (2011). A persistent oxygen anomaly reveals the fate of spilled methane in the deep Gulf of Mexico. *Science, 331*(6015), 312–315.

Khudzari, J., et al., (2016). Effect of C/N ratio and salinity on power generation in compost microbial fuel cells. *Waste Manag., 48*, 135–142.

Kim, J. R., et al., (2007). Power generation using different cation, anion, and ultrafiltration membranes in microbial fuel cells. *Environ. Sci. Technol., 41*, 1004–1009.

Kondash, A. J., et al., (2017). Quantity of flowback and produced waters from unconventional oil and gas exploration. *Sci. Total Environ., 574*, 314–321.

Kostka, J. E., et al., (2011). Hydrocarbon-degrading bacteria and the bacterial community response in Gulf of Mexico beach sands impacted by the Deepwater Horizon oil spill. *Appl. Environ. Microbiol., 77*(22), 7962–7974.

Kumar, R., et al., (2015). Exoelectrogens in microbial fuel cells toward bioelectricity generation: A review. *Int. J. Energy Res., 39*(8), 1048–1067.

Kuntke, P., et al., (2012). Ammonium recovery and energy production from urine by a microbial fuel cell. *Water Res., 46*(8), 2627–2636.

Leavitt, M., et al., (2004). *Technical Support Document for the 2004 Effluent Guidelines Program Plan* (p. 126). U.S. Environmental Protection Agency: Washington, D. C., 2004/08//.

Lee, H. S., et al., (2009). Fate of H_2 in an up flow single-chamber microbial electrolysis cell using a metal-catalyst-free cathode. *Environ. Sci. Technol., 43*(20), 7971–7976.

Lester, Y., et al., (2013). Can we treat hydraulic fracturing flowback with a conventional biological process? The case of guar gum. *Environ. Sci. Technol. Lett., 1*(1), 133–136.

Lewis, A. J., & Borole, A. P., (2016). Understanding the impact of flow rate and recycle on the conversion of a complex biorefinery stream using a flow-through microbial electrolysis cell. *Biochem. Eng. J., 116*, 95–104.

Lewis, A. J., & Borole, A. P., (2017). Adapting microbial communities to low anode potentials improves performance of MECs at negative potentials. *Electrochim. Acta, 254*(10), 79–88.

Lewis, A. J., et al., (2015). Hydrogen production from switchgrass via an integrated pyrolysis-microbial electrolysis process. *Bioresour. Technol., 195*, 231–241.

Lewis, A. J., et al., (2018). Unravelling biocomplexity of electroactive biofilms for producing hydrogen from biomass. *Microb. Biotechnol., 11*(1), 84–97.

Li, X., et al., (2014). Extended petroleum hydrocarbon bioremediation in saline soil using Pt-free multianodes microbial fuel cells. *RSC Adv., 4*(104), 59803–59808.

Li, X., et al., (2015). Sand amendment enhances bioelectrochemical remediation of petroleum hydrocarbon contaminated soil. *Chemosphere, 141*, 62–70.

Liu, H., et al., (2005). Power generation in fed-batch microbial fuel cells as a function of ionic strength, temperature, and reactor configuration. *Environ. Sci. Technol., 39*(14), 5488–5493.

Liu, W. F., & Cheng, S. A., (2014). Microbial fuel cells for energy production from wastewaters: The way toward practical application. *J. Zhejiang Univ. Sci. A, 15*(11), 841–861.

Logan, B. E., (2009). Exoelectrogenic bacteria that power microbial fuel cells. *Nat. Rev. Microbiol., 7*(5), 375–381.

Logan, B. E., et al., (2006). Microbial fuel cells: methodology and technology. *Environ. Sci. Technol., 40*, 5181–5192.

Logan, B. E., et al., (2008). Microbial electrolysis cells for high yield hydrogen gas production from organic matter. *Environ. Sci. Technol., 42*(23), 8630–8640.

Logan, B. E., et al., (2015). Assessment of microbial fuel cell configurations and power densities. *Environ. Sci. Technol. Lett., 2*(8), 206–214.

Lovley, D. R., (2012). Electromicrobiology. *Annu. Rev. Microbiol., 66*, 391–409.

Lu, L., et al., (2014a). Microbial metabolism and community structure in response to bioelectrochemically enhanced remediation of petroleum hydrocarbon-contaminated soil. *Environ. Sci. Technol., 48*(7), 4021–4029.

Lu, L., et al., (2014b). Enhanced bioremediation of hydrocarbon-contaminated soil using pilot-scale bioelectrochemical systems. *J. Hazard. Mater., 274*, 8–15.

Lu, L., et al., (2016). Active H_2 harvesting prevents methanogenesis in microbial electrolysis cells. *Environ. Sci. Technol. Lett., 3*(8), 286–290.

Lu, Y., et al., (2015). When bioelectrochemical systems meet forward osmosis: Accomplishing wastewater treatment and reuse through synergy. *Water, 7*(1), 38–50.

Lutzu, G. A., & Dunford, N. T., (2019). Algal treatment of wastewater generated during oil and gas production using hydraulic fracturing technology. *Environ. Technol., 40*(8), 1027–1034.

Majumder, D., et al., (2014). Electricity generation and wastewater treatment of oil refinery in microbial fuel cells using *Pseudomonas putida*. *Int. J. Mol. Sci., 15*, 16772–16786.

Marcus, A. K., et al., (2007). Conduction-based modeling of the biofilm anode of a microbial fuel cell. *Biotechnol. Bioeng., 98*(6), 1171–1182.

Marone, A., et al., (2017). Coupling dark fermentation and microbial electrolysis to enhance bio-hydrogen production from agro-industrial wastewaters and by-products in a bio-refinery framework. *Int. J. Hydrog. Energy, 42*(3), 1609–1621.

McCurdy, R. (2021). *Underground Injection Wells for Produced Water Disposal*. Chesapeake Energy Corporation: Oklahoma City, OK. https://www.epa.gov/sites/default/files/documents/21_McCurdy_-_UIC_Disposal_508.pdf (accessed on 20 December 2021).

Miceli, III. J. F., et al., (2012). Enrichment and analysis of anode-respiring bacteria from diverse anaerobic inocula. *Environ. Sci. Technol., 46*(18), 10349–10355.

Mohan, S. V., & Chandrasekhar, K., (2011). Self-induced bio-potential and graphite electron accepting conditions enhances petroleum sludge degradation in bio-electrochemical system with simultaneous power generation. *Bioresour. Technol., 102*(20), 9532–9541.

Mohanakrishna, G., et al., (2018). Enhanced treatment of petroleum refinery wastewater by short-term applied voltage in single chamber microbial fuel cell. *Bioresour. Technol., 253*, 16–21.

Monzon, O., et al., (2014). Microbial fuel cells under extreme salinity: Performance and microbial analysis. *Environ. Chem., 12*, 293–299.

Monzon, O., et al., (2016). Quorum sensing autoinducers enhance biofilm formation and power production in a hypersaline microbial fuel cell. *Biochem. Eng. J., 109*, 222–227.

Monzon, O., et al., (2017). Microbial fuel cell fed by Barnett Shale produced water: Power production by hypersaline autochthonous bacteria and coupling to a desalination unit. *Biochem. Eng. J.,* 117(Part A), 87–91.

Morris, J. M., & Jin, S., (2007). Feasibility of using microbial fuel cell technology for bioremediation of hydrocarbons in groundwater. *J. Environ. Sci. Health A, 43*, 18–23.

Morris, J. M., & Jin, S., (2012). Enhanced biodegradation of hydrocarbon-contaminated sediments using microbial fuel cells. *J. Hazard. Mater., 213, 214*, 474–477.

Morris, J. M., et al., (2009). Microbial fuel cell in enhancing anaerobic biodegradation of diesel. *Chem. Eng. J., 146*(2), 161–167.

Moussavi, G., & Ghorbanian, M., (2015). The biodegradation of petroleum hydrocarbons in an upflow sludge-blanket/fixed-film hybrid bioreactor under nitrate-reducing conditions: Performance evaluation and microbial identification. *Chem. Eng. J., 280*, 121–131.

Neff, J., et al., (2011). Produced water: Overview of composition, fates, and effects. In: Lee, K., & Neff, J., (eds.), *Produced Water* (pp. 3–54). Springer, New York.

Ogden, B. L., (2005). *Sulfide Stress Cracking- Practical Application to the Oil and gas Industry*. Texas Tech University: Southwest Petroleum Short Course.

Oh, S. E., et al., (2009). Effects of applied voltages and dissolved oxygen on sustained power generation by microbial fuel cells. *Water Sci. Technol., 60*, 1311–1317.

Ostenberg, R., (2009). Optimization of water usage at petroleum refineries. In: Ostenberg, R., (ed.), *Water/Energy Sustainability Forum*. Salt Lake City, Utah.

Pandey, P., et al., (2016). Recent advances in the use of different substrates in microbial fuel cells toward wastewater treatment and simultaneous energy recovery. *Appl. Energy, 168*, 706–723.

Pannell, T. C., et al., (2016). Effect of fed-batch vs. continuous mode of operation on microbial fuel cell performance treating biorefinery wastewater. *Biochem. Eng. J., 116*, 85–94.

Pant, D., et al., (2010). A review of the substrates used in microbial fuel cells (MFCs) for sustainable energy production. *Bioresour. Technol., 101*(6), 1533–1543.

Pant, D., et al., (2012). Bioelectrochemical systems (BES) for sustainable energy production and product recovery from organic wastes and industrial wastewaters. *RSC Adv., 2*(4), 1248–1263.

Paque, M., (1997). *Use of Injection Wells for Refinery Waste Disposal*. DE-FG22-95BC14828; Ground Water Protection Council: Oklahoma City, OK, 1997/01//.

Paul, V. G., et al., (2014). Ability of a haloalkaliphilic bacterium isolated from Soap Lake, Washington to generate electricity at pH 11.0 and 7% salinity. *Environ. Technol., 35*(8), 1003–1011.

Popat, S. C., & Torres, C. I., (2016). Critical transport rates that limit the performance of microbial electrochemistry technologies. *Bioresour. Technol., 215*, 265–273.

Rabaey, K., et al., (2010a). *Bioelectrochemical Systems: From Extracellular Electron Transfer to Biotechnological Application*. IWA Publishing: London, UK.

Rabaey, K., et al., (2010b). High current generation coupled to caustic production using a lamellar bioelectrochemical system. *Environ. Sci. Technol., 44*(11), 4315–4321.

Rakoczy, J., et al., (2013). Benzene and sulfide removal from groundwater treated in a microbial fuel cell. *Biotechnol. Bioeng., 110*(12), 3104–3113.

Ramasamy, R. P., et al., (2008). Impact of initial biofilm growth on the anode impedance of microbial fuel cells. *Biotechnol. Bioeng., 101*(1), 101–108.

Reguera, G., et al., (2005). Extracellular electron transfer via microbial nanowires. *Nature, 435*(7045), 1098–1101.

Ren, L., et al., (2013). Treatability studies on different refinery wastewater samples using high-throughput microbial electrolysis cells (MECs). *Bioresour. Technol., 136*, 322–328.

Rimboud, M., et al., (2015). Multi-system Nernst-Michaelis-Menten model applied to bioanodes formed from sewage sludge. *Bioresour. Technol., 195*, 162–169.

Roach, R. W., et al., (1993). *An Assessment of Produced Water Impacts in the Galveston Bay System, 993*, 38.

Rogers, J. D., et al., (2017). Inhibition of biodegradation of hydraulic fracturing compounds by glutaraldehyde: Groundwater column and microcosm experiments. *Environ. Sci. Technol., 51*(17), 10251–10261.
Rousseau, R., et al., (2014). Correlation of the electrochemical kinetics of high-salinity-tolerant bioanodes with the structure and microbial composition of the biofilm. *ChemElectroChem, 1*(11), 1966–1975.
Roustazadeh, S. P., et al., (2017). Investigation of different configurations of microbial fuel cells for the treatment of oilfield produced water. *Appl. Energy, 192*, 457–465.
Rozendal, R. A., et al., (2008). Towards practical implementation of bioelectrochemical wastewater treatment. *Trends Biotechnol., 26*(8), 450–459.
Rozendal, R. A., et al., (2009). Efficient hydrogen peroxide generation from organic matter in a bioelectrochemical system. *Electrochem. Commun., 11*(9), 1752–1755.
Sabina, K., et al., (2014). Microbial desalination cell for enhanced biodegradation of waste engine oil using a novel bacterial strain *Bacillus subtilis* moh3. *Environ. Technol., 35*(17), 2194–2203.
Saeed, H. M., et al., (2015). Microbial desalination cell technology: A review and a case study. *Desalination, 359*, 1–13.
Salminen, J. M., et al., (2004). Potential for aerobic and anaerobic biodegradation of petroleum hydrocarbons in boreal subsurface. *Biodegradation, 15*, 29–39.
Satinover, S. J., Schell, D., & Borole, A. P. (2020). "Achieving high hydrogen productivities of 20 L/L-day via microbial electrolysis of corn stover fermentation products." *Applied Energy, 259*, 114126.
Sekar, N., & Ramasamy, R. P., (2013). Electrochemical impedance spectroscopy for microbial fuel cell characterization. *J. Microb. Biochem. Technol., S6*(004).
Sevda, S., et al., (2015). Microbial desalination cells as a versatile technology: Functions, optimization and prospective. *Desalination, 371*, 9–17.
Sevda, S., et al., (2017). Bioelectricity generation from treatment of petroleum refinery wastewater with simultaneous seawater desalination in microbial desalination cells. *Energy Convers. Manag., 141*, 101–107.
Sharma, M., et al., (2014). A critical revisit of the key parameters used to describe microbial electrochemical systems. *Electrochim. Acta, 140*, 191–208.
Shehab, N. A., et al., (2017). Enrichment of extremophilic exoelectrogens in microbial electrolysis cells using Red Sea brine pools as inocula. *Bioresour. Technol., 239*, 82–86.
Sleutels, T., et al., (2011). Effect of mass and charge transport speed and direction in porous anodes on microbial electrolysis cell performance. *Bioresour. Technol., 102*(1), 399–403.
Sleutels, T., et al., (2012). Bioelectrochemical systems: An outlook for practical applications. *ChemSusChem, 5*(6), 1012–1019.
Sleutels, T., et al., (2016). Low substrate loading limits methanogenesis and leads to high coulombic efficiency in bioelectrochemical systems. *Microorganisms, 4*(7), 1–11.
Srikanth, S., et al., (2016). Electro-biocatalytic treatment of petroleum refinery wastewater using microbial fuel cell (MFC) in continuous mode operation. *Bioresour. Technol., 221*, 70–77.
Stoll, Z. A., et al., (2015). Shale gas produced water treatment using innovative microbial capacitive desalination cell. *J. Hazard. Mater., 283*, 847–855.
Stringfellow, W. T., et al., (2017). Identifying chemicals of concern in hydraulic fracturing fluids used for oil production. *Environ. Pollut., 220*(Pt A), 413–420.
Sure, S., et al., (2015). Inquisition of *Microcystis aeruginosa* and *Synechocystis* nanowires: Characterization and modelling. *Antonie Van Leeuwenhoek, 108*(5), 1213–1225.

Thiel, G. P., et al., (2015). Energy consumption in desalinating produced water from shale oil and gas extraction. *Desalination, 366*, 94–112.

Tong, K., et al., (2013). Treatment of heavy oil wastewater by a conventional activated sludge process coupled with an immobilized biological filter. *Int. Biodeterior. Biodegradation, 84*, 65–71.

Torres, C. I., et al., (2009). Selecting anode-respiring bacteria based on anode potential: Phylogenetic, electrochemical, and microscopic characterization. *Environ. Sci. Technol., 43*(24), 9519–9524.

Ullery, M. L., & Logan, B. E., (2014). Comparison of complex effluent treatability in different bench scale microbial electrolysis cells. *Bioresour. Technol., 170*, 530–537.

Venkidusamy, K., et al., (2016). Enhanced removal of petroleum hydrocarbons using a bioelectrochemical remediation system with pre-cultured anodes. *Sci. Total Environ., 539*, 61–69.

Von, W. F., et al., (2001). *Bordetella petrii* sp. nov., isolated from an anaerobic bioreactor, and emended description of the genus Bordetella. *Int. J. Syst. Evol. Microbiol., 51*(4), 1257–1265.

Wang, H., et al., (2015). Bioelectrochemical system platform for sustainable environmental remediation and energy generation. *Biotechnol. Adv., 33*(3), 317–334.

Wang, X., et al., (2012). Bioelectrochemical stimulation of petroleum hydrocarbon degradation in saline soil using U-tube microbial fuel cells. *Biotechnol. Bioeng., 109*(2), 426–433.

Wei, M., et al., (2014). Harvesting electricity from benzene and ammonium-contaminated groundwater using a microbial fuel cell with an aerated cathode. *RSC Adv., 5*, 5321–5330.

Wu, C. H., et al., (2013). Generation of power by microbial fuel cell with ferricyanide in biodegradation of benzene. *Clean (Weinh), 41*(4), 390–395.

Wu, C. H., et al., (2014). Enhancement of power generation by toluene biodegradation in a microbial fuel cell in the presence of pyocyanin. *J. Taiwan Inst. Chem. Eng., 45*(5), 2319–2324.

Xie, Z., et al., (2013). Influence and mechanism of dissolved oxygen on the performance of ammonia-oxidation microbial fuel cell. *Int. J. Hydrog. Energy, 38*, 10607–10615.

Yan, F., & Reible, D., (2015). Electro-bioremediation of contaminated sediment by electrode enhanced capping. *J. Environ. Manage., 155*, 154–161.

Zhang, F., et al., (2014). Treating refinery wastewaters in microbial fuel cells using separator electrode assembly or spaced electrode configurations. *Bioresour. Technol., 152*, 46–52.

Zhang, T., et al., (2010). Stimulating the anaerobic degradation of aromatic hydrocarbons in contaminated sediments by providing an electrode as the electron acceptor. *Environ. Microbiol., 12*(4), 1011–1020.

Zhang, X., et al., (2018). Simultaneous removal of organic matter and iron from hydraulic fracturing flowback water through sulfur cycling in a microbial fuel cell. *Water Res., 147*, 461–471.

Zhao, R., et al., (2013). Energy consumption in membrane capacitive deionization for different water recoveries and flow rates, and comparison with reverse osmosis. *Desalination, 330*, 35–41.

Annex: Resources for Environmental Pollution and Hydrocarbon Biodegradation

BOOKS

- **Handbook of Hydrocarbon and Lipid Microbiology. Springer, 2020**
  Editor-in-chief: Timmis, Kenneth N.
  Series Editors: Boll, M., Geiger, O., Goldfine, H., Krell, T., Lee, S.Y., McGenity, T.J., Rojo, F., Sousa, D.Z., Stams, A.J.M., Steffan, R., Wilkes, H.

- **Oilfield Microbiology.** CRC Press/Taylor & Francis, 2019, Torben Lund Skovhus and Corinne Whitby, ed.
- **Microbial Metabolism of Xenobiotic Compounds.** Springer, 2019, Arora, Pankaj Kumar, ed.
- **Introduction to Petroleum Biotechnology.** Gulf Professional Publishing, An Imprint of Elsevier, 2018, James G. Speight and Nour Shafik El-Gendy.
- **Petroleum Microbiology: Concepts, Environmental Implications, Industrial Applications.** Editions Technip, 2008, Jean-Paul Vandecasteele, ed.
- **Microbial Biodegradation and Bioremediation.** Elsevier, 2014, Surajit Das, ed.
- **Biosurfactants: Production and Utilization-Processes, Technologies, and Economics.** CRC Press, Taylor & Francis, 2015, Naim Kosaric and Fazilet Vardar-Sukan, ed.
- **Biocatalysis and Biodegradation: Microbial Transformation of Organic Compounds.** ASM Press, 2001, Lawrence P. Wackett and C. Douglas Hershberger.

JOURNALS

- Applied and Environmental Microbiology
- Applied Microbiology and Biotechnology
- International Biodeterioration and Biodegradation
- Environmental Microbiology
- Bioresource Technology
- Frontiers in Microbiology
- Microbial Biotechnology
- Biodegradation
- Biocatalysis and Biotransformation
- Current Opinion in Biotechnology

INTERNET RESOURCES

- Biocatalysis/Biodegradation Database (http://eawag-bbd.ethz.ch/)
- BioTransformer: A computational tool for the prediction of small molecule metabolism, and metabolite identification (https://bitbucket.org/djoumbou/biotransformerjar/src/master/)
- U.S. Geological Survey (https://www.usgs.gov/science-explorer-results?es=biodegradation)
- U.S. Environmental Protection Agency (https://www.epa.gov/)
- MetaCyc Metabolic Pathway Database (https://metacyc.org/)
- KEGG Pathway Database (https://www.genome.jp/kegg/pathway.html)

Index

B

E

F

L

M

O

P

T

U

V